National Audubon Society®
Field Guide to
North American Fossils

A Chanticleer Press Edition

National Audubon Society®
Field Guide to
North American Fossils

Ida Thompson,
Associate Research Professor,
Center for Coastal and Environmental
Studies, Rutgers University

With photographs by
Townsend P. Dickinson

Visual Key by
Carol Nehring

Alfred A. Knopf, New York

I dedicate this book to my children,
Matthew, Alice, and Carla

This is a Borzoi Book
Published by Alfred A. Knopf

www.aaknopf.com

Prepared and produced by
Chanticleer Press, Inc., New York.

Color reproductions by Nievergelt Repro
AG, Zurich, Switzerland. Typeset in
Garamond by Dix Type, Inc., Syracuse,
New York. Printed and bound by Toppan
Leefung Printing Limited, China.

Published September 1982
Twenty-first printing, October 2016

Library of Congress Cataloging-in-
Publication Number: 81-84772
ISBN: 978-0-394-52412-2

CONTENTS

11 Introduction
23 Life and Geologic Time
24 Geologic Time Chart
87 Fossil-bearing Rocks

Part I Color Plates
99 How to Use This Guide
103 Key to the Maps
Maps
137 *Color Plates*
Fossiliferous Rocks and Outcroppings
1–33
149 Key to the Color Plates
150 Thumb Tab Guide
Scallop-shaped Fossils 34–102
Clam-shaped Fossils 103–165
Snail-shaped Fossils 166–216
Gastropods 217–267
Arthropods 268–315
Cystoids, Blastoids, and Crinoids 316–336
Cone-shaped Fossils 337–357
Disc-shaped Fossils 358–375
Sponges, Corals, and Bryozoans 376–456
Trace Fossils and Graptolites 457–471
Vertebrates 472–486
Plants 487–507

Part II Text
339 Sponges
353 Cnidarians
389 Mollusks
537 Annelids

541 Small Tubes of Uncertain Affinities
547 Arthropods
597 Bryozoans
625 Brachiopods
687 Echinoderms
723 Hemichordates
731 Trace Fossils
738 Plants
751 Insects
755 Vertebrates

Part III Appendices
771 Phylum Illustrations
779 How to Collect and Preserve Fossils
789 Glossary
805 List of Major Fossil Displays
813 List of Geological Surveys
817 List of Specimen Localities
828 Picture Credits
829 Index

NATIONAL AUDUBON SOCIETY

The mission of NATIONAL AUDUBON SOCIETY, *founded in 1905, is to conserve and restore natural ecosystems, focusing on birds, other wildlife, and their habitats for the benefit of humanity and the earth's biological diversity.*

One of the largest, most effective environmental organizations, AUDUBON has nearly 550,000 members, numerous state offices and nature centers, and 500+ chapters in the United States and Latin America, plus a professional staff of scientists, educators, and policy analysts. Through its nationwide sanctuary system AUDUBON manages 160,000 acres of critical wildlife habitat and unique natural areas for birds, wild animals, and rare plant life.

The award-winning *Audubon* magazine, which is sent to all members, carries outstanding articles and color photography on wildlife, nature, environmental issues, and conservation news. AUDUBON also publishes *Audubon Adventures,* a children's newsletter reaching 450,000 students. Through its ecology camps and workshops in Maine, Connecticut, and Wyoming, AUDUBON offers nature education for teachers, families, and children; through *Audubon Expedition Institute* in Belfast, Maine, AUDUBON offers unique, traveling undergraduate and graduate degree programs in Environmental Education.

AUDUBON sponsors books and on-line nature activities, plus travel programs to exotic places like Antarctica, Africa, Baja California, the Galápagos Islands, and Patagonia. For information about how to become an AUDUBON member, subscribe to *Audubon Adventures,* or to learn more about any of our programs, please contact:

AUDUBON
225 Varick Street, 7th Floor
New York, NY 10014
(212) 979-3000
(800) 274-4201
www.audubon.org

THE AUTHOR

Ida Thompson is Associate Research Professor at the Center for Coastal and Environmental Studies of Rutgers University. Dr. Thompson received her A.B., M.S., and Ph.D. from the University of Chicago, where she was a Danforth Scholar. She worked at the Field Museum of Natural History in Chicago and, as a member of the faculty of Princeton University, taught courses in paleoecology, historical geology, invertebrate zoology, paleontology, and oceanography. Dr. Thompson is the author of numerous articles on terrestrial and marine fossils, and the recipient of two major grants from the National Oceanographic and Atmospheric Administration to study marine bivalves.

ACKNOWLEDGMENTS

Many people helped in the preparation of this field guide. I am especially grateful to the following paleontologists who reviewed the manuscript and photographs: J. Keith Rigby, Brigham Young University (sponges); Roger L. Batten, American Museum of Natural History (gastropods); Harold E. Vokes, Tulane University (Tertiary bivalves); Walter C. Sweet, Ohio State University (cephalopods); Frederick R. Schram, San Diego County Museum (crustaceans); Roger J. Cuffey, Pennsylvania State University (bryozoans); James C. Brower, Syracuse University (cystoids, blastoids, and crinoids); Alfred G. Fischer, Princeton University (echinoids); William B. N. Berry, University of California, Berkeley (graptolites); Erling Dorf, Princeton University (plants); Charles L. Remington, Yale University (insects); and Donald Baird, Princeton University (vertebrates and sedimentary rocks). In addition to reviewing the section on trilobites, Niles Eldredge provided invaluable advice.

For allowing us to photograph their fossil collections, thanks are due to the National Museum of Natural History, Smithsonian Institution; Pennsylvania

State University; Princeton University; and Yale University. The following individuals were especially helpful: at Princeton, Donald Baird and Erling Dorf; at Yale, Jean S. Lawless, Copeland MacClintock, David E. Schindel, and Karl M. Waage; at the National Museum of Natural History, Frederick J. Collier and Raymond T. Rye II; and at Pennsylvania State University, Roger J. Cuffey.
Maps kindly provided by the United States Geological Survey formed the basis for the ones that we have used. Canada's National Museum of Natural Sciences was also helpful.
In the preparation of the manuscript and numerous other tasks, I am grateful to Joelene Bergonzi, Andrew Ranicki, and Joelle Collot. I am also greatly indebted to David Stager, Geology Librarian of Princeton University, for his help and encouragement.
Several people, including some old friends, provided information and advice. My thanks to Shirley Albright, George R. Clark II, William Cobban, Ken McKinney, Matthew Nitecki, Eugene S. Richardson, Jr., Paul Strother, Franklin Van Houten, Ellis Yochelson, and Keith Young.
Particular thanks go to Paul Steiner and the staff of Chanticleer Press. I am especially grateful to Carole Slatkin and Jill Hamilton for their wise advice and intelligent editing, and to Townsend P. Dickinson, who took most of the photographs. Thanks also go to John Farrand, Jr., who developed the idea for the book and edited parts of the text; to Helga Lose and John Holliday, who saw the book through production; and to Carol Nehring, who helped solve the design problems, supervised the layout, and developed the visual key.

INTRODUCTION

Fossils, the traces of creatures that lived ages ago, have intrigued mankind throughout history. Primitive peoples were mystified by these strange forms of plants and animals in stone, and wore them as amulets, attributing supernatural powers to them. In past centuries, fossils have been variously regarded as vestiges of the Great Flood, as objects placed in rocks by the Devil to deceive the unfaithful, and as the remains of unsuccessful forms of life discarded by the Creator.

Today we know that fossils are a record of life as it has evolved through the long span of geologic time. They are invaluable to scientists in reconstructing prehistoric environments, identifying animals and plants that flourished millions of years ago, and tracing the evolution of living things, including Man, during three and a half billion years.

But fossils are not objects displayed only in museums and appreciated only by paleontologists: they are abundant in many parts of North America and can be discovered by an amateur fossil-hunter. Along the western shore of Chesapeake Bay in Maryland, fossil shells and bones wash out of the cliffs and mingle with ordinary seashells on the beach. In the bottoms of glens in upstate New York, delicately sculptured brachiopods, more than 350 million years old, can be found in the crumbling shale of streambeds. Thousands of corals, dating from more than 300 million years ago, have been collected along the Ohio River near Louisville, Kentucky. Beautifully coiled pearly ammonites, each one encased in a sphere of stone and more than 100 million years old, can be discovered in sea cliffs along the coast of California and Oregon. At thousands of places across North America, fossils can be found and collected by anyone. To split open a fragment of shale and find embedded within it a creature that lived and died millions of years ago, and that lay in darkness until you brought it to light, is a thrill as great as that experienced by any bird-watcher, shell-collector, or wildflower enthusiast. The purpose of this guide is to make fossil-collecting as easy and enjoyable as possible. Take it into the field with you—and welcome to the ever-growing ranks of fossil-hunters.

Geographical Scope:
This field guide covers fossils found in North America north of Mexico. All of the fossils included are commonly found in some part of the continent. Fossils that are too small to be studied easily by most collectors, or so scarce that they are seldom found by the average fossil-hunter, have not been included.

What Is a Fossil?
Fossils are the remains of ancient life that have been buried in the earth or under the sea for anywhere from thousands of years to hundreds of millions of years. They include not only the bones of huge dinosaurs and other vertebrates, but the shells and skeletons of marine and freshwater invertebrates, the fragile remains of insects, and the preserved leaves, buds, bark, and other parts of plants. Since fossils form most

readily in sediments laid down on the floor of the sea, more than 90% of all known fossils are marine animals. For this reason, the great majority of fossils described in detail in this book are of this kind. The remains of plants are common, but generally consist of single leaves or other fragments and are very difficult to identify, while insect fossils are fragile, and consequently very rare. Vertebrate fossils are seldom found by the average collector, and usually consist of single bones that cannot be identified except by a specialist. These 3 kinds of fossils—plants, insects, and vertebrates—are therefore described in general terms, rather than in detail. Paleontologists disagree about how old a specimen must be before it can be called a fossil. Some require that it merely be older than historic time. Others stipulate that it must date from before the start of the geological epoch known as the Recent, or from at least 11,000 years ago. If a plant or animal is extinct, its remains are usually considered a fossil even if they were buried only 100 years ago. If such remains are relatively recent and have not yet been mineralized, they are sometimes called subfossils. The oldest known fossils, tiny cells that look like bacteria, occur in rocks about 3.4 billion years old.

How Fossils Are Formed:
Fossils are formed in a variety of ways. In rare instances, the whole organism has been preserved, as in the case of the well-known frozen mammoths of Siberia, whose flesh, when found, was still edible. But most fossils are only partial remains of animals or plants—generally the hard parts of the organism that were buried quickly and thus preserved from decay. In animals, these are usually shells, bones, or teeth. If these hard parts are not very old, they may be virtually unchanged from their condition when the animal was alive.

Molds: Given enough time, however, even the hard parts of a fossil may dissolve. Very often, when a shell or bone has been buried in mud or sand, the sediment hardens, and the material of the shell or bone dissolves, leaving a cavity in the rock. This cavity is called a mold. The outside of a shell leaves an imprint, called an external mold (Plate 159), while the imprint of the inside of the shell is called an internal mold (Plate 149).

Steinkerns: If the interior of a shell fills with mud that hardens and becomes free of the surrounding matrix, the resulting internal mold is called a steinkern (Plate 255). Steinkerns from snails are particularly common, but bivalves and brachiopods may also be preserved in this way. Since a steinkern is formed in the interior of a shell, it does not show the outer form or surface features.

Casts: If the cavity left by the whole shell has filled in with a mineral such as calcite, silica, or pyrite, the result is called a cast—or a natural cast, to distinguish it from an artificial cast made by someone studying the fossil.

Petrifactions: Often the original material is replaced by minerals, which gradually seep into the fossil from the surrounding matrix. When this happens, the resulting fossil is called a petrifaction. Wood (Plate 487) and bone (Plate 484) are often preserved in this way.

Films: Leaves are frequently preserved as carbon films on rock surfaces, left after the other substances that made up the original leaf have been dissolved away. Only the more stable carbon remains, usually as a black or brown film that records the exact shape and venation of the leaf (Plate 496). This process is known as distillation. Occasionally animal fossils are also preserved as films, but the material that remains may not be carbon, even when it is black. Usually a leaf or animal preserved by distillation leaves an impression in the rock as well. The carbon film may be washed away before it is collected, so that all that remains is this impression (Plate 503).

Trace Fossils: Sometimes a fossil is not part of the animal itself, but a trace of the animal's activities. Such traces may record the movement of the animal as it travelled from one place to another, as in the case of dinosaur tracks (Plates 472–474). The trace may be of feeding activity (Plates 464, 465), or the burrow once occupied by an animal (Plates 470, 471). Another kind of trace fossil is animal feces, called coprolites (Plate 483). Besides leaving a record of the presence of the animal, a coprolite may contain the partially digested remains of its food, giving us a glimpse of the animal's ecology.

Life and Geologic Time: During the immense span of time that has elapsed since life first evolved on Earth at least 3.4 billion years ago, environmental conditions have changed and different groups of plants and animals have appeared, flourished, and disappeared. Geologic time is divided into basic units known as eons, eras, periods, and epochs, each one characterized by different environmental conditions and by specific kinds of life that flourished. The story of these geologic ages, their changing conditions, and the rise and fall of different groups of animals and plants, is told in the section entitled "Life and Geologic Time" (pp. 23–85).

Fossil-bearing Rocks: Nearly all fossils are found in rock formed from sand, silt, or other fine sediments that have settled to the bottom of a body of water and have gradually hardened into stone. There are several kinds of such sedimentary rock, each one formed in a different way or made up of particles of different sizes. By knowing how to identify the various kinds of sedimentary rock, a fossil-collector can tell where he is likely to find fossils; and when he has found one, he can understand something about its environment ages ago. The different kinds of sedimentary rock are described and discussed in the section entitled "Fossil-bearing Rocks" (pp. 87–96), and are illustrated in a section of the color plates.

Collecting and Preserving Fossils: Valuable information on where to find fossils, how to remove them from the rock, and how to prepare and preserve them, as well as a discussion of the simple equipment useful in collecting fossils, are given in the section entitled "How to Collect and Preserve Fossils" (pp. 779–788).

Classifying Fossils: Fossils are classified in the same way as are living things. Biologists divide both plants and animals into major groups called phyla. Whether a phylum contains thousands of species—as does the phylum Arthropoda, which includes the insects, spiders, crustaceans, and the ancient trilobites —or only a handful of species, all of its members share certain basic characteristics and differ strikingly from the members of other phyla. Phyla may be divided into classes, smaller categories that differ from one another but possess the fundamental features of the phylum. The extinct trilobites, for example, are placed in the class Trilobita, readily distinguishable from the insects, crustaceans, and other classes of the phylum Arthropoda. Classes are further divided into orders; one of the major orders of trilobites is the order Phacopida. Orders, in turn, are divided into families; the order Phacopida, for example, contains the family Phacopidae.

For very large and complex phyla, these basic categories may still contain many species, making it convenient to recognize additional categories. Classes can be grouped into superclasses and divided into subclasses; orders may be combined into superorders and split into suborders; and families can be arranged in superfamilies and broken up into subfamilies.

Within a family, or subfamily, closely related species are placed together in the same genus (plural genera), and it is the name of the genus, always italicized, that forms the first part of the scientific name of a species. One of the important genera in the trilobite family Phacopidae is the genus *Phacops,* which includes perhaps the best-known of all trilobite species, *Phacops rana.* This trilobite was given the species name *rana* (Latin for "frog") because its bulging eyes give it a froglike appearance.

Biologists usually define a species as an interbreeding population of plants or animals that cannot interbreed with other such populations. While this definition works well for living organisms, it is difficult to apply to fossils, which are not alive and therefore give us no clue as to whether or not they were able to interbreed with other similar creatures. So paleontologists prefer to define a species as a group of very similar fossils that show no more variation among themselves than is found within a typical living species.

Since there are many thousands of common fossil species, but only about 1000 common fossil genera in North America, this field guide concentrates on identifying fossils to the level of the genus. A collector using this book has a good chance of finding the correct generic name for a fossil. If the guide had tried to describe fossil species, the odds of finding and identifying the correct one would have been exceedingly small.

How to Identify Fossils: Because fossils are diverse, numerous, and often fragmentary, identifying them may seem difficult at first. For that reason this book employs a new approach—one that uses color photographs of fossils and arranges them by readily distinguishable categories of shape. Similar fossils are pictured side by side, enabling the reader to note their differences and thus to identify most specimens quickly. The photographs emphasize the features most readily observed in the field, those needed to identify a fossil to genus, and at the same time allow the reader to enjoy the fossils' natural beauty.

When you find a specimen, turn to the map section and consult the detailed map of the area in which you are collecting, to determine the age of the rock in which you found your specimen. Then, after locating the photograph of the fossil most closely resembling your specimen, check the text description of the genus to confirm that your fossil comes from rock of the same age. Then compare the features described in the text with those of your fossil. Fragments are worth collecting too, because they often show enough features to enable you to determine what an intact specimen looked like, and they also furnish clues about ecological conditions that prevailed.

Organization of the Color Plates: The color plates are arranged in three major sections: a series of geologic maps, a section of photographs of fossil-bearing rocks and outcroppings, and the photographs of the fossils themselves. These sections complement one another in helping to identify a fossil and in providing a glimpse of the ancient environment in which the fossil lived.

The Maps: This section contains 15 detailed geologic maps, preceded by an overall map of North America. These maps show the distribution of fossil-bearing rock deposits of the various geological ages. Before searching for fossils in an area, consult these maps to determine the age of the rock strata there. Knowing the age of a deposit will enable you to anticipate which fossils you are likely to find in it, and to narrow an identification to just those fossils that occur in rocks of that age. Once you have identified a fossil and know the age of the rock that contained it, the Geologic Time Chart and the text section on "Life and Geologic Time" will tell you what other plants and animals flourished millions of years ago, when your fossil was alive.

Fossil-bearing Rocks and Outcroppings: These photographs are a field guide to the kinds of fossil-bearing rocks, one that will enable you to identify a rock you are holding in your hand, as well as rocks seen from a distance, as they appear as part of the landscape. Since each type of fossiliferous rock was formed under different conditions, the kind of rock in which a fossil is found tells us much about the habitat in which the ancient animal lived. Each photograph is keyed by number to an entry in the text section on "Fossil-bearing Rocks," where more information on the origin and formation of each rock type may be found.

The Fossils: The photographs of the fossils are arranged according to the shapes of specimens a collector will find in the field. For example, the section called "Snail-shaped Fossils" contains, in addition to the fossils of snails, those of certain bivalves, cephalopods, and annelid tubes whose shells coil in a snail-like fashion. Similarly, the section called "Disc-shaped Fossils" includes, among others, the rounded forms and starlike surface patterns of sponges, corals, and sand dollars. Within each section, the photographs have been arranged so that similar fossils are close together, allowing the reader to determine at a glance the differences between them. The color plates of the fossils are in the following order:

Scallop-shaped Fossils
Clam-shaped Fossils
Snail-shaped Fossils
Gastropods
Arthropods
Cystoids, Blastoids, and Crinoids
Cone-shaped Fossils
Disc-shaped Fossils
Sponges, Corals, and Bryozoans
Trace Fossils and Graptolites
Vertebrates
Plants

Thumb Tab
Guide:
The grouping of the color plates is explained in a table preceding them. A silhouette of a typical fossil in each group appears on the left. Silhouettes of fossils within that group are shown on the right. For example, the silhouette of a crinoid represents the group Cystoids, Blastoids, and Crinoids. This representative silhouette also appears on a thumb tab at the left edge of each double page of color plates devoted to that group of fossils.

Captions:
The caption under each fossil photograph gives the plate number, genus name, and maximum measurement in inches of the fossil depicted, and the page number of its description in the text. The measurement in inches indicates the largest dimension of the animal depicted. As a cross reference, the color-plate number is repeated at the beginning of each text description.

Organization of
the Text:
The text is divided into 14 main sections, 11 containing genus entries with each animal described in detail, and 3 consisting of more general descriptions of plants, insects, and vertebrates. Each of the first 11 sections

begins with a description of the group: its form, geologic history, and preservation. Phyla are broken down into classes and, where necessary, into orders and other categories, and the genera are arranged within these categories. In the sections on plants and vertebrates, only the most important divisions are discussed. Trace fossils are grouped according to the activity of the animal that formed the trace.

The classification presented in each phylum is generally the one used in the *Treatise on Invertebrate Paleontology,* published by the University of Kansas Press. For a few groups we have used a sequence different from that in the *Treatise on Invertebrate Paleontology,* on the advice of a specialist.

Scientific Name: The scientific name of the genus opens the text account. Since almost no fossils have vernacular English names, none is used in the heading. If there is a vernacular name, it is given in the "Comments" section.

Description: Each description begins with measurements of the specimen illustrated in the color plate. When more than one color plate accompanies an entry, measurements are given for the specimen in each color plate, and are distinguished accordingly. If more than one specimen appears in a color plate, the measurements refer to the largest specimen shown. These measurements may not be of whole specimens, since some fossils are fragments. Bear in mind that these measurements belong to a single specimen of one species, and a genus may contain hundreds of species with a wide variation in size. Measurements of fossils illustrated by marginal drawings only are taken from the *Treatise on Invertebrate Paleontology.*

The size is followed by a description of the genus. This is generally not an exact description of the specimen illustrated, because that may be only one species among many in a genus. Rather, the description indicates the characteristics shared by all species in the genus. The descriptions are often supplemented by marginal drawings, especially when the specimen in the plate is fragmentary. These drawings are often of an entire fossil, and are based on many specimens. They may be enlargements of part of the fossil, or show a cut section of the fossil.

Age: Here the geologic periods during which a genus lived is given. We have followed the practice of the *Treatise on Invertebrate Paleontology* by subdividing these ages according to the terms "Lower," "Middle," and "Upper," which refer to the relative positions of rock strata, rather than using the terms "Early," "Middle," and "Late," which have to do with time.

Distribution: This section indicates where in North America a genus has been found. When the distribution is given simply as "Widespread in North America," the fossil is apt to occur wherever sedimentary rocks of the proper age appear at the surface. When the distribution is given more precisely, it is always possible that the genus also occurs in states not listed, since not all occurrences have been reported. The distribution is given only for North America, but most genera occur on other continents as well.

Comments: The Comments include notes on the habits and ecology of the animal before it was fossilized and any living relatives. There may be a discussion of the photographed specimen and how it was preserved. Other similar genera may also be described. Where necessary, former names of a genus are given. In recent decades, the names of many fossils have been changed as paleontologists shifted species from one genus to another because of new ideas about relationships.

LIFE AND GEOLOGIC TIME

Since its primitive beginnings over 3.4 billion years ago, life on this planet has evolved from a few simple, 1-celled organisms to the great array of complex animals and plants we see around us today. Fossils tell us the story of this long and unfolding evolution of living things, as well as of changes in the earth's environment that have influenced this evolution. The purpose of this chapter is to survey the past history of life on earth. The chapter begins with an introduction to some basic concepts, such as time itself, the way in which fossil-bearing rocks are named and classified historically, the profound effects that environmental changes have had on living things, the mass extinctions that mark the boundaries between geologic periods, and some of the theories that paleontologists and geologists have suggested in piecing together this fascinating story.

Geologic Time Chart: The Geologic Time Chart provides a summary of the geologic ages, and the history of life in graphic form. The geologic periods are arranged in this chart just as the rocks representing them are positioned in the earth, with the oldest periods at the bottom and the most recent periods at the top.

Time Period	Succession of Life
Cenozoic Era	Angiosperms, gastropods, bivalves, teleosts, mammals abundant.
Quaternary Period Recent Epoch Present to 11 thousand years ago	
	Mass extinction of large terrestrial mammals.
Pleistocene Epoch 11 thousand to 1.8 million years ago	First *Homo sapiens*.
Tertiary Period Pliocene Epoch 1.8 million to 5 million years ago	First australopithecines.
Miocene Epoch 5 million to 26 million years ago	Largest sharks and whales. First hominids.
Oligocene Epoch 26 million to 37.5 million years ago	First grasses, anthropoids.
Eocene Epoch 37.5 million to 54 million years ago	First marine and large terrestrial mammals.
Paleocene Epoch 54 million to 65 million years ago	Mammals diversify.

Time Period	Succession of Life
Mesozoic Era	Cycads, conifers, gastropods, bivalves, ammonoids, reptiles abundant.
Cretaceous Period 65 million to 136 million years ago	Mass extinction of ammonoids, belemnoids, flying reptiles, dinosaurs, marine reptiles. Angiosperms, scleractinian corals, gastropods, bivalves, ammonoids, belemnoids, bryozoans, teleosts, dinosaurs abundant. First angiosperms, primates.
Jurassic Period 136 million to 190 million years ago	Cycads, conifers, ammonoids, belemnoids, dinosaurs abundant. First frogs, salamanders, teleosts, crocodiles, flying reptiles, birds.
Triassic Period 190 million to 225 million years ago	Mass extinction: tabulate corals extinct; conodonts disappear; ammonoids, amphibians, reptiles decimated. Gastropods, bivalves, ammonoids abundant. First scleractinian corals, turtles, lizards, dinosaurs, marine reptiles, mammals.

Time Period	Succession of Life
Paleozoic Era	Spore-bearing plants, sponges, corals, trilobites, brachiopods, crinoids, fishes abundant.
Permian Period 225 million to 280 million years ago	Mass extinction: rugose corals, trilobites, eurypterids, blastoids eliminated; ammonoids, bryozoans, brachiopods, crinoids, amphibians, reptiles decimated. Foraminiferans, ammonoids, insects, bryozoans, productid brachiopods, reptiles abundant. First cycads, mammal-like reptiles.
Pennsylvanian Period 280 million to 325 million years ago	Trilobites decline. Gastropods, bivalves, insects, productid brachiopods, amphibians abundant; coal swamps flourish. First conifers, insects, reptiles.
Mississippian Period 325 million to 345 million years ago	Rugose corals, trilobites decline. Foraminiferans, bryozoans, crinoids, blastoids abundant. First seed ferns, belemnoids.
Devonian Period 345 million to 395 million years ago	Mass extinction: corals, ammonoids, trilobites, fishes decimated; graptolites decline. Land plants, sponges, corals, brachiopods, sharks, bony fishes abundant. First ammonoids, arachnids, sharks, bony fishes, amphibians.
Silurian Period 395 million to 430 million years ago	Rugose and tabulate corals, eurypterids, bryozoans, crinoids, graptolites abundant. First land plants, blastoids, jawed fishes.

Time Period	Succession of Life
Ordovician Period 430 million to 500 million years ago	Mass extinction: trilobites, brachiopods decimated. Gastropods, cephalopods, trilobites, brachiopods, graptolites abundant. First rugose and tabulate corals, eurypterids, bryozoans, crinoids, starfishes, vertebrates.
Cambrian Period 500 million to 570 million years ago	Mass extinction: sponges, trilobites decimated. Stromatolites abundant. First radiolarians, stromatoporoids, gastropods, bivalves, cephalopods, trilobites, ostracodes, brachiopods, cystoids, graptolites, conodonts.
Precambrian Eon	
.57 billion to .70 billion years ago	Sponges, jellyfishlike and sea-penlike cnidarians, annelids, primitive arthropods diversify.
.70 billion to 1 billion years ago	First multicellular plants and animals; stromatolites, sponges.
1 billion to 1.4 billion years ago	First advanced cells.
1.4 billion to 2 billion years ago	Primitive cells diversify.
2 billion to 3 billion years ago	First stromatolites.
3 billion to 3.4 billion years ago	First primitive cells.
3.4 billion to 4.6 billion years ago	Solar system formed.

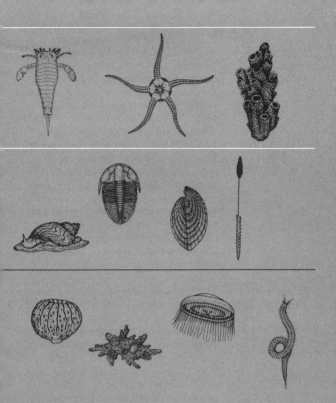

Time: Time is one of the most important concepts in both geology and paleontology. But until the beginning of this century, scientists could only determine the relative ages of rocks and fossils, and not their absolute ages in years. The scale of relative time was based on the study of fossil-bearing rocks. These layered rocks seemed to form natural units, which were separated by erosional surfaces, or distinguished by different kinds of fossils, or both. The relative ages of the layers were judged by their positions where they were found undisturbed. Older layers were assumed to have been laid down first, and to lie beneath younger layers.

Measurement of Absolute Time: The first successful attempts to measure absolute geologic time—the ages of rocks and fossils in years—came after the discovery of radioactive decay in 1896. Scientists learned that many elements "decay" by losing particles from their atomic nuclei, giving off electromagnetic radiation in the process. As particles are lost, new isotopes or different elements are formed. Geologists quickly realized that radioactive decay occurs in rock crystals. Crystals have a constant and predictable proportion of different elements in them when they form, and it is possible to measure the amount of a new isotope or element that has accumulated since the formation of a crystal. The rate of radioactive decay of the elements is also known, and so the absolute age of a rock crystal can be determined. Uranium, for example, decays to lead at a known rate and in a series of stages; the relative proportions of uranium and its decay products in a crystal thus reveal the age, in absolute time, of the crystal. In the last few decades, many measurements have been made on rocks from throughout geologic time, and fairly accurate dates have been calculated for all the boundaries between the natural rock units whose ages were previously known only in relative terms, although in each case there is still a possible error of 2–10 million years.

Eons and Eras: All of geologic time is divided into 2 eons: the Precambrian Eon, which began with the formation of the Earth and ended when fossils became abundant in rocks about 570 million years ago, and the Phanerozoic Eon, which began with the close of the Precambrian and has lasted until now. Although it is far longer than the Phanerozoic, the Precambrian is not subdivided in this book. The Phanerozoic Eon, however, is divided into 3 eras, the Paleozoic, Mesozoic, and Cenozoic.

Periods: Each era is divided into periods: the Paleozoic Era is broken into the Cambrian, Ordovician, Silurian, Devonian, Mississippian, Pennsylvanian, and Permian periods; the Mesozoic into the Triassic, Jurassic, and Cretaceous periods; and the Cenozoic into the Tertiary and Quaternary periods.

Epochs: In turn, all of these periods are divided into epochs. The periods of the Paleozoic and Mesozoic are divided into "Early," "Middle," and "Late" (or simply into "Early" and "Late") epochs, while the epochs of the Cenozoic Era are given distinctive names. The 5 epochs of the Tertiary Period are called the Paleocene, Eocene, Oligocene, Miocene, and Pliocene, while the 2 epochs of the Quaternary are called the Pleistocene and the Recent, or Holocene.

Rock-Unit Names: Fossil-bearing rocks are given the same names as the geologic periods they represent. All rocks of the Cambrian

Period, for example, are called the Cambrian System, while those of the Early, Middle, and Late Cambrian epochs make up the Lower, Middle, and Upper Cambrian series, respectively.

Plate Tectonics: The theory of plate tectonics is another concept that is important in understanding the history of life. According to this theory, the crust of the Earth, which is 8–280 miles (13–450 km) thick, consists of about 10 large, rigid plates that move over the surface of the earth. The continents are carried on the backs of these plates. The plates move at a rate of 1 to 6 cm per year—either drifting apart, coming together, or sliding past one another. When plates drift apart, molten rock from below the crust flows up to fill the space between them, producing a ridge of new crust. These ridges are present only in ocean basins and are called mid-ocean ridges.

When 2 plates come together, one of them slides beneath the other, and a trench is formed on the ocean bottom where this descent, or subduction, occurs. The plate that slides down into the Earth's mantle melts, along with its load of sediment. Some of this material rises again as magma—molten rock—and erupts in volcanoes at the surface. Along the western coast of South America, where the Eastern Pacific plate is sliding under the American plate, there are high volcanoes just inland, and a deep trench just off shore.

If both of the plates moving toward each other carry a continent, the continents will eventually collide. Then the sediment that has accumulated near the margins is squeezed, folded, and pushed up into a mountain range.

Two plates may also slide past each other. This causes faulting, earthquakes, and volcanism, as along the West Coast of North America where the Eastern Pacific plate is moving northward past the American plate. The relative positions of the continents have been changing continually since well back in the Precambrian. When continents collide, they may stay together, forming a new, larger continent, or plate movements may reverse, reopening the suture or making a new rift somewhere else.

Plate Tectonics and Life: Because of plate tectonics, many phenomena concerning the history of life can now be readily explained. For example, paleontologists had been unable to account for the movements of organisms from 1 continent to another without postulating "land bridges" between them. A knowledge of the movements of continents has made such hypothetical land bridges unnecessary. Another problem that has plagued paleontologists has been to explain how rocks of the same age and located very close to one another can have very different fossils. Now that previously separate continents are known to have drifted together, such questions can be answered.

Changes in Sea Level: Estimates of sea level relative to the continents have greatly improved during the last few years. We now know that during the entire Phanerozoic Eon there have been long-term trends of gradually rising or falling sea level, as well as very sudden and sometimes severe drops in sea level. We tend to think of sea level as a constant, but it has been clear to geologists ever since they started studying rocks that the oceans have varied enormously in their level relative to the continents. It is also clear that sea level is a very important factor to both marine and terrestrial life. Most marine species live in the shallow seas that border the continents. These shallow seas lie over the continental shelves, the underwater extensions of the continents. At the edge of these shelves, the water becomes much deeper very quickly until the abyssal depths of the ocean bottom are reached. Three factors determine how much of the continents are covered by seas at any particular time: (1) the volume of the ocean basins; (2) the volume of water in the ocean basins; (3) the height of the continents. One of the most important factors influencing the volume of the ocean basins is the height of the mid-ocean ridges. When ridges are high, the oceans hold less water and parts of the continents are flooded. When they are low, water drains off the continents. The factor controlling the height of the ridges is the rate at which the crustal plates are separating at the ridges. The faster 2 plates are traveling, the higher the ridge. There is good evidence that the rate of spreading has varied considerably through time. But this factor can only very gradually raise and lower sea level, and only explain long-term changes in sea level.

Changes in the volume of sea water are affected, as far as we know, only by continental glaciation. Glaciation withdraws huge amounts of water from the oceans. In the Pleistocene Epoch, sea level varied by about 500′ (150 m) between maximum glaciation and minimum glaciation over a period of only 10,000 years. If sea level fell 500′ now, the edges of the continental shelves would be exposed. Changes in the height of the continents have both short-term and long-term aspects. In the short term, mountain ranges rise and are eroded down to sea level again, all in a period of a few million years. In the long term, the processes of plate tectonics add light crustal material to continents, and continents have gradually become higher through time.

Mass Extinctions of Life: Six times during the Phanerozoic Eon between 25% and 50% of all fossil families have become extinct in a geologically short time. Such mass extinctions have occurred at the close of the Cambrian, Ordovician, Devonian, Permian, Triassic, and Cretaceous periods. Since each family usually contains many species, and only 1 species need survive in order to keep a family from becoming extinct, the rate of extinction at the species level is considerably higher. It has been estimated that a 50% reduction in the number of families is equivalent to a reduction of 90% or more in the number of species. The causes of these mass extinctions, whose occurrences mark major boundaries between geologic periods, have been the subject of much debate.

Precambrian Eon

The Precambrian Eon is the largest
division of Earth history. It begins with
the formation of the Earth and lasts
until fossils become abundant at the
beginning of the Cambrian Period. It
represents about 88% of Earth time, or
4 billion years. The Phanerozoic—
which includes all time after the
Precambrian—is less than 600 million
years long, but contains most of what
we know of Earth history.

Oldest Rocks: The oldest Earth rocks that have been
dated by radioactive decay are only
about 3.8 billion years old, but
meteorites yield radiometric dates that
fall between 4.5 and 4.7 billion years.
Astronomers agree that the solar system
formed all at once, so the Earth and the
meteorites should be about the same
age. The oldest rock samples from the
moon found so far are 4.66 billion years
old. A commonly accepted estimate of
the age of the Earth is 4.6 billion
years.

First Life: The first unequivocal evidence of life is
found in rocks 3.4 billion years old.
These rocks are chert, or silicon
dioxide, and come from a formation
called Fig Tree in southern Africa. The
fossils can be seen when the chert is
sliced into very thin, transparent
sheets. The fossils are tiny, only a few
microns in diameter. Some are
spherical, some cylindrical; all look like
varieties of living bacteria. Bacteria are
examples of the simple cell type called
a procaryote, which has no nucleus or
other organelles. Instead, the genetic
material is scattered throughout the
cell. Among living organisms, only
bacteria and blue-green algae have this
kind of cell organization. All other
plants and animals have a more
advanced cell type, called eucaryote.
The bacteria-like procaryotes may have

fed like most living bacteria, by absorbing nutrients through their cell walls, but the source of these nutrients cannot have been the same then as today—animal and plant tissues. In the early Precambrian, nutrients may have formed spontaneously from compounds in the atmosphere and ocean such as methane, ammonia, and carbon dioxide. Without oxygen in the atmosphere, organic compounds would not have been oxidized and would probably have begun to accumulate. Sedimentary rocks formed at this time contain unoxidized minerals, so the absence of atmospheric oxygen seems to be a reasonable hypothesis.

Fig Tree Procaryotes: The first organisms may have subsisted on nutrients dissolved in water, but eventually the supply would have diminished; organisms had to begin to make their own food. There is some evidence that the Fig Tree procaryotes were capable of photosynthesis—that is, that they could combine carbon dioxide and water, using sunlight as an energy source, to make simple sugars as modern plants do. These sugars, in turn, can be burned to provide the energy to make all the other chemicals a cell needs. In older rocks, just below the Fig Tree Formation, the proportion of carbon isotopes in the rock changes. This change could be explained if some of the carbon was generated by photosynthesis.

Stromatolites: The first certain photosynthesizers appeared about 3 billion years ago. These were blue-green algae, whose activities built stromatolites (Plate 488). Stromatolites are mounds of layered rock, usually a few cm to 1 m across, that form in shallow water. A stromatolite begins to grow when a layer of filaments of blue-green algae binds sediment together to form a feltlike mat. The blue-green algae also deposit calcium carbonate over the mat. The filaments, growing up through the mat, form a new layer that, in turn, binds more sediment. Stromatolites are present in rocks throughout the rest of the Precambrian and on into the Cambrian. They still occur today in very limited areas.

First Eucaryotes: The first known fossil eucaryote cells occur in rocks in the Beck Spring area of California, and are about 1.4 billion years old. Other floras with eucaryotic cells similar to those of modern single-celled green algae are fairly common around this time. One of the first to be discovered, and still the most varied, is the Bitter Springs flora of Australia, about 900 million years old. The presence of these floras probably means that oxygen, given off by photosynthesis, had begun to accumulate in the atmosphere. Such atmospheric accumulation could only occur after sufficient free oxygen had been released to oxygenate all the unoxidized materials on the Earth's surface. That this process was well advanced 2 billion years ago is shown by the formation then of red beds, that is, sedimentary rocks stained red by oxidized iron.

First Multicellular Organisms: In rocks of almost the same age as those containing the first eucaryote organisms, the earliest multicellular plants have been discovered. Worm burrows in Nevada, in rocks that have been dated at 1 billion years, represent other evidence for multicellular life at this time.

Diverse Multicellular Animals: Between the initial multicellular life of around 1 billion years ago and a diversified multicellular fauna, about 300 million years elapsed. In the Ediacran Hills of southern Australia, in rocks about 680 million years old, occurs one of the most interesting of all fossil faunas. This is a group of fossils, all made by soft-bodied animals, that contains over 34 species from 4 or more phyla. There are jellyfishes, worms, sea pens, sponges, coral-like animals, primitive arthropods, and various tracks and trails.

The fossils are molds and casts that formed when these animals died on a shallow sea bottom and were covered over by sand. The bodies decayed, leaving cavities that filled with sediment. This sandy sediment ultimately hardened to form sandstone, but the impression left in the sea floor by each animal remained as a plane of weakness; when the sandstone split, impressions were preserved on the lower layer and casts on the upper layer.

Many of these animals were several inches long. It is likely that such large animals needed an atmosphere rich in oxygen, perhaps as much as 10% of the amount now present.

Elements of the Ediacran fauna are found in several other parts of the world, including North America. There is a diverse fauna of this age with many similar forms in eastern Newfoundland.

Cambrian Period

The Phanerozoic Eon begins with the Paleozoic Era and its first subdivision, the Cambrian Period. The boundary between the Precambrian and the Cambrian has been dated at about 570 million years ago. The Cambrian lasted 70 million years and is one of the longest of all Phanerozoic periods. By its end, almost all the major groups of living things had appeared on Earth.

Position of Continents: By and large, the continents of the Cambrian were widely separated by deep oceans and were smaller than they are today. Most of the present major coastal mountain belts were still to be built, since these would result either from the subduction of a plate beneath a continent or from collisions between continents later in the Phanerozoic. These had not yet occurred. For example, the eastern coast of North America lacked the Appalachian Mountains, which did not yet exist; they did not begin to form until the European plate collided with North America in the Ordovician Period. The southeastern United States and Mexico were located near the South Pole and were attached to South America and Africa. The equator passed through northern Canada and the western United States. During the Cambrian, the oceans between continents became smaller as the continents moved closer together.

The Cambrian Boundary: Originally, the base of the Cambrian System was defined as those rocks that first bear fossils. When this definition was proposed, it was thought that the Precambrian was totally devoid of fossils. Now we know that this is not true. Not only are microscopic fossils common in some areas, but large fossils like jellyfish and worms occur toward the end of the Precambrian. Some geologists recommend moving the Cambrian boundary back to the time of the Ediacran fauna, about 680 million years ago, but the majority prefer to keep the boundary where it is and to redefine it as the first appearance of numerous fossil shells. These first shelly fossils appear suddenly and in great abundance, making it possible for the first time to correlate rocks all over the Earth on the basis of their fossil content.

Trilobites: In the Cambrian, the most common and important guide fossils for correlating rocks of different ages are the trilobites. These segmented, 3-lobed animals had calcified skeletons and are well preserved in a variety of rock types. Each of the 3 divisions of the Cambrian—the Early, Middle, and Late—has its characteristic trilobite genera. Within these 3 main divisions, smaller time units, called zones, can be distinguished by the ranges of particular trilobite species.

Other Invertebrates: The trilobites, while the most abundant and most interesting of Cambrian fossils, were by no means alone on Earth. Even during the Early Cambrian, almost all the main groups of animal life were present. Sponge spicules are common, indicating that these simple animals had evolved. The archaeocyathids, an unusual group of organisms that may have been related to sponges, are common in some areas, such as Nevada, and have such distinctive skeletons that they are very good indicators of Lower and Middle Cambrian rocks. Stromatolites were still numerous, building large reefs in many places. Hyolithids (Plate 347) left their small tubes in great abundance in many formations. Brachiopods were well represented, first by the small, chitinous inarticulates. Among the mollusks, the first to

appear were monoplacophorans like
Scenella (Plate 236), to be followed
almost immediately by small
gastropods that are rare as fossils.
Toward the end of the Cambrian, some
small cephalopods appeared, which may
have been the first predators. Bivalves
are known from the Early Cambrian,
but they are very scarce as fossils until
the Ordovician.

Echinoderms were also present but are
very rare as fossils; the major classes did
not appear until the Ordovician.
Absent from the Cambrian were
bryozoans and corals. Cnidarians
without skeletons were important
members of the late Precambrian
Ediacran fauna, so these must have
been present in the Cambrian. In many
Lower Cambrian rocks, the contact
between the Precambrian and the
Cambrian is marked by a coarse sand
full of the burrows of the trace fossil
Skolithos (Plate 470). The wormlike
animals that inhabited these tubes
sought protection from the turbulent
environment of the open shoreline, like
the many worms that build similar
burrows on modern beaches.

Terrestrial Life: Almost certainly the land was barren
during the Cambrian, as it had been
since the origin of the Earth. It seems
probable that as soon as atmospheric
oxygen levels were high enough for the
upper atmosphere to develop a layer of
ozone that could shield the surface of
the Earth from deadly ultraviolet
radiation, life moved out onto land.
But this did not happen until late in
the following period, the Ordovician.

Mass Extinction: The Cambrian ended with a mass
extinction of almost 75% of trilobite
families, 50% of all sponge families,
and many brachiopods and gastropods.
The cause of this extinction is
unknown.

Ordovician Period

The Ordovician Period lasted about 70 million years. At its beginning, most life was still primitive in comparison to modern forms. By the end of the Ordovician, all the common invertebrate fossil groups were present, and even the vertebrates had appeared.

Cambrian-Ordovician Boundary: At the end of the Cambrian, sea level fell and much erosion occurred in North America. Withdrawal of seas from a continent must have disrupted life in the shallow marine waters and thus caused extinctions. When the seas returned, new organisms evolved to take over the newly created niches. When the sea level rose again, it flooded most of North America.

Plate Movements and Mountain-Building: The continents were moving together during the Ordovician, and in the northeastern part of the North American continent, mountain-building began. This appears to have resulted from the collision of North America with Greenland and Europe. This mountain-building episode, which was the first stage in the formation of the Appalachian Mountains, lasted throughout the Ordovician, leaving a profound mark on the structure and sediments of the whole eastern half of the continent. While the eastern half of the United States experienced mountain-building—with its concomitant heavy erosion and the deposit of detrital sediments in deep basins—the rest of the continent was quiet. The dominant sediment was lime mud, which later formed limestone and dolomite.

Invertebrate Life: Most Ordovician rocks are rich in fossils. Brachiopods are the dominant forms and are particularly abundant and well preserved in Middle Ordovician rocks of the Cincinnati

region of Ohio, Indiana, and
Kentucky. Almost none of the
Cambrian trilobite genera survived the
Cambrian, and the Ordovician was
populated with new forms.

Bryozoans make their first appearance
in the Lower Ordovician. Some deposits
from the Middle Ordovician are
composed almost entirely of the
skeletons of bryozoans.

Other newcomers to the Ordovician
world were the still-living classes of
echinoderms—starfishes, brittle stars,
echinoids, and crinoids. Crinoids were
still small in the Ordovician, but they
were abundant. Starfishes also thrived,
probably as predators on the gastropods
and bivalves. While gastropods had
been small and rare in the Cambrian,
in the Ordovician they diversified
greatly and became quite large. The
first bivalves appeared in the Cambrian,
but they were not a conspicuous part of
the fauna. In the Ordovician, bivalves
became larger, more numerous, and
more varied. Bivalves were less diverse
and abundant than brachiopods, but
they were able to move about. This
seems to have given them an advantage
over the sedentary brachiopods in soft-
bottomed areas near shore.

The other major class of mollusks, the
cephalopods, showed a similar increase
in size and diversity. All other animals
in the Ordovician, with the exception
of starfishes, appear to have been either
filter-feeders, grazers, or detritus-
feeders.

Corals (phylum Cnidaria) appeared for
the first time in the Early Ordovician.
The tabulates came first, and in the
Middle Ordovician, rugose corals
appeared.

The tiny shelled arthropods called
ostracodes were an important group in
the Ordovician, although they were
very small, mostly microscopic.

Graptolites—strange, weedlike animals
probably belonging in the phylum

Hemichordata—had appeared in the Cambrian but did not become abundant in North America until the Ordovician. They often appear in black shale where little else is preserved, and are enormously useful for correlating rocks on different continents.

First Vertebrates: Of all the organisms swimming in the warm, shallow seas of the Ordovician, the most fascinating were a group of small, strange-looking creatures with heavy shields of bone around their flattened heads and with little, fishlike tails. These intriguing first vertebrates are found along the front range of the Rocky Mountains in Middle Ordovician rocks. The fossils are little more than small flakes of bone, and in themselves would not tell us much about the organisms that left them; but the flakes are true bone, made of calcium phosphate. They belong to the class Agnatha, which includes the living lampreys and hagfishes. The mouth was a small hole, without jaws, and flanked by rows of small gill slits. These primitive vertebrates, called ostracoderms, probably fed by picking up mud from the bottom and straining out bits of food. Surviving until the Devonian, these were the ancestors of all other vertebrates.

Life on Land: In the meantime, some activity on land might have begun. A few plant fossils have been found—bits of apparently vascular tissue and some spores—that may be from terrestrial plants. But no well-preserved land plants occur until the Silurian.

Mass Extinction: The Ordovician ended with another mass extinction. About 25% of all families perished. As at the end of the Cambrian, the trilobites suffered most, losing over 50% of all families. Sponges, brachiopods, echinoids, and fishes were also decimated.

Silurian Period

The Silurian Period lasted about 35 million years. During this time, the northern continents were joined together into a supercontinent, called Laurasia, and the southern continents were united to form Gondwanaland. These 2 supercontinents were moving closer to each other. Most of North America was part of Laurasia, but Mexico and the southeastern United States were moving toward Laurasia on the Gondwanaland supercontinent. Europe had collided with the northeastern coast of North America during the Ordovician, and tectonic activity was still high. A great deal of volcanism, generating lava flows and ash, persisted, as did the erosion of highlands and the deposition of thick sediment in sinking basins.

Sea Level: There appears to have been another lowering of sea level at the Ordovician-Silurian boundary, with erosion over most of the continent; so Silurian strata are deposited over an unconformity. Most of North America was covered by warm, shallow, clear seas. Most of the rocks deposited in these seas were limestone. Some extensive, thick coral reefs were deposited in Michigan, Indiana, and Illinois. Late in the Silurian, the seas withdrew, except from basins centered in Michigan and New York. When these basins became partially landlocked, thick deposits of salt formed at their bottoms because of the high evaporation rate and little influx of fresh water. North America was moving south, and these areas were very close to the equator.

Life in the Seas: No new major groups of organisms appeared during this time. The old groups flourished, with some exceptions. Trilobites were on the decline but still locally abundant.

Graptolites were also declining, but can still be used for correlation in the black shales that were being deposited in the basins along the East Coast. Cephalopods were not as conspicuous as in the Ordovician, and eurypterids became the giants of the Silurian. Eurypterids of the genus *Pterygotus* grew to 9′ (3 m).

Vertebrates were still very rare in North America. Fragments of ostracoderms have been discovered, and some excellently preserved ostracoderms from Norway and Sweden give us a good idea of how these animals looked. Certain spines and scales in Upper Silurian rocks possibly belonged to a group of fishes called acanthodians. These were probably the first fishes to develop jaws, and they became common in the Devonian.

Life on Land: During the Silurian, evidence of life on land is sparse. From the Upper Silurian of Wales, there are some well-preserved vascular plants—small forms, called psilophytes, with a few branches but no leaves. Psilophytes were transitional between marine algae and true terrestrial plants, and probably inhabited marshes. Some millipedelike fossils have been interpreted as land animals, but may have been aquatic.

Devonian Period

The Devonian Period lasted about 50 million years, ending 345 million years ago. Life in the seas abounded, and may have been more varied than at any other time during the Paleozoic Era. Modern fishes made their first appearance, land plants flourished, invertebrates successfully colonized the land, and the first amphibians appeared.

Plate Movements: The Devonian saw a continuation of trends from the Lower Paleozoic. Europe and North America were still drifting together and the northern Appalachians were being completed; the Acadian Orogeny in the Middle Devonian was the last great movement before this part of the world became stable. The rest of North America was quiet. The equator ran through Canada, approximately from northern British Columbia to Newfoundland. North America was largely tropical or subtropical.

Marine Invertebrates: Among the marine invertebrates, there was little change from the Silurian. Corals and stromatoporoids were the principal reef builders. In western New York, glass sponges became abundant. Trilobites continued to decline, but the largest trilobites lived at this time and attained a length of 28″ (70 cm). The brachiopods reached their maximum abundance and diversity. The most numerous were the spiriferids.

Freshwater and Terrestrial Invertebrates: New environments were colonized by invertebrates at this time. The bivalves moved into fresh water, where they can still be found, and gastropods moved onto land. At about the same time or possibly in the Late Silurian, arthropods developed tracheae, or breathing passages, in their tissues, and became terrestrial. These were the millipedes and scorpions. The first colonizers must have been plant-eaters, but predators were soon to follow, such as the primitive spiders that have been found in the Early Devonian of Germany.

First Jawed Fishes: Fishes, meanwhile, became the dominant animals of the oceans. The first jawed fishes, the acanthodians, which had appeared at the very end of the Silurian, were abundant by the Devonian, and a new group, the placoderms, had appeared. This latter group contained some monsters. *Dunkleosteus* reached a length of 30' (9 m) and had an enormous mouth. There were no true teeth in the jaw, but the bones were as sharp as knives and the jaws opened so wide that *Dunkleosteus* must have been able to swallow whole anything that lived in the Devonian oceans.

The acanthodians and placoderms were to become extinct by the end of the Paleozoic, but 2 other groups of fishes appeared, both still living and very successful—Chondrichthyes and Osteichthyes.

Chondrichthyes are the sharks, rays, and skates. They have cartilaginous skeletons that are uncalcified.

The Osteichthyes are an even more important group. These are the bony fishes, and include all living fishes except the Chondrichthyes, and the lampreys and hagfishes. In the Devonian, the bony fishes split into 2 main groups, the ray-fins and the lobe-fins. These differed in the structure of the paired fins of the ventral side. The ray-fins had no muscular base to their fins, which were thin and supported by a system of bony rays. This group is dominant today and includes the teleosts—95% of living fishes.

The lobe-fins had a muscular base and a system of bones in the fins. These fish had lungs, outpocketings of the throat with a vascular lining for absorbing oxygen from the air. Lobe-fins must have lived in small, stagnant pools where there was little oxygen in the water. Their lungs made it possible for them to breathe air, and their strong fins allowed them to travel short distances across land to find new pools when the old ones dried up. The lobe-fins are divided into 2 groups: the Dipnoi, including lungfishes that survive today in the tropics, and the Crossopterygii, the direct ancestors of amphibians.

First Amphibians: The first amphibian fossils occur in the Late Devonian in Greenland. The bones in these skeletons are so little changed from the bones of a crossopterygian that there can be no doubt that amphibians are descended from this group. The early amphibian skull can be matched bone for bone with that of the crossopterygians. Adaptations can also be seen in the rest of the skeleton. The bones in the fins have changed proportion, with those near the base growing longer and heavier and those near the tips forming toes. The vertebrae have lengthened lateral spines forming ribs to help support the viscera, and the tail has become longer and more whiplike. The result is an animal that could maneuver reasonably well on land or in water—in other words, an amphibian.

Why did the crossopterygians begin to live partly on land? There are at least 3 possible answers: (1) to eat, since there must have been quite a selection of terrestrial insect life by the end of the Devonian; (2) to escape aquatic predators in the water; and (3) to seek new water holes, as the ones they inhabited dried up. But like living amphibians, these Paleozoic amphibians were still dependent on water. They had to return to it to lay their eggs, which had no covering and would have dried out on land. Not until reptiles appeared in the Pennsylvanian did vertebrates become fully terrestrial.

Early Terrestrial Plants:
Like the amphibians, Early Devonian plants were dependent on water for reproduction. These early plants were the lycopods, sphenophytes, and ferns. Their spores had to fall into water in order to germinate, producing the sexual plants that would then yield fertilized eggs from which the large asexual generation grew.

Seed Plants:
By the late Devonian, plant forms had developed that could reproduce in the absence of water—the seed plants. The oldest known seeds are from the Upper Devonian of Pennsylvania. They might have belonged to the seed ferns, a group of plants abundant in the Pennsylvanian, with foliage like that of true ferns. Seeds arose as the culmination of a process which began when some plants retained their spores on the leaves or on small stalks in closed structures where they could be kept moist. Some of the spores developed into female plants, producing eggs; other spores developed into males, producing pollen. Each pollen grain was a sperm cell enclosed in an airtight cover designed to be carried by the wind until it came to rest against the female, into which a tube then grew. The sperm passed down this tube and fertilized the egg— now a developing seed.

Mass Extinction:
The Devonian period ended with a mass extinction similar in magnitude to the one that ended the Ordovician. About 25% of all families became extinct. Ammonoids, fishes, and amphibians lost almost all their families. Corals and trilobites lost over 50% of their families, and brachiopods, bryozoans, ostracodes and crinoids suffered important but less severe losses.

Mississippian Period

In Europe, the Mississippian Period and the period immediately following it, the Pennsylvanian, are combined into a single period known as the Carboniferous. The European Carboniferous Period can be divided naturally into 3 epochs, but in North America this whole span of time can only be divided into 2 parts, separated by a withdrawal of the seas from the North American continent. American geologists regard this division as a major boundary, and accord the Mississippian and Pennsylvanian the rank of periods, rather than epochs. In some texts, the Mississippian and Pennsylvanian are termed the "Lower Carboniferous" and "Upper Carboniferous" respectively. The Mississippian Period was relatively short—lasting only 20 million years—and its fossil record is almost entirely marine. Shallow, clear, warm seas covered much of North America. This is the last time that widespread seas would deposit limestone, the dominant Mississippian rock, over this continent.

Age of Crinoids: The Mississippian is known as the Age of Crinoids, and many Mississippian limestones are rich with the remains of crinoids as well as blastoids, a related group of animals that first appeared in the Silurian. Crinoids flourished because they are filter-feeders, and most of the particles in the clear oceans would have been bits of food. They also needed warm water to produce their elaborate skeletons, since warm water can hold more dissolved calcium carbonate than can cold, making it easier to precipitate.

Fenestrate Bryozoans: The lacy fenestrate bryozoans are another group of very successful Mississippian animals. Like the crinoids and blastoids, they had much calcium in their skeletons, and removed tiny particles of food from the water with their microscopic tentacles. They could not have fed if there had been much sediment in the water. The fenestrates were so common in some areas that bedding planes are covered with their skeletons.

Foraminiferans: One other group of Mississippian marine invertebrates deserves mention. In the Ordovician, the first foraminiferans had appeared, but they were inconspicuous. In the Mississippian, however, these tiny, single-celled animals became abundant enough in places to produce whole strata composed almost entirely of their skeletons. The genus *Endothyra,* about the size of a pinhead; was so common in parts of Illinois and Missouri that the Salem Limestone is mostly composed of its globular shells. *Endothyra* lived on the bottom of shallow seas and picked up bits of food with its pseudopods in the manner of its close relatives, the amoebas.

Life on Land: There are few Mississippian fossils of terrestrial plants and animals, a situation unlike that of the next period, the Pennsylvanian. Either conditions in North America were unfavorable for land life or unsuitable for its preservation. Coal did not form during this time in North America as it did in Europe.

Pennsylvanian Period

The Pennsylvanian lasted 45 million years and was a time of mountain-building. The ocean to the south of North America finally closed, and Gondwanaland pushed Mexico and the southeastern United States against ancestral North America. All the continents came together to form the supercontinent Pangea. The southern Appalachians began to rise, as did mountains to the west, the remnants of which are the Arbuckle and Wichita Mountains. The Ancestral Rockies rose in Colorado, Wyoming, and Utah, and another chain of mountains rose along what is now the New Mexico–Arizona border. The seas withdrew from the interior, and long, narrow troughs bordered the southern mountain ranges. Volcanism and erosion caused enormous volumes of sediments to pour into these troughs.

Life in the Sea: The loss of the warm, shallow seas dramatically changed marine life. The crinoids permanently declined, and so did the fenestrate bryozoans. Mobile bivalves and gastropods flourished on the muddy bottoms where sedentary brachiopods were smothered. Spiriferid brachiopods declined, but the spiny productids became more abundant.

Coal Swamps: During most of the Pennsylvanian, large areas of North America supported lush, swampy forests, and the fossils of these forests make up almost all the important coal deposits of central and eastern North America.
The lycopods, or scale trees, were the most primitive and also the largest, reaching heights of 100' (30 m). The sphenophytes were abundant, although they only grew to 30' (9 m). Ferns grew to 50' (15 m).
Two groups of gymnosperms were also present in the coal swamps. The

cordaites had broad, straplike leaves instead of needles and bore their seeds in long clusters instead of cones. The other group was the seed ferns, which had evolved during the Devonian. These plants may have been the ancestors of the flowering plants.

Terrestrial Invertebrates: The invertebrate life of the coal swamps included the largest insects that ever lived. The giant of all was a dragonfly with a wingspan of 29" (74 cm). Cockroaches reached 4" (100 mm). There was a foot-long centipede, many kinds of spiders, scorpions, and other primitive insects. Land snails have been found in hollows of fossil tree stumps in Nova Scotia, and freshwater bivalves were common.

Terrestrial Vertebrates: Terrestrial vertebrates were also abundant and diversified. Amphibians were common, and looked much like modern salamanders. Some were giants, as shown by tracks in sandstone from near Lawrence, Kansas, that are 5" (125 mm) long, with a stride of 30" (76 cm).

First Reptiles: The first reptile fossils are found in the Lower Pennsylvanian in Nova Scotia. The distinguishing characteristic of a reptile is the amniote egg, with a strong, watertight covering, a large yolk for nourishing the embryo, and with the yolk surrounded by a water-filled membrane, the amnion. With such an egg, reptiles no longer had to return to the water to reproduce. The amniote egg freed reptiles from their dependence on standing water, and enabled them to colonize lands closed to amphibians.

Permian Period

All during the Permian, which lasted 55 million years, the continents were welded together into the supercontinent Pangea. Greenland, Europe, and Africa were pressed against North America, Africa also abutted South America, and Australia, India, and Antarctica were attached to southern Africa. There was pressure at the eastern and southern borders of North America, and the southern Appalachians continued to rise. Permian marine rocks are found only along the West Coast, in the Southwest, and in a few other areas. The western interior was hot and dry, and red beds, deposits of sandstone stained red by iron, formed. Deposits of salt formed in Kansas, Texas, and New Mexico as sea water evaporated, and sand dunes were widespread. The West Coast experienced volcanism, and deep troughs were filled with sediment.

Marine Invertebrates: For the fauna of the seas, it was a time of specialization. Huge foraminiferans, called fusulines, were so numerous in some areas that rocks were deposited that are formed mostly of their shells. Spiny brachiopods experienced a great expansion; areas such as Texas yield fossil evidence of the vast reefs formed by their intertangled spiny shells. As in the Mississippian, bryozoans of the fragile, lacy types flourished, as did crinoids and blastoids. Echinoids were locally abundant, and bivalves and gastropods were diversifying.

Plant Life: At the beginning of the Permian, plant life on land was much as it had been in the Pennsylvanian. But by the end of the period, probably as a consequence of falling sea levels and an increase in area of higher, drier land, the vegetation changed. The spore-bearers and the primitive seed plants declined, many becoming extinct by the end of the period. Ginkgos and cycads appeared for the first time. Conifers became more abundant, probably because they could colonize dry land. Conifers reproduce with cones, their pollen carried by the wind. Cycads are still living, but are rare and restricted to Australia, southern Africa, and the American tropics.

Insects: Insects continued to flourish. At Insect Hill, near Elmo, Kansas, 20 orders of insects have been described. Insects were smaller, on the average, than they had been in the Pennsylvanian, perhaps reflecting the generally more arid conditions of the Permian.

Amphibians: Amphibians have left a good fossil record from this period. Especially in the red beds of the Southwest, abundant material has been found. Amphibians clustered around water holes and streams, and were generally quite large.

Reptiles: Similarly, there is an abundant fossil record of reptiles from the Permian. The reptilian body tends to have longer limb bones, and fewer bones in the wrists and ankles, than does the amphibian body. The head is deeper, with fewer bones in the skull, and there is only 1 occipital condyle instead of 2, where the skull joins the first vertebra. Reptiles have 2 sacral vertebrae, instead of 1. These early reptiles varied in size. Some were carnivores, some herbivores.

Pelycosaurs: Among the most important and spectacular reptiles of the Permian were the pelycosaurs. They developed high, sail-like extensions of their backs, supported by spines from the vertebrae. Explanations of the function of these sails are various, but they most likely served as heat regulators—reptiles must regulate their body temperature by behavioral means, since they have no automatic internal thermoregulation as mammals and birds do. A pelycosaur could use its sail both to absorb energy from the sun's rays when the air temperature was cool, and to radiate heat from the sail if its blood became overheated.

Therapsids: Mammals are in a direct line of descent from the pelycosaurs. Late in the Permian, some pelycosaur descendants developed into the ancestors of mammals: the therapsids, or mammal-like reptiles. Therapsids were a varied group, mostly fairly small; like the pelycosaurs, they were meat-eaters. Their skulls were more mammal-like than reptilian, with fewer bones. Their teeth were differentiated rather than all alike, and separate nasal passages opened on the tops of their skulls so they could eat and breathe at the same time. Their legs became more efficient for running, with longer limbs and knees drawn inward, instead of bent outward as in amphibians and earlier reptiles. The therapsids dominated the vertebrate faunas of the Late Permian, and spread to areas not previously colonized by other land vertebrates. In fact, some of these areas were so close to the poles that it seems likely that some therapsids were warm-blooded; they may even have had fur. Therapsids became extinct in the Triassic, but not before giving rise to the mammals.

Mass Extinction: The Permian Period ended with the most severe of all mass extinctions; as many as 96% of all species were lost. Many classes, orders, and families disappeared. The effects were felt equally in the water and on land.

Extinctions in All marine invertebrate groups
the Sea: suffered. Among the cnidarians, the rugose and tabulate corals vanished.

The 2 most abundant orders of Paleozoic bryozoans, the trepostomes and cryptostomes, perished; these included the lacy fenestrate bryozoans that had flourished in the Mississippian and Permian. Almost all brachiopods became extinct, and they have bordered on complete extinction ever since. Many mollusk families disappeared; the ammonoids lost all but 1 family. Among the arthropods, the trilobites and the eurypterids became extinct. All the blastoids and all but 1 family of crinoids vanished. The echinoids, which had been abundant, came within 1 genus of total extinction. Among the invertebrates, it was the small, unspecialized animals that survived. These were usually detritus- and sediment-eaters, not large, specialized carnivores or filter-feeders.

Extinctions on Land: Life on land fared no better. Fully 75% of all amphibian families and 80% of all reptile families became extinct. In addition, many groups of land plants perished, including most of those that had composed the great forests of the Pennsylvanian.

The Cause of the Permian Mass Extinction: The most widely accepted explanation of this mass extinction is that there was a sudden, drastic drop in sea level and the area of shallow seas was reduced virtually to nothing. But there are difficulties with this theory. The most recent studies of coastal sediments indicate that at several other times during the Phanerozoic Eon the sea level has dropped as much or more than it did at the end of the Permian, but there were few or only minor extinctions associated with these other drops.

Other explanations have attempted to tie the extinctions to the positions of the continents. Pangea, the supercontinent, formed during the late Paleozoic, and this might have had profound effects on life forms. For one thing, each of Pangea's constituent continents would have carried its own species. When the continents collided, these species would have spread from one land mass to another and would have begun competing. Many would have become extinct. This explanation also presents problems. The continents did not suddenly come together at the end of the Permian. They had been in the process of uniting all during the Paleozoic. If continental positions were an important factor, then extinctions should have begun in the Cambrian and continued all through the Paleozoic, with few new extinctions in the Permian.

There is no evidence that climate deteriorated during the Permian, which is another often-heard explanation. If anything, climate improved, insofar as a general warming trend can be considered an improvement.

Extraterrestrial Events:
From time to time, geologists have proposed such extraterrestrial events as supernova explosions as reasons for mass extinctions. These explanations have been criticized on the ground that land animals and plants would suffer more during such a cataclysm than would marine life, since ultraviolet and cosmic radiation is screened out by only a few meters of water. But there is now increasing evidence that the extinctions at the end of the Cretaceous were indeed caused by a particular type of extraterrestrial event: the impact of a large meteorite on the Earth. There is as yet no direct evidence for such an occurrence at the end of the Permian, but such an explanation can no longer be dismissed.

Triassic Period

The extinctions at the end of the Permian were so severe that when the Triassic began, life forms were very different. This was evident to the first geologists in the 18th century, and they called the next 160 million years of history the Mesozoic Era, or Time of Middle Life. The Mesozoic is divided into 3 periods—the Triassic, the Jurassic, and the Cretaceous—with relatively minor extinctions between them. The first period, the Triassic, lasted for 30–35 million years.

Plate Movements: Most North American Triassic rocks are in the West. There was a seaway running down through western Canada and the western United States; mountain-building, volcanism, and the deposit of terrestrial and marine sediments were concentrated in these areas. Toward the end of the Triassic, however, tensional forces developed that were related to the breakup of Pangea and the reopening of the Atlantic Ocean. The result was that from Nova Scotia to Florida rift valleys developed where great blocks of the Earth's crust sank hundreds of feet. The valleys filled with red sandstones, siltstones, and shales, interbedded with volcanic rock. During the Triassic, the equator passed through central Mexico; much of North America lay in the northern "horse latitudes," a climatic zone about 30° north of the equator where dry high-pressure air masses tend to descend, bringing very little rainfall. Most of the world's deserts today are located in the horse latitudes.

Marine Invertebrates: At the beginning of the Triassic, there was very little marine life left. The first group to recover from the mass extinctions at the end of the Permian appears to have been the ammonoids. These coiled cephalopods had barely survived the Permian, but in the Triassic they diversified rapidly to 400 genera.

Eventually the other surviving invertebrate groups recovered too, except for brachiopods. The first modern corals, the scleractinians, appeared in the Middle Triassic. In the Triassic and Jurassic, North American fossil corals are rare, except for some non-reef-building corals from the West Coast.

Land Plants: The fossil record of Triassic land plants is poor. Some plants are preserved in the rift valleys that developed along the East Coast. These are remains of swamp-dwelling ferns and sphenophytes (scouring rushes), and of cycads and conifers that probably grew on nearby uplands. Whole tree trunks are preserved in the Petrified Forest of Arizona. These trunks are conifers, but some foliage in the deposits is that of ferns and cycads.

The Age of Reptiles: The Mesozoic Era is known as the Age of Reptiles. Eventually reptiles became so diversified that they were to dominate the land, the sea, and even the air.

Thecodonts: One of the first new groups of reptiles to appear in the Early Triassic were the thecodonts. They are the earliest of the archosaurs, or ruling reptiles, which include the dinosaurs, phytosaurs, birds, and crocodiles. At first, thecodonts were small, lizardlike, and meat-eaters. Some of them walked on 2 legs, using their long, heavy tails for balance.

Dinosaurs: In the Late Triassic, the first dinosaurs appeared. These were the saurischians. The first saurischians also walked on 2 legs and had short forelimbs and stabbing teeth. They did not become abundant until the Jurassic, when the other group of dinosaurs, the ornithischians, appeared.

Phytosaurs and Crocodiles: Another important group of archosaurs were the phytosaurs, which show up in the fossil record in the Late Triassic. They looked like crocodiles, but were not closely related to them. Phytosaurs became extinct by the end of the Triassic. Crocodiles, which are still living, appeared a little later, near the Triassic-Jurassic boundary, and are the only archosaurs besides birds to have survived the end of the Cretaceous.

Lizards and Turtles: The living groups of lizards and turtles also originated in the Triassic. Lizards stayed small until the Cretaceous, but turtles were fairly large throughout the Mesozoic.

Ichthyosaurs: One of the most successful groups of reptiles during the Triassic was the ichthyosaurs. These were fully marine reptiles, shaped like sharks or dolphins, which had completely lost their legs. Ichthyosaurs quickly became the largest animals, terrestrial or marine, of the Triassic. Ichthyosaur vertebrae from Nevada indicate that some of these creatures were 30′ (9 m) long. They had elongated jaws full of uniform stabbing teeth.

Plesiosaurs: The plesiosaurs were another group of large marine reptiles. They had small heads, their legs were modified into flippers, and—except for their longer necks—their overall body form was similar to that of living marine turtles. Some plesiosaurs had teeth for crushing shellfish; others had stabbing teeth for catching fish.

Therapsids: The therapsids, the mammal-like reptiles that had evolved during the Permian, continued to be common during the Triassic, but became extinct early in the Jurassic.

Mammals: The first mammal fossils are found just below the Triassic-Jurassic boundary. These mammals were about the size of a mouse, and are thought to have eaten insects.

Mass Extinction: At the end of the Triassic, there was a mass extinction in which perhaps 25% of all animal families disappeared. Particularly affected were the ammonoids, which seem to have been reduced to only 1 genus. The brachiopods were also devastated again, and the conodonts, microscopic, toothlike fossils that had first appeared in the Cambrian, were completely eliminated. Land vertebrates were depleted, too; almost all amphibian and reptile families were lost.

At the same time as this biological crisis, Pangea was beginning to split up, and there was a sudden, worldwide drop in sea level. The only mechanism known to cause such a precipitous drop is glaciation, but there is no indication of glaciation in rocks of this age.

It is tempting to try to account for the Triassic-Jurassic extinctions in the same way that the Cretaceous-Tertiary extinction has been explained—as the result of a huge meteorite, or a series of meteorites, hitting the Earth. How could this have affected the drop in sea level? Dust thrown into the atmosphere by meteoritic impact could have reduced sunlight appreciably for a number of years, cooling the Earth sufficiently to form thick glaciers over northern latitudes—thereby dropping sea level. The absence of glacial sediments might be due to the short duration of these glaciers; if they melted after a few years, when the atmosphere cleared and the Earth warmed, traces of their presence would be very scarce.

Jurassic Period

The Jurassic, which lasted for about 55 million years, was named for the Jura Mountains on the border between France and Switzerland, where rocks of this age are well exposed.

Plate Movements:
While Pangea was in existence, the East Coast of North America was bordered by Greenland in the north and Africa in the south. During the Late Triassic or Early Jurassic, Africa and North America began to drift apart. This caused faulting along the East Coast and on the coast of northwestern Africa. During the Jurassic, these fault valleys continued to deepen and fill with sediment. At about the same time, Gondwanaland started to pull away from the southern border of North America. Later in the Jurassic, Africa began separating from South America, and Greenland from North America. There are very few outcrops of Jurassic rocks in North America. In the East, there are terrestrial sediments in the fault valleys, but there are no marine outcrops.

The West Coast was a geologically active zone because the Pacific plate was still sliding under it. There were volcanoes, troughs, and thick deposits of sediments. There are some Jurassic outcrops along the West Coast now, but the area was compressed toward the end of the Jurassic and most of the rocks have been highly folded and metamorphosed and their fossils destroyed. In the interior of North America, both marine and terrestrial rocks occur. One formation contains a rich marine fauna, but the best-known Jurassic formation is the Morrison Formation (Plate 20), which covers much of the Western Interior. This is a terrestrial deposit of sandstone, siltstone, and shale covering millions of square miles. Almost all the large dinosaurs found in North America come from this formation. Along with the dinosaurs, some plant fossils and primitive mammals are preserved here.

Sea Level and Climate: Sea level was low but rising during the Early Jurassic, and high throughout the Middle and Late Jurassic. Climate, in general, was mild, warm, and more equable than it is today, since the North American continent lay farther to the south than it does now.

Marine Invertebrates: The record for marine invertebrates in the Jurassic is sparse in North America, but from the record on other continents it is clear that most groups were doing well. The ammonoids recovered from their third near-extinction and continued to flourish and diversify until the end of the Mesozoic, when they finally disappeared. Reefs of many different kinds of organisms were widespread, particularly in the Mediterranean region. Oysters had appeared, as well as crabs, lobsters, and shrimplike crustaceans. Sea urchins were diversifying. The belemnoid cephalopods reached their climax during this period. These squidlike animals were as long as 5–6′ (1.5–1.8 m). In general, all of the Jurassic and Cretaceous was a time of diversification and proliferation of marine animals, with no major biological crises.

Land Plants: Worldwide, many coal deposits formed during the Jurassic. In North America, there are major deposits in British Columbia and the Western Interior. The forests that produced this coal were dominated by conifers; ginkgos and tree ferns were also important. In drier areas, cycadeoids, which are ancient relatives of living cycads, were abundant. Cycadeoids, like cycads, were short trees crowned with palmlike branches. If there were flowering plants at this time, they have left no clear fossil record.

Land Reptiles: Although reptiles had suffered the extinction of many families at the end of the Triassic, they remained the dominant animals on land. The saurischian dinosaurs grew larger, and some reverted to walking on 4 legs. In North America, the largest saurischian fossils are found in Upper Jurassic rocks. Predatory dinosaurs were still smaller than they were to become in the Cretaceous. Another order of dinosaurs appeared in the Jurassic—the ornithischians, or bird-hipped dinosaurs. These may have evolved independently from the thecodonts or, as some paleontologists think, from an early saurischian. Most still walked on 2 legs, and all were plant-eaters. Among the ornithischians that reverted to 4 legs were the stegosaurs, of which *Stegosaurus* is the best-known example. *Stegosaurus* had 2 rows of triangular plates down its back. These plates probably functioned in heat regulation.

Marine Reptiles: The ichthyosaurs were still the largest animals in the Early Jurassic seas. They had suffered during the extinctions at the end of the Triassic, but by the middle of the Jurassic some were 50' (15 m) long. Then ichthyosaurs waned in size and abundance, and by the end of the Jurassic the role of large marine predators passed to the plesiosaurs. These animals were much larger now— as long as 42' (13 m) by the end of the Jurassic. They continued to dominate the seas until the Late Cretaceous, when mosasaurs and sharks took over.

Pterosaurs: Flying reptiles appeared during the Jurassic. The pterosaurs differed from birds in that they lacked feathers and walked on all fours when on the ground. Only 1 finger was elongated to support the wing; the other digits were used as claws. In the Jurassic, they ranged in size from a few inches to species with wingspans of 3–4′ (0.9–1.2 m) and became considerably larger in the Cretaceous.

Birds: The first fossil birds are from the Upper Jurassic Solnhofen Formation in southern Germany. These birds, called *Archaeopteryx,* were similar to the pterosaurs in that their teeth were basically reptilian and they still used some of their digits as claws. But they had feathers and could have been the ancestors of later birds. Some paleontologists think that these first birds could not fly because their breastbones were not properly shaped to support flight muscles. They may only have glided from tree to tree.

Mammals: Most of the Jurassic mammal fossils come either from England or from the Morrison Formation in the Western Interior of North America. Mammals were still quite small, but had become more diverse: 6 orders are known from the Jurassic.

End of the Jurassic: Although there was no major mass extinction at the end of the Jurassic, certain groups, notably the ammonoids and some of the dinosaurs, did die out, to be replaced by more advanced forms in the Cretaceous. The Jurassic and Cretaceous boundary is further marked by unconformities in most places, caused by erosion following a drop in sea level.

71

Cretaceous Period

There are more rocks from the Cretaceous than from any other period. There are 3 reasons for this. First, the Cretaceous is a relatively recent period, and erosion has had less time to remove these rocks than those from earlier periods. The Cretaceous was also one of the longest periods, lasting about 70 million years. Finally and most importantly, the seas stood higher on the continents during the Cretaceous than ever before or since, and thick deposits were laid down. A rise in sea level that affected all the continents began in the Jurassic and continued with minor fluctuations until the end of the Cretaceous, when sea level began to drop again, slowly falling throughout the Tertiary. This pattern can most easily be explained by changes in the volume of the ocean basins caused by plate movements. When crustal plates are moving rapidly, the areas where they separate become high mid-ocean ridges that displace water onto the continents. When plates are moving slowly, these ridges cool and sink, causing water to drain off the continents. Radiometric dating of ocean crust formed during this time confirms that the plates were moving rapidly throughout the Jurassic and Cretaceous, and more slowly later, during the Tertiary.

Cretaceous Rocks: Chalk, so characteristic of the Cretaceous, is a limestone that is mostly composed of the skeletal remains of plants and animals. The White Cliffs at Dover are composed almost entirely of the tiny, round shells of calcareous algae. In North America, there are 2 large chalk formations, the Selma Chalk of the Gulf Coast and the Niobrara Chalk in Kansas. Marine Cretaceous rocks, rich in fossils, occur along the eastern North American coastal plains from New Jersey south to Florida and along the coast of the Gulf of Mexico. Near the coast, these rocks tend to be covered by younger deposits from the Tertiary. Beginning early in the Cretaceous, a seaway opened up from the Gulf of Mexico, eventually extending north through Canada and Alaska. Toward the end of the Cretaceous, this sea withdrew and swamps took its place; thus, the marine rocks are covered by terrestrial deposits containing fossils of land vertebrates and plants. Much coal formed in this terrestrial sediment.

Formation of the Rocky Mountains: The Rocky Mountains began to rise during the Cretaceous. Severe compressional forces operated through Idaho, Wyoming, and Utah, causing great sheets of rock to be moved horizontally and folded. Vast quantities of sediments eroded from these early Rocky Mountains and poured into the seaway, leaving deposits of conglomerate rock 10,000' (3100 m) thick in some places.

In the Middle Cretaceous, molten rock was forced into the crust along the West Coast, forming great masses of granite hundreds of miles long.

Marine Invertebrates: In the vast continental seas of the Cretaceous, it is not surprising that marine life flourished. Many groups reached the height of their abundance and diversity, never to recover after the Cretaceous-Tertiary extinction.

Corals: The scleractinian corals were far more various in the Cretaceous than they are now. In North America, reefs formed in Mexico and along the Gulf Coast.

Mollusks: The rudists, a group of bivalves that resembled in form the rugose corals of the Paleozoic, built reefs that rivaled those of the corals. Rudists had 1 valve modified into a cone, with the other valve serving as a lid. They cemented themselves together to form large, strong structures, and they built upward, just as corals do. A few of these rudist reefs developed along the Gulf Coast and in the West Indies, where 1 species grew to 5' (1.5 m) tall, but the rudists were best developed in Europe. *Exogyra* (Plate 211) was another common Cretaceous bivalve, and is found in almost all marine rocks. *Exogyra,* the rudists, and their close relatives became extinct at the end of the Cretaceous. Ammonoids continued to be diverse and abundant, and are very useful as Cretaceous guide fossils. The most advanced of the shelled marine gastropods, the neogastropods, appeared in the Cretaceous. The mesogastropods were also numerous in the Cretaceous, although they had first appeared in the Ordovician.

Crustaceans: Among the crustaceans, lobsters and shrimps were present in the Cretaceous, although the number of fossils is small. Crabs, important as predators on gastropods and bivalves, had appeared in the Jurassic and now became more abundant.

Teleost Fishes: The dominant modern group of fishes, the teleosts, appeared in the Jurassic, but it is not until the Cretaceous that they outnumber the previously dominant holosteans in marine rocks. A teleost differs from earlier fishes in that the internal skeleton is fully ossified, and there are bones not present in earlier fishes. The backbone does not extend into the tail, which is composed of fin-rays. The scales are thin, flexible, and overlapping or, in some teleosts, lost altogether.

Marine Reptiles: Among the marine reptiles, the record is scanty for the Early Cretaceous. Ichthyosaurs were present but small. By the Middle Cretaceous, the plesiosaurs reappeared, as large as they had been in the Late Jurassic; but the genera are different. Soon the plesiosaurs were replaced as the largest marine animals by the lizardlike mosasaurs. Mosasaurs had flippers instead of legs, and long, lizardlike tails. Their teeth were those of reptilian predators. The largest mosasaurs were over 56' (17 m) long. Many of their bones are found in the Niobrara Chalk of Kansas. This formation also contains marine turtle skeletons. The giant was *Archelon,* 12' (3.7 m) across and 11' (3.4 m) long.

Pterosaurs: Flying reptiles, the pterosaurs, became larger and more specialized than they had been in the Jurassic. Their fossilized bones are more common than bird bones, and are found in the Niobrara Chalk. One genus had a wingspan of 25' (7.6 m), but a body no bigger than that of a goose. Its legs were so weak that it probably could not have walked on land, and must have spent all its time at sea.

Birds: The only common bird fossil from the North American Cretaceous is the aquatic *Hesperornis,* which could not fly. It was similar to *Archaeopteryx* in that, unlike living birds, it had teeth.

Dinosaurs: Between the Jurassic and the Cretaceous, some dinosaurs disappeared and new types emerged, but diversity remained until their sudden extinction at the end of the period. Among the saurischians was *Tyrannosaurus,* the largest land predator ever to live. It was 45' (14 m) long and stood 20' (6 m) high. There were also many smaller predators, some no larger than a rooster or a dog. Among the plant-eating ornithischians, the stegosaurs became extinct early in the Cretaceous and were replaced by ankylosaurs, which looked something like large armadillos with heavy armor and clublike tails. Another 4-legged group was the ceratopsians. Most genera had 3 horns on their skulls, strictly for defense, since these dinosaurs were plant-eaters. The most common ornithischian fossils are the duck-billed dinosaurs. These animals walked on 2 legs, and their snouts were flattened and bill-shaped. They probably ate aquatic vegetation.

Mass Extinction: The close of the Cretaceous is also the end of the Mesozoic Era and is marked by the second most severe of all mass extinctions. At this time, about half of all animal families disappeared. All of the dinosaurs became extinct. Marine reptiles and pterosaurs vanished as well. Among the invertebrates, the ammonoids and belemnoids were entirely wiped out. Many families of bivalves and gastropods disappeared. Two-thirds of the corals became extinct; the group was never to recover its former diversity. Bryozoans lost about 25% of their families, as did the echinoids. Crinoids suffered, and the number of sponge families was reduced by half. Planktonic foraminiferans, tiny shelled animals that float in the ocean, were almost extinguished; only 3 species are known to have survived. A few groups seem to have escaped obvious injury. Land plants were largely unaffected, as were fishes. The brachiopods and amphibians lost only a few families. Mammals show little effect also, but this may be because the fossil record of mammals in the Mesozoic is so poor and obviously incomplete.

Meteorite Impact: At present, the only explanation for the mass extinction that seems plausible is that a large meteorite hit the earth 65 million years ago. Certain elements, most notably iridium, which are known to occur in meteorites but are very rare in the Earth's crust, are concentrated at the boundary between the Cretaceous and Tertiary periods. There appears to be no way to explain the presence of these elements except as the debris from a very large meteorite, perhaps 6 miles (10 km) across. The extinctions would have been caused by the deterioration of the climate following the ejection of much dust into the upper atmosphere. Estimates vary as to how dark the Earth might have become, and for how long, but most likely the effects would have appreciably reduced sunlight and the ability of plants to photosynthesize for years. With plant growth halted, many animal species would simply have starved to death.

Some geologists think that the impact from a large meteorite could cause increased volcanism and changes in the rate of movement and direction of movement of crustal plates. Volcanism would further reduce sunlight and add to the food crisis. The dust from volcanoes and the meteorite impact could cool the Earth to the point where glaciers would begin growing, causing the sudden drop in sea level that is often observed at the time of a mass extinction. The combination of reduced photosynthesis, cooling climate, and lowering sea level might well cause a mass extinction.

Tertiary Period

The era following the Mesozoic is the Cenozoic, or "Time of Recent Life." The Cenozoic includes the Tertiary Period, which lasted about 63 million years and is divided into 5 epochs, as well as the Quaternary Period, which is broken down into 2 epochs, together lasting less than 2 million years. From earliest to latest, the 5 Tertiary epochs are called the Paleocene, Eocene, Oligocene, Miocene, and Pliocene.

Sea Level and Climate: Sea level fell gradually during the Tertiary. On the average it was low, with only 3% of present land areas covered by seas. After a worldwide rise during the Cretaceous, temperatures gradually fell in the Tertiary. Until the end of the Oligocene, the climate was humid and subtropical as far north as the Dakotas. During the latter part of the Tertiary, subtropical vegetation retreated southward. These gradual drops in sea level and temperature seem to have had little effect on the diversity of life. Indeed, there is every indication that diversity increased throughout the Tertiary and into the Pleistocene. Today there are probably more kinds of plants and animals living on Earth than ever before.

Plate Movements: The continents continued to move apart during the Tertiary. The Pacific plate continued to slide under the West Coast of North America, and this area remained geologically active throughout the Cenozoic. In the Early Tertiary, Australia and Antarctica started to separate; Alaska and Siberia moved closer together, beginning to close off the Arctic Ocean from the Pacific Ocean. In the Late Tertiary, India collided with Asia and the Himalayas began to rise; they are still rising now, as these 2 land masses push against each other.

Tertiary Rocks: In North America, almost all the marine rocks of the Tertiary Period are confined to the coastal margins. These rocks are mostly unconsolidated sands and clays. In the East, the Tertiary coastal sediments are partly material eroded from the Appalachians. These mountains had been eroded almost to sea level in the Cretaceous, but in the Tertiary the mountain roots rose, to be eroded once again into the characteristic series of parallel ridges we see today.

Something similar happened to the Rocky Mountains. After mountain-building ceased in the Eocene, they too were eroded, and the valleys and basins between peaks were filled with sediment. Then, following uplift, these sediments began to erode away. Much was carried eastward by streams, to be deposited in thick sheets over the Great Plains. This Tertiary sediment has been weathered into the characteristic topography of badlands in several locations. Many of the North American mammal fossils are found in places like the badlands east of the Black Hills in South Dakota.

Marine Invertebrates: Fossils contained in the Tertiary coastal beds are modern in appearance; most of the genera and many of the species are still living. Gastropods and bivalves are dominant in these beds, as they are in coastal waters today. Sea urchin and bryozoan fossils are present, but these forms were more fragile and easily destroyed. Solitary corals are common, but reef-building corals are now restricted to Florida and Mexico.

Fishes: The teleost fishes, which first appeared in the Jurassic, continued to diversify. They were common in fresh water as well as in the ocean. Many intact skeletons of teleosts have been recovered from the Green River and Florissant shales. Sharks also flourished.

They began to get larger in the Eocene, reaching 36' (11 m), and by the Miocene some *Carcharodon* were 60' (18 m) long (Plate 478).

Age of Mammals:
The Tertiary is the Age of Mammals. The first mammals to appear in the Tertiary were still small, with correspondingly small brains. They had 5 toes on each foot and walked on their soles. They probably ate insects, and their teeth were unspecialized. Some, like shrews and moles, have remained small insect-eaters, but most began a trend toward increased size, with specializations of the feet for running, for grasping prey, or for swinging in trees. Teeth became specialized for predation—large canines and shearing cheek teeth—or adapted for grinding up coarse vegetation—broad, flat cheek teeth.

First Horses:
The first horse, *Hyracotherium* (or *Eohippus*, the "dawn horse") appeared in North America in the Early Eocene. It was about the size of a fox terrier. It had 4 toes on its front feet and 3 toes on its hind feet, so it had already begun to specialize for running. Through a series of genera, the horses show a gradual loss of all but 1 toe on each foot, and an increase in size. Their teeth changed too, from unspecialized cheek teeth to very long cheek teeth with raised grinding surfaces and convoluted folds of enamel and cement for increased hardness. Grass has a high silica content and is therefore more abrasive than other kinds of plant food; it probably appeared in the Oligocene, when horses and other plant-eaters developed high-crowned, folded teeth. Horses were abundant in North America until near the end of the Pleistocene, when this population died out. Fortunately, some had previously spread to Asia and Europe, and they survived there.

Titanotheres: Another group of the early plant-eaters were the rhinoceroslike titanotheres, the largest land animals in North America in the Oligocene. *Brontotherium* stood 8' (2.4 m) high at the shoulder. On its nose it had 2 horns, which were composed of hair compressed into hard masses. *Brontotherium* fossils are abundant in the badlands of South Dakota. All titanotheres disappeared in the middle of the Oligocene.

Camels: Camels were abundant animals on the plains of North America from the Eocene until they vanished from the continent at the end of the Pleistocene. Their evolution was similar to that of the horses. They first appeared as very small browsers, the size of large rabbits, with 4 toes on each foot. In the Oligocene, they were the size of sheep, and the number of toes on each foot had been reduced to 2. Camels became larger and more diverse in the Miocene. Before all North American camels became extinct, they had spread to South America, where the alpaca and llama are their living representatives, and to Asia.

Oreodonts: The oreodonts were another group of browsers and grazers abundant on the western plains in mid-Tertiary times. These undistinctive goat-sized animals had proportions somewhat like those of a hog. They first appeared in the Eocene, became most abundant in the Oligocene—when their bones accumulated in vast numbers in the badlands of South Dakota—and died out in the Pliocene.

Proboscideans: Elephants are the only living members of the order Proboscidia, named for the greatly extended proboscis, or nose, of its members. Proboscidians probably evolved in North Africa, where the first fossils are found in the Early Oligocene

One group of extinct proboscidians, the mastodons, crossed from Asia into North America over the Siberian-Alaskan land bridge in the Miocene and were common on this continent. About 7000 years ago they became extinct. Another proboscidian that was common in North America during the Pleistocene was the woolly mammoth.

Carnivores: As soon as some mammals became specialized for eating plants, other mammals began specializing to eat the plant-eaters. The first of these carnivores, the creodonts, appeared in the Paleocene. They varied in size and shape, but almost all had small brains and unspecialized teeth. Carnivores radiated in the Eocene and Miocene until there were many different types. By the end of the Pliocene in North America, there were wolves, biting cats (like living cats, with fairly small canine teeth), and stabbing cats (with very long canines that were used to stab the prey to death).

Hominids: Humans are the only living species of the family Hominidae, but several fossil genera have been placed in this group. Hominids are mammals that walk on 2 legs, have teeth similar to human teeth, and hands capable of manipulating objects. The Hominidae are thought to have had their origins among members of the family Ramapithecidae, ape-like animals that flourished in the Old World during the Miocene. The earliest definite hominid is *Australopithecus,* of which several species flourished in Africa between 4 million and 1 million years ago.

Quaternary Period

The Quaternary is divided into 2 epochs, the Pleistocene and the Recent (or, as it is sometimes called, the Holocene). The beginning of the Pleistocene is defined as the start of the cooling trend that began about 1.8 million years ago and resulted in continental glaciations. The Pleistocene ended, and the Recent Epoch began, about 11,000 years ago, midway in the warming up of the oceans after the last glacial retreat.

Glaciation: During the Pleistocene, glaciers advanced and retreated 4 times across North America, Europe, and Asia. As much as 30% of all land area was covered with ice—3 times the area ice covers today. This glacial ice was 10,000–13,000' (3,000–4,000 m) thick. In addition to the major advances, there were many small fluctuations in the extent of ice cover. Between the glacial stages, the climate was often warmer than it is at present.

Sea Level: At the height of each glacial stage, the sea level was probably about 430' (130 m) lower than it is today. Since the average depth of water at the edge of the continental shelf is now 450' (137 m), this means that sea level fell almost to the edge of the shelf 4 times. If all the ice still piled up on Greenland and Antarctica were to melt, the sea level would rise about 210' (65 m).

Causes of Glaciers may have begun to grow at the
Glaciation: beginning of the Pleistocene because of crustal plate movements. These movements closed the Arctic Ocean off from the Pacific Ocean in the late Tertiary, and also caused Antarctica to move to the South Pole. Heat was then distributed less evenly over the globe, since ocean water could not flow over the poles to warm them. Consequently,

the Arctic Ocean froze permanently and thick ice sheets grew over Antarctica. This may have been a crucial factor in causing the Pleistocene glaciations.

Milankovitch Cycles: An explanation for the pattern of glacial advance and retreat was suggested in 1932 by the Yugoslavian meteorologist M. Milankovitch, who thought that cycles in the solar system had caused the oscillations. With information on past ocean temperatures from sea shells in deep sea cores, variations in climate are known fairly precisely. Milankovitch's predictions fit the temperature curves very well. Milankovitch suggested that there were 3 important cycles in the relation of the Earth to the sun, and that in those parts of the cycles when summers were cooler than usual, glaciation would increase. The first of these cycles is caused by changes in the shape of the Earth's orbit about the sun; the second results from the degree of tilt of the Earth's axis; and the third involves variations in the direction of tilt of the Earth's axis. Maximum glaciations should take place when the axis is closest to vertical, the orbit most elliptical, and summer occurs when the Earth is farthest from the sun.

Hominids: The Pleistocene is rich in the fossils of early hominids. *Australopithecus,* first found in the late Pliocene, survived in Africa in the Pleistocene. *Australopithecus* is known to have used tools as early as 2.5 million years ago.

Homo: In the earliest Pleistocene, the first fossils of a more advanced hominid, *Homo habilis,* appear in Africa. Another species, *Homo erectus,* is first found in deposits 1.6 million years old. Both *Homo habilis* and *Homo erectus* coexisted with highly specialized species of *Australopithecus.* By the middle of the Pleistocene, *Homo* *erectus* was widespread in the Old World. Charcoal is preserved at some of the sites, indicating that *Homo erectus* used fire.

Homo sapiens: The oldest known fossils of our own species, *Homo sapiens,* are of Neanderthal man—*Homo sapiens neanderthalensis*—found in Europe, Africa, and Asia in deposits dating from 150,000 to 35,000 years ago. Neanderthal man had a braincase as large as that of modern man, a massive frame, brow ridges, and large jaws. These fossils are found in association with many tools, including needles. Fully modern humans, *Homo sapiens sapiens,* replaced Neanderthals in Europe about 35,000 years ago; by 30,000 years ago, they were widely distributed.

Humans in North America: It is clear that hominids evolved in the Old World, which was cut off from North and South America during the Pliocene and most of the Pleistocene. Some time during the late Pleistocene, humans crossed from Asia on the Bering Strait when the sea level was low because of glaciation. The first abundant evidence of human habitation in North America is from sites about 11,000–12,000 years old. But there is some evidence that humans may have reached the New World thousands of years before this.

Migrations of Mammals: As the ice sheets advanced and retreated, mammals migrated south and north. In North America during the cold glacial intervals, there was a fauna of mastodons, woolly mammoths, and woolly rhinoceroses. In the warm interglacials, the fauna consisted of giant ground sloths, saber-toothed cats, wolves, lions, bison, cattle, camels, and 1-toed horses. These mammals seem to have adjusted to the fluctuating climate, and there were very few extinctions until the very end of the last glaciation. Then—between 10,000 and 5,000 years ago—about 32 genera of North American mammals died out.

Large Mammal Extinctions: These extinctions of large mammals are a mystery to paleontologists. These same genera had survived other glacial and interglacial intervals. And there are no comparable extinctions in the oceans.

Paul S. Martin, an anthropologist at the University of Arizona, has proposed that early human hunters became so efficient that they exterminated all the large plant-eating mammals. Predators like wolves and saber-toothed cats would also have vanished when their natural prey was gone.

Other experts have disagreed with this concept. They argue that if large mammals became rare, their human predators also would have decreased in numbers. A balance would have been maintained, with neither large mammals nor the humans becoming extinct.

Future: There is no reason to suppose that the glacial advances of the Pleistocene have ended. We are almost certainly still living in the Ice Age, and geologists were overly optimistic to have concluded that a new epoch began 11,000 years ago. If the Milankovitch hypothesis is correct, the next glacial advance should begin within a few thousand years.

FOSSIL-BEARING ROCKS

Learning to identify the principal fossil-bearing rocks is not difficult, and after a little practice the fossil-hunter will find it easy to distinguish them at a distance, or even from a car speeding past an outcrop along a highway. This section provides information on the major kinds of fossiliferous rocks—how they were formed and how they may be recognized.

Geologists classify rocks in 3 broad categories based on how they were formed. Igneous rocks such as granite and basalt are formed by the cooling and hardening of molten materials, either deep inside the earth or, in the case of lava released by volcanoes, on the surface. Sedimentary rocks are formed by the hardening of deposits of sand, silt, and other fine particles, or from substances dissolved in water that have precipitated and settled on the bottom. Finally, metamorphic rocks are formed from igneous or sedimentary rocks that have been subjected to pressure and high temperatures deep under the earth's surface, causing their original structure to change.

Fossils are extremely rare in igneous rocks; those found usually consist of a rough mold formed when a plant or animal was trapped in lava. Very few fossils survive the intense heat and pressure that produce metamorphic rock, and those that do are generally badly damaged and deformed. Nearly all fossils are found in sedimentary rocks or in sediments that have not yet hardened into stone. Fortunately for the fossil-collector, sedimentary rocks cover about 75% of the land surface of the earth.

Formation of Sedimentary Rock: Almost all sedimentary rocks are derived from the weathering or erosion of other rocks, although a few kinds, coal among them, are formed directly from plant or animal material. The products of weathering are carried, either as particles or dissolved in water, to a place of deposition on the floor of the ocean, on a lake bottom, or on a floodplain. Rock formed mainly of particles is called clastic rock. Sandstones and shales, for example, are clastic rocks. If the weathering products are dissolved, they may be precipitated chemically or by evaporation to form a non-clastic rock. These dissolved substances may be derived from rocks, but some are derived from the shells of animals. Non-clastic rocks include limestone, dolomite, and a group known as evaporites—gypsum, salt, anhydrite, phosphates, and others. Often sedimentary rocks are a mixture of both clastic and non-clastic materials. For example, a sandstone may be formed when calcium carbonate, a dissolved substance, seeps among the grains of sand and precipitates, cementing the grains together. Since the different kinds of sedimentary rock grade into one another, they are classified on the basis of their most abundant component.

Fossils in Sedimentary Rocks: Sedimentary rocks vary in their fossil content. Most fossils are found in shales, sandstones, and the rocks composed of calcium carbonate, called limestones. Very coarse sedimentary rocks like conglomerate, composed of grains more than .08″ (2 mm) in diameter, rarely contain fossils, although the bones of large mammals like mastodons are occasionally found in gravel pits. The evaporites are not fossiliferous.

Shales

1, 2, 3, 4, 9, 20, 21, 22, 23, 24, 26, 29, 30, 31, 32, 33

The most abundant sedimentary rocks are shales. They make up about half of all the sedimentary rocks on the continents and, if sediments on the ocean floor are included, about 80% of all sedimentary rocks. Strictly speaking, shale is a hardened rock that splits along bedding planes and at least 25% of which consists of clay-sized particles less than .00008″ (.002 mm) in diameter. Certain kinds of rocks differ slightly from this definition, but are included here as being closely related to true shale.

Formation of Shale:

The more rapidly a current of water is flowing, the larger the particles it can carry. As a current gradually slows, particles carried in it begin to settle to the bottom. Very fine particles like those that form shale settle out of very slowly moving streams or out of the still waters of lakes, bays, and larger bodies of water. A deposit of shale, therefore, is an indication that a site was once the floor of a slow-moving or still body of water.

Recognizing Shale:

Shales may be recognized at a distance by their obviously layered structure (Plates 32, 33), and in the hand by their very fine texture, as opposed to the gritty feel that distinguishes sandstones. When they weather, shales tend to fracture and to split along bedding planes, producing sharp, angular edges and fragments (Plate 20). Sandstones commonly weather by disintegrating, producing deposits of loose sand (Plate 26). Deposits of limestone and other finely textured non-clastic rocks often weather by dissolving, so they generally have more rounded contours. Unlike limestones, shales generally contain little calcium carbonate or calcite and hence do not react with acid when it is applied to them.

Color of Shales: The color of shale varies, and is an important indication of its origin and composition. Very light-colored shale is composed of many particles of quartz and other minerals, and contains little or no organic material such as the decayed remains of plants. The darker a specimen of shale, the more organic material it contains. Much shale is black (Plate 1) because of its high organic content, and coal (Plate 31) is composed almost entirely of decomposed plant material. The layer of coal shown in Plate 31 is bounded above and below by deposits of sandstone. Well-preserved plant fossils are common in the gray shale that often lies just above a deposit of coal. Many shales are red or brown (Plates 22, 23, 29), the result of oxidized iron in the rock. Newly exposed red or brown surfaces indicate that the iron was oxidized before the rock formed; but if a weathered surface is red or brown, and a fresh surface is gray, then the iron is being oxidized, as the rock is exposed to the air. The lake shale in Plate 29, for example, is brown only on the surface of the outcrop. On the other hand, the soft sediments in the badlands shown in Plates 22 and 23 were deposited on floodplains, exposed to the air or to oxygen-rich water, so that the iron was oxidized before and during deposition. Some shales with unoxidized iron are green.

Paper Shale: Some shales are very finely layered, as in Plate 4, and are called paper shales. They often form in fresh water, and occur in many parts of the western United States in ancient lakebeds. The ease with which they can be split indicates a low content of lime and silica, soluble substances that would have cemented the particles together, and often a high content of organic material. Paper shales are often rich in fossils, and some of the finest plant fossils are found in them. Some paper shales are black.

Mudstone and Siltstone: If a rock has the same texture and particle size as true shale but does not split along bedding planes, it is called mudstone (Plate 3). Siltstone (Plate 9) is a rock slightly gritty to the touch and composed of larger particles; 50% of the particles in siltstone are .00016–.002″ (.004–.06 mm) in diameter. If this rock splits along bedding planes, it is sometimes called silty shale.

Calcareous Shale: Although shales are composed mainly of clay and quartz, small amounts of calcium carbonate may also be present. If the amount of calcium carbonate is more than 6%, the rock is called calcareous shale (Plate 2). The addition of calcium carbonate makes the rock harder and less likely to split along bedding planes.

Slaty Shale: If shale is subjected to heat and pressure, the particles are often realigned, and the rock becomes harder and more brittle. If the heat and pressure are great enough, the rock becomes crystalline and turns into slate, a metamorphic rock. If the rock has not turned all the way to slate, new cleavage planes may be developed, but fossils in the original shale will not have been destroyed; such rock is called slaty shale (Plate 30).

Badlands: In many places in western North America where rainfall is scanty, soft freshwater sediments deposited during the last 65 million years erode into badlands (Plates 21–24, 26). Badlands are generally a mixture of rock types. Shale and siltstone are the dominant rocks, but sandstone is usually present, commonly in lens-shaped deposits. Most terrestrial vertebrate fossils come from badlands.

Volcanic Ash: A material often found in shale is volcanic ash, minute bits of igneous rock that have been spewed out by erupting volcanoes. It settles out of the air slowly and is washed into streams along with other materials. It may form beds in lake and river sediments. Ashfalls were fairly frequent in the western part of North America during the Tertiary; the freshwater Tertiary sediments shown in Plates 21 and 24 contain large amounts of ash.

Concretions: Many shales and sandstones contain concretions, rounded nodules harder than the surrounding rock. These concretions may contain fossils. The composition of the concretions may differ from that of the shale, containing more calcium carbonate or iron carbonate. The bedding planes of the enclosing rock pass through the concretions, showing that the concretions formed within the rock and after the sediment was deposited.

5, 7, 8, 10, 14,   ## Sandstones
17, 18, 20, 22,
25, 26, 31   Sandstone is composed of particles the size of sand grains, that is, .002–.08″ (.06–2 mm) in diameter. These grains are most often composed of quartz, but sandstones may also contain grains of feldspar, magnetite, garnet, hornblende, mica, and other minerals. Frequently, water seeps among the sand grains, and if calcium carbonate or silica are dissolved in the water, these substances may precipitate, cementing the grains together and filling the tiny spaces between them, a process known

as consolidation. Unconsolidated sandstones still have spaces between the grains, and the water seeping through often dissolves any fossils that may have been present. Sandstones vary in the readiness with which they weather; those that are consolidated are quite often resistant to erosion, while unconsolidated sandstones erode quite easily, leaving more resistant layers projecting from an exposure (Plates 26, 31).

Formation of Sandstone:
Sandstones are deposited by faster-moving water than that which produces deposits of shale. Sand accumulates in the beds, deltas, and floodplains of rivers, on beaches, and also on land in sand dunes. Very old sandstones seldom contain original fossil material because water has seeped through them, dissolving fossils and leaving molds (Plate 5). More recent deposits of sandstone may preserve entire fossils, as in the unconsolidated sandstones of the Calvert Cliffs in Maryland (Plates 17, 25) or the California sandstone in Plate 18. Because they are formed on land or by more rapidly moving water, sandstones tend to have fewer fossils than do shales or limestones.

Recognizing Sandstone:
Sandstone may be distinguished by the size of the particles, which can usually be seen with the naked eye. The particles give sandstone a gritty feel and can usually be rubbed off the rock mass. Softer, more easily eroded deposits of sandstone often have apronlike piles of loose sand beneath them (Plate 26).

Greensand:
Glauconite is a green mineral that is often found in sandstone, giving it a greenish color if the surface is fresh (Plate 10), or a greenish-yellow, buff, or brown color if the surface has weathered (Plates 7, 14).
Unconsolidated sands rich in glauconite

are called greensands, and are usually fossiliferous. Some greensands contain as little as 1% sand, but most are about half sand and half glauconite. Glauconite is composed mainly of iron, potassium, and silica, and is usually present as small granules about .016″ (.4 mm) in diameter. The sandstone in Plate 7 has a salt-and-pepper appearance because it is a mixture composed mainly of sand, with some glauconite. Greensands are common sedimentary deposits on the coastal plain of the eastern United States. The greensand in Plate 14 contains well-preserved brachiopod fossils.

Freshwater Sandstone: Much sandstone in the western United States was deposited in fresh water, usually on floodplains or in river channels. These deposits tend not to be extensive, and grade into siltstones and shales. Lens-shaped deposits of sandstone occur in what is mainly shale in the formations shown in Plates 20, 22, and 26.

Beach Sandstone: Ancient beaches are sometimes preserved as sandstone. Characteristic fossils of many of these beach deposits are vertical burrows called *Skolithos* (Plate 8). These vertical burrows are a sign of shallow water, since they were presumably formed by animals digging downward to protect themselves from being dislodged by currents and wave action or drying out at low tide.

6, 11 ,12, 13, 15, 16, 19, 27, 28   **Limestones**

Limestones are rocks that contain at least 50% calcium carbonate. The calcium carbonate may be organic, derived from calcareous shells, or it may be inorganic in origin. Most limestones are organic; even a very fine-grained limestone may be composed of small crystals from algae or from the disintegration of animal shells.

Formation of Limestone: Limestones are formed either by the precipitation of calcium carbonate dissolved in water or by the accumulation of undissolved particles or fragments of calcium carbonate derived from the shells of animals such as bivalves, brachiopods, or corals. As in the case of shale, a deposit of fine-textured limestone is an indication that the water was quiet when the sediment was laid down. Coarser limestones can form in more turbulent water that would keep the fine particles in suspension. Limestones of all kinds are readily cemented, or consolidated, because of the abundant calcium carbonate present.

Fossils are generally common and well preserved, although shells may dissolve and leave only molds. The limestone in Plate 19 is full of molds of bivalves. The fossils in the limestone in Plate 13 are partially dissolved. After fossils in limestone dissolve, they may be replaced by silica; such casts can be freed from the surrounding matrix by dissolving it with acid.

Recognizing Limestone: Fine-textured limestone resembles shale, but it often occurs in thicker deposits and is less obviously layered. Since it contains much calcium carbonate, it reacts with acids and will fizz if a few drops of acetic acid or hydrochloric acid are applied to it. From a distance, limestone may be distinguished from shale by the way it weathers: instead of splitting and fracturing, limestone dissolves, so that a natural outcrop of limestone tends to have rounded surfaces (Plate 16). A fresh exposure of limestone may fracture, but it usually appears to be made up of larger blocks (Plate 28), rather than smaller, more angular fragments, as in shale. Limestones of coarser texture are usually recognizable by the shells or shell fragments they contain (Plate 19).

Dolomite: Besides calcium carbonate, limestones commonly contain another mineral, magnesium carbonate. If the amount of magnesium carbonate exceeds 50%, the rock is called dolomite (Plate 12). Often dolomite forms from limestone when some of the calcium is replaced by magnesium. Fossils are often dissolved during this process, leaving molds. The example of dolomite in Plate 12 contains the mold of a receptaculid alga. Dolomite is not as reactive with acids as is limestone, and will only fizz if the acid has been heated or if the dolomite has been pulverized.

Siliceous Limestone: A common component of some limestones is silica. The silica may be present as sand grains or may have precipitated along with the calcium carbonate. Siliceous limestones (Plate 6) are very hard, but split readily into layers. The surfaces of these layers often contain intact skeletons of fishes and crustaceans; soft-bodied animals may also be found.

Chalk: Chalk is a very pure limestone, white, porous, fine-grained, and usually uncemented. It is composed of tiny fragments of calcareous shells. The chalk shown in Plate 11 is made up of the microscopic shells of foraminiferans, single-celled animals related to amoebas.

Coquina: Another kind of limestone composed chiefly of fossil shells is coquina. It is named for a small bivalve called a coquina (*Donax*), which is an abundant component of some limestones of this kind. Coquina is forming today along some beaches, where shells accumulate and water percolates through them, depositing calcium carbonate and cementing the shells together. Pieces of this coquina, called beachrock, wash ashore along many Florida beaches. The coquina in Plate 27 is nearly pure, while that in Plate 15 contains a high percentage of silt and sand.

Part I
Color Plates

HOW TO USE THIS GUIDE

Example 1
Iridescent, coiled shell in a stream bank

Wading near a stream bank in western South Dakota, you come upon a glistening, iridescent, coiled shell. You note that the outer coil does not touch the inner ones, and that the shell's surface is covered with a series of ribs. On the surface, where the outer layer has broken off, you notice a complex pattern of curved lines.

1. To discover the age of the rocks in which you are hunting, turn to the index map of North America. You will find each state labelled with the abbreviation of the state name, and keyed to a particular color. Note the color in which South Dakota appears, and turn to the thumb tab of that color in the section of individual geologic maps that follows the index map. The abbreviation for South Dakota will also appear on the thumb tab. The geologic map of South Dakota indicates that you are in an area of Cretaceous rock.

2. Next, turn to the Thumb Tab Guide preceding the color plates and look for the silhouette that most resembles your fossil. In the group called Snail-shaped Fossils you find a silhouette similar to your specimen. It refers you to color plates 181–204 and 207–209.

3. Checking these color plates, you see that 2 fossils look like your shell: *Scaphites* and *Sphenodiscus*, color plates

194, 198, and 199. The captions refer you to pages 529 and 535.

4. The text tells you that both specimens do occur in Cretaceous rock, but you eliminate *Sphenodiscus* because all its whorls overlap and it does not have ribs.

5. Comparing the drawings of the fossils' sutures—the pattern of curved lines that you noticed below the shell's surface—you confirm that the fossil you have found is *Scaphites*.

Example 2
Round fossil from a quarry

You are hunting through rock debris in an old quarry in upstate New York and pick up a round fossil, about 2″ high and 2″ wide. You notice that it is composed of many tiny tubes with openings at their ends. Using a hand lens, you see 4 tiny platelike structures projecting inward from the walls of some of the tubes.

1. Find the geologic map of New York State by locating the color of its thumb tab on the index map. You learn from the geologic map that the rocks in the quarry date from the Ordovician Period.

2. In the Thumb Tab Guide you find a silhouette resembling your fossil in the group called Sponges, Corals, and Bryozoans. It refers you to color plates 377, 379, 389, 390, 401, and 402.

3. When you check the color plates, you find 2 fossils with similar tiny tubes and openings: *Chaetetes* and *Tetradium,* color plates 401 and 402. The captions tell you the sizes and refer you to entries on pages 378 and 379.

4. Reading the text, you eliminate *Chaetetes* because its tubes do not contain platelike structures and because it did not occur in the Ordovician. The text on *Tetradium* states that it is found in Ordovician rocks and that, like your fossil, it has 4 plates in some tubes. The marginal drawing shows an enlargement of the tube openings. An additional photograph of a different

form of the fossil (Plate 419) also illustrates the tube openings. Your fossil is *Tetradium*.

Example 3
Black fossil in a slab of shale

While splitting shale on a mountainside in British Columbia, you discover a black fossil, about 2″ long and 1½″ wide, in one of the slabs. The fossil is oval and flattened, has a raised section down the middle, and is transversely grooved except at one end, which is broad and shieldlike. You notice that the middle transverse segments seem to end in small spines, and that a few pairs of short spines also project from one end of the specimen.

1. Turn to the geologic map of British Columbia. You note that you are in an area of Cambrian rock.

2. In the Thumb Tab Guide you see a silhouette of an oval, transversely segmented fossil similar to your specimen, and discover that it is a trilobite. It refers you to plates 268–305 and 307.

3. In examining the color plates you find 2 fossils that look like yours: *Olenoides* and *Ogygopsis,* color plates 274, 279, and 284. The captions refer you to text pages 558 and 560.

4. Consulting the text, you learn that both trilobites occur in the Middle Cambrian. You then discover that the transverse segments of *Ogygopsis* end in points but do not bear spines, and that *Ogygopsis* has no tail spines; you thus rule out *Ogygopsis.* The description of *Olenoides* confirms your identification

Map Key

C	Carboniferous
$\mathcal{C}$	Cambrian
CD	Carboniferous and Devonian
D	
D_1	Devonian
D_2	
D$\mathcal{C}$	Devonian, Silurian, Ordovician, and Cambrian
DS	Devonian and Silurian
DSO	Devonian, Silurian, and Ordovician
J	Jurassic marine and continental
J$\bar{R}$	Jurassic and Triassic
K	Cretaceous
KJ	Cretaceous and uppermost Jurassic
Kl	
Kl_1	Lower Cretaceous
Kl_2	
Ku	Upper Cretaceous
Kuc	
Kuc_1	Upper Cretaceous continental
Kuc_2	
Kum	
Kum_1	Upper Cretaceous marine
Kum_2	
Kum_3	
M	
M_1	Mississippian
M_2	
Mz	Mesozoic sedimentary
O	
O_1	Ordovician
O_2	
O$\mathcal{C}$	Ordovician and Cambrian
P	Permian marine and continental
P_1	

P_2	Permian marine and continental
$I\!P$	
$I\!P_1$	Pennsylvanian
$I\!P_2$	
PC	Permian and Carboniferous
pKm	Pre-Cretaceous metamorphic and sedimentary
Pz	Paleozoic
Qc	Quaternary continental
Qm	Quaternary marine
S	
S_1	Silurian
S_2	
SO	Silurian and Ordovician
T	
Tc	Tertiary
Tm	
Tec_1	
Tec_2	Eocene and Paleocene continental
Tepc	
Tpac	
Tem	
Tepm	Eocene and Paleocene marine
$Tepm_1$	
Tmc	Miocene continental
Tmm	
Tmm_1	Miocene marine
Tmm_2	
Tmm_3	
Toc	Oligocene continental
Tom	Oligocene marine
Tpc	Pliocene continental
Tpm	Pliocene marine
$\bar{R}$	Triassic marine and continental
$\bar{R}P$	Triassic and Permian

The Maps

The maps on the following pages show the distribution and geological age of nearly all North American fossil-bearing rock strata. The first, a map of the whole continent, is divided into 15 sections, each in a different color and covering a region containing important fossiliferous areas. Each of these 15 regions is shown in larger scale on a double-page map following the index map. At the left edge of each detailed map is a thumb tab of the same color as the region on the index map, and bearing abbreviations of the states or provinces included. Since there is little or no fossil-bearing rock in New England and parts of northern and eastern Canada, detailed maps of these areas are not included.

By locating the fossil-bearing rocks from each period, and by helping to determine the age of a rock under examination, the maps provide an additional shortcut to identification: they narrow the choice to those fossils found in rocks of that age. The Geologic Time Chart and the section on "Life and Geologic Time" may also be consulted, in conjunction with the maps, to find out about other plants and animals that lived at the same time as the fossil you have identified. Fossil-bearing rocks of the various geological periods are indicated on the maps by a letter code; a key to this code is given on the facing page. With each of the regional maps, a key is given only for those fossil-bearing rocks found in the area covered by that map, and not for any volcanic rocks.

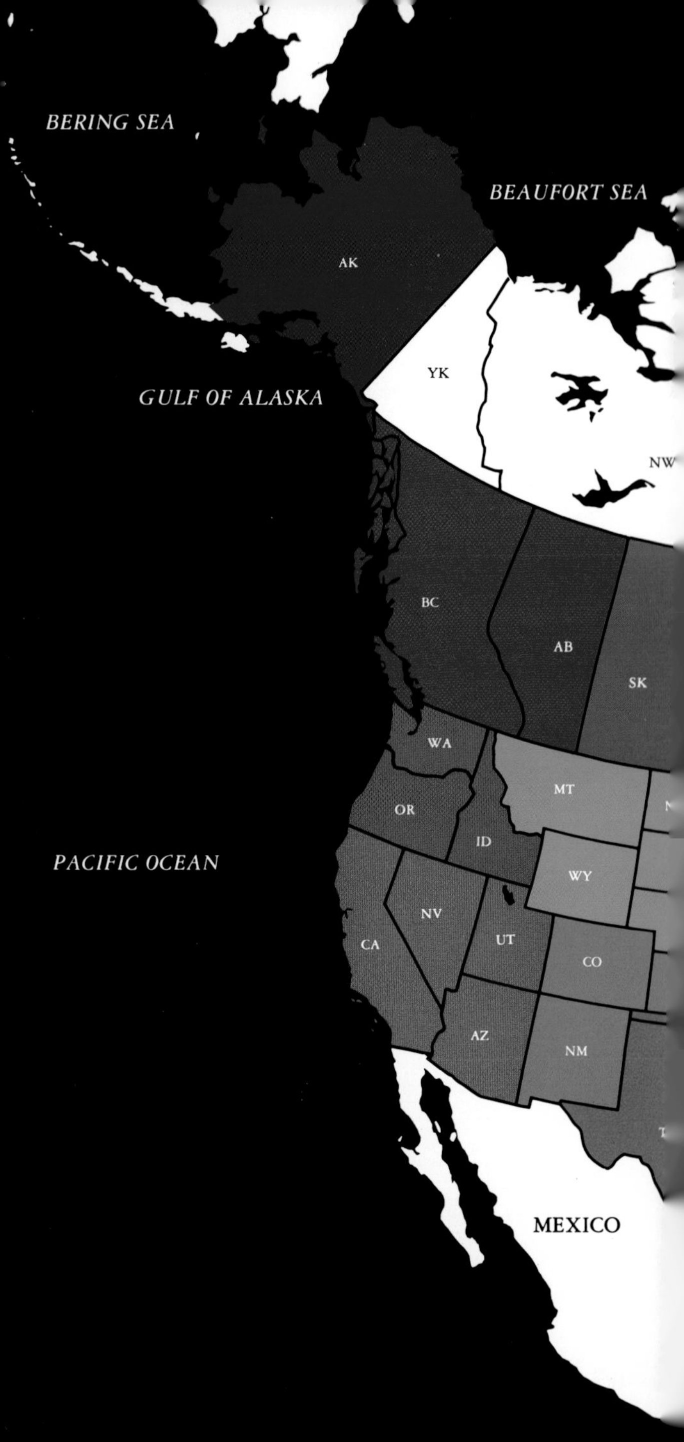

BERING SEA

BEAUFORT SEA

AK

YK

GULF OF ALASKA

NW

BC

AB

SK

WA

MT

N

OR

ID

PACIFIC OCEAN

WY

NV

UT

CA

CO

AZ

NM

T

MEXICO

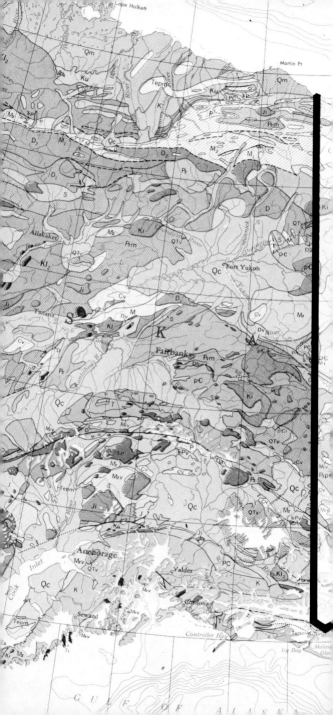

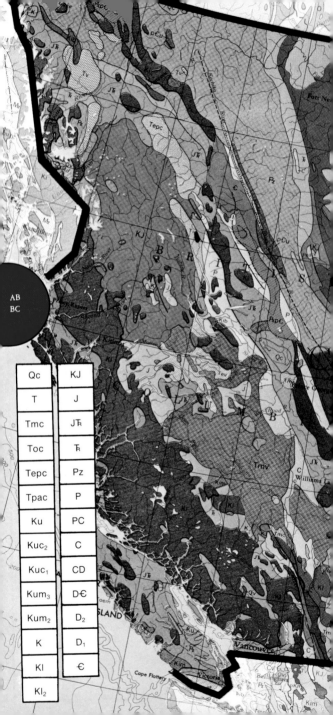

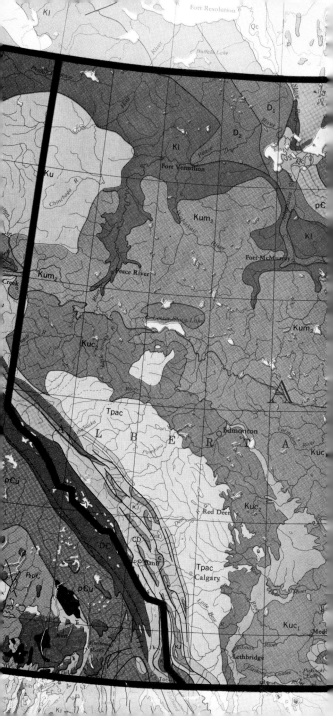

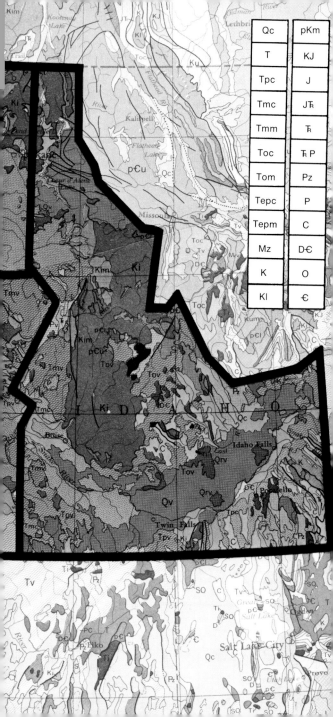

Qc	pKm
T	KJ
Tpc	J
Tmc	JⱤ
Tmm	Ⱡ
Toc	ⱤP
Tom	Pz
Tepc	P
Tepm	C
Mz	DϾ
K	O
Kl	Ͼ

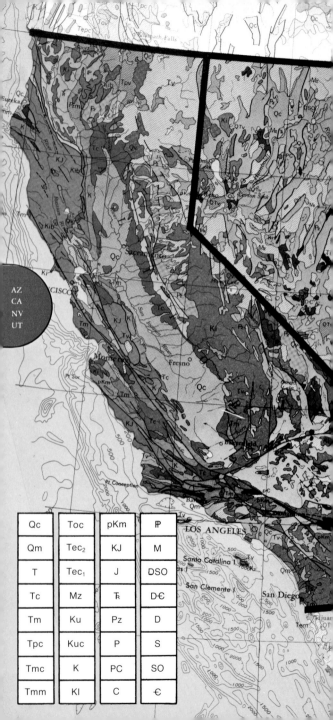

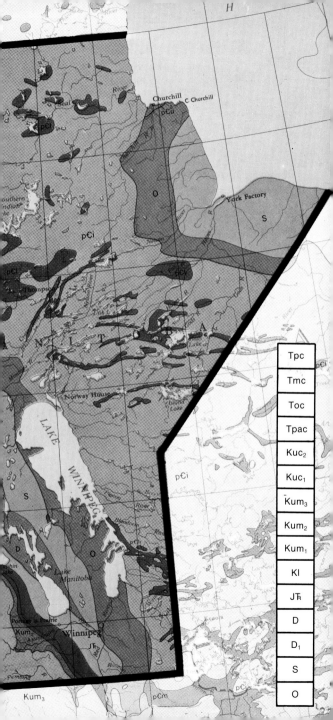

Churchill

C Churchill

pCu

York Factory

S

O

pCi

pCi

pCl

pCl

Thompson

pCl

N I T O B A

Norway House

Island
Lake

LAKE WINNIPEG

S

D

O

Lake
Manitoba

Portage la Prairie

Kum

Winnipeg

JTr

Pembina

Kum₃

pCm

pCi

Tpc

Tmc

Toc

Tpac

Kuc₂

Kuc₁

Kum₃

Kum₂

Kum₁

Kl

JTr

D

D₁

S

O

MT
ND
NE
SD
WY

Qc	Tec₁	Kum₃	JŦ	₱₂
T	Tpac	Kum₂	Ŧ	₱₁
Tpc	Ku	Kum₁	P	M
Tmc	Kuc	Kl	PC	D-€
Toc	Kuc₂	KJ	C	D
Tec₂	Kuc₁	J	₱	

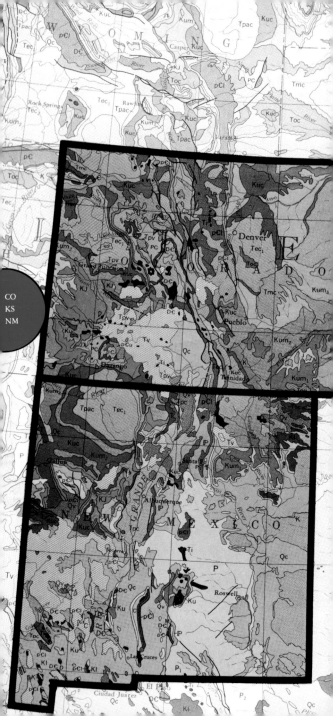

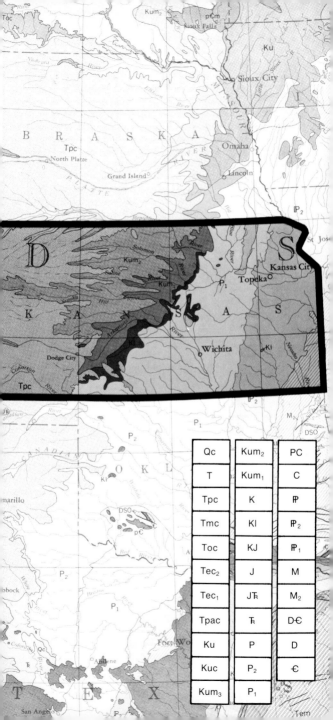

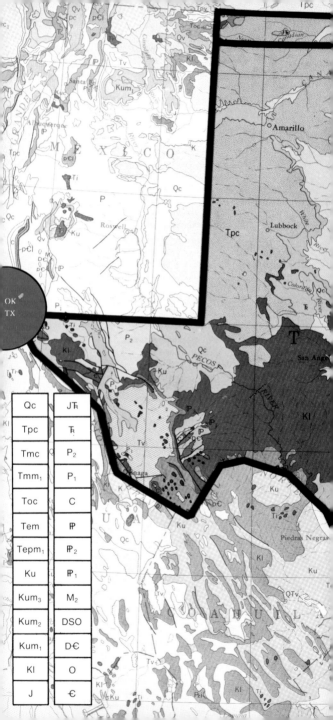

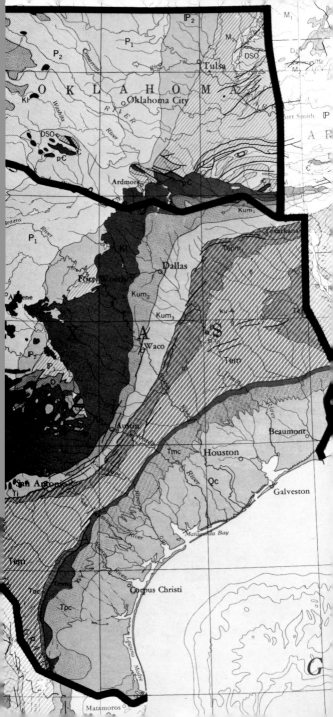

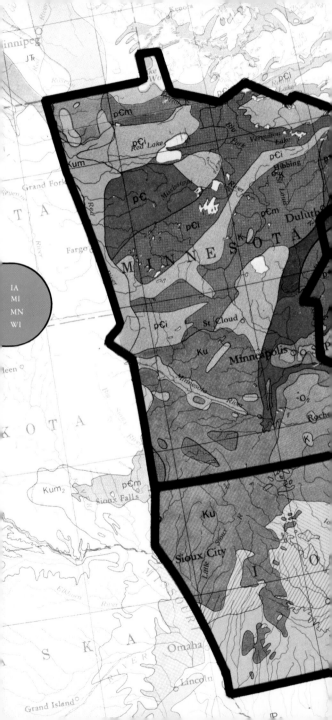

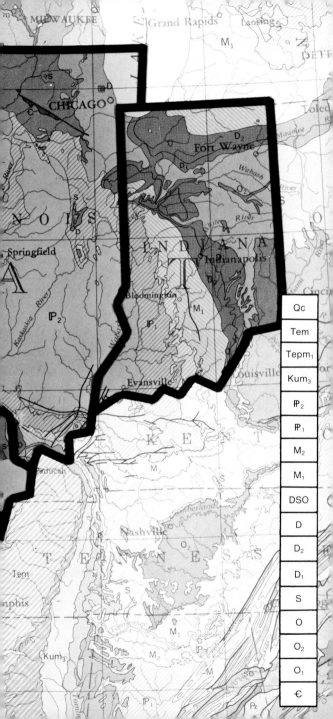

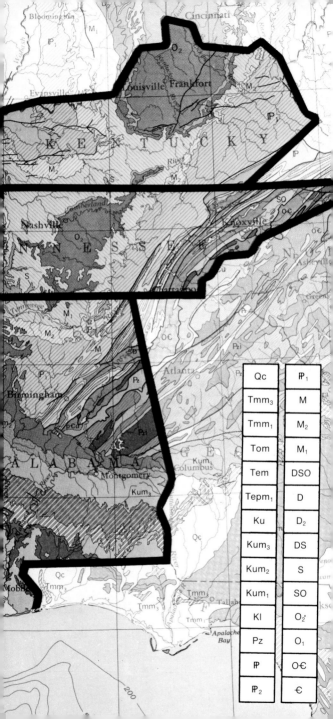

Qc	IP_1
Tmm_3	M
Tmm_1	M_2
Tom	M_1
Tem	DSO
$Tepm_1$	D
Ku	D_2
Kum_3	DS
Kum_2	S
Kum_1	SO
Kl	O_2
Pz	O_1
IP	O€
IP_2	€

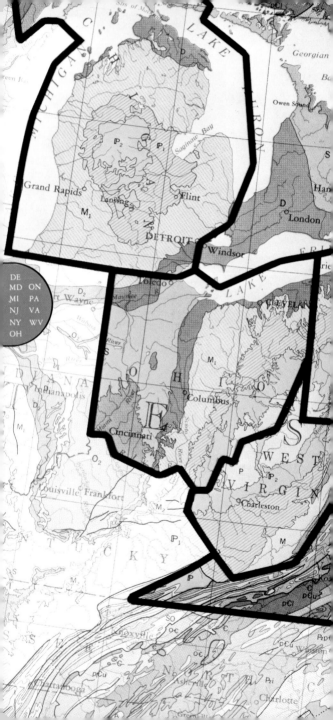

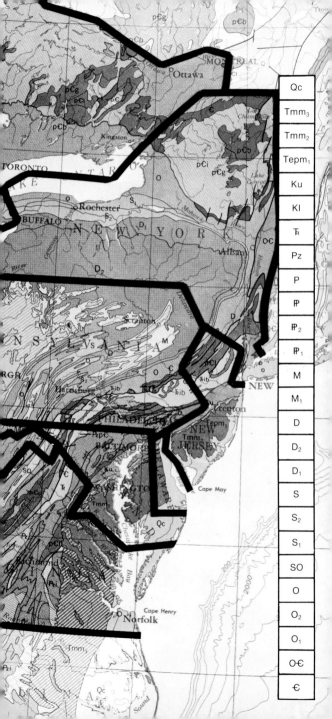

| Qc |
| Tmm₃ |
| Tmm₂ |
| Tepm₁ |
| Ku |
| Kl |
| Ŧr |
| Pz |
| P |
| ₱ |
| ₱₂ |
| ₱₁ |
| M |
| M₁ |
| D |
| D₂ |
| D₁ |
| S |
| S₂ |
| S₁ |
| SO |
| O |
| O₂ |
| O₁ |
| O€ |
| € |

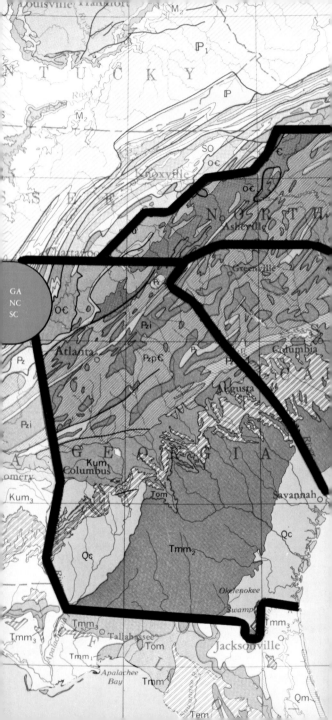

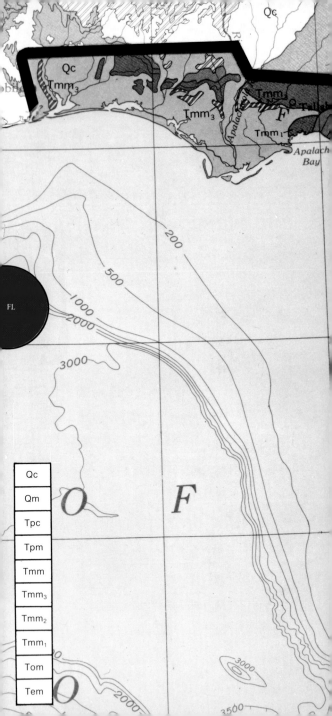

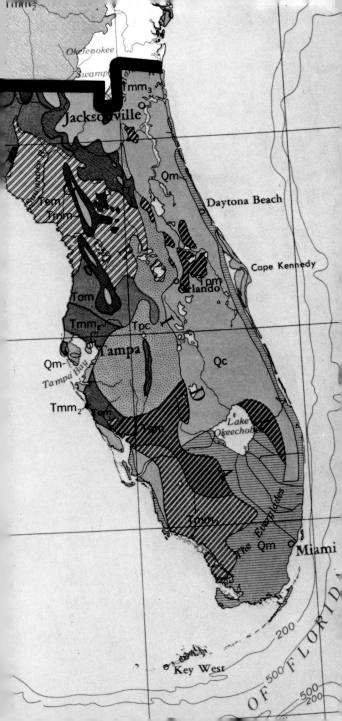

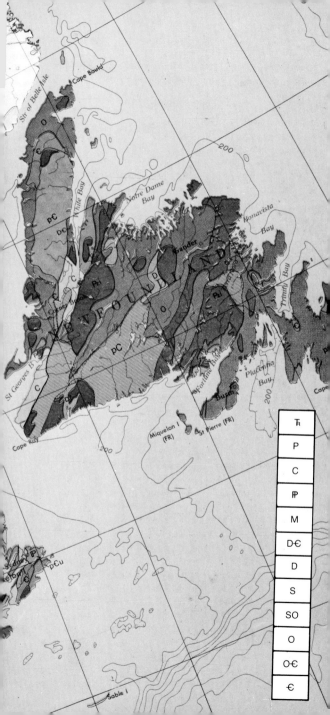

Fossiliferous Rocks and Outcroppings

The photographs in this section show various rocks and outcroppings in which fossils are commonly found. Plates 1–15 show specimens of shales, sandstones, and limestones, to assist the collector in finding and identifying sedimentary rocks that might contain fossils. Plates 16–33 show these rocks as they appear in the landscape, often interspersed with other rock types.

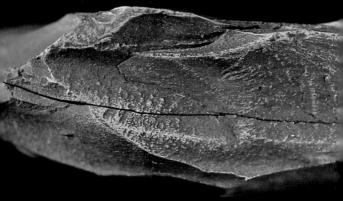

1   Black Shale, *p. 89*

2   Calcareous Shale, *p. 89*

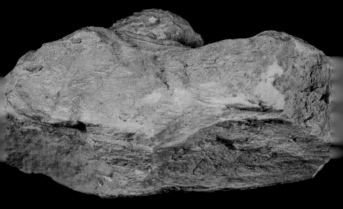

3   Mudstone, *p. 89*

4   Paper Shale, *p. 89*

5   Fossiliferous Sandstone, *p. 92*

7   Sandstone with Glauconite, *p. 92*

8   Beach Sandstone, *p. 92*

9   Siltstone, *p. 89*

10 Greensand, *p. 92*

11 Chalk, *p. 94*

13   Fossiliferous Limestone, *p. 94*

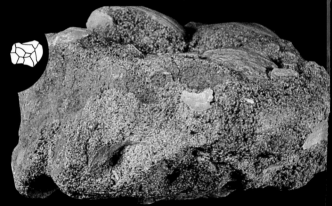

14   Greensand, *p. 92*

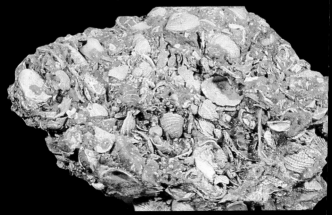

15   Coquina, *p. 94*

16 Limestone, *p. 94*

17 Sandstone, *p. 92*

18 Sandstone, *p. 92*

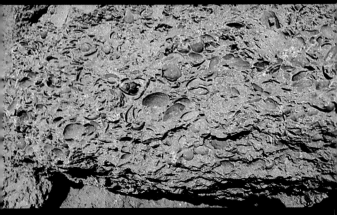

19   Fossiliferous Limestone, *p. 94*

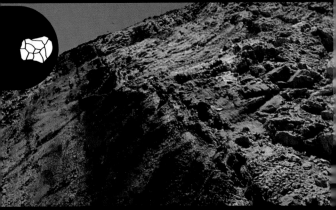

20   Sandstone, Shale, *pp. 89, 92*

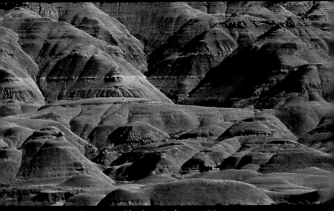

21   Freshwater Shale, Ash, *p. 89*

22   Freshwater Shale, Sandstone, *pp. 89, 92*

23   Freshwater Shale, *p. 89*

24   Freshwater Shale, Siltstone, Ash, *p. 89*

25 Sandstone, *p. 92*

26 Freshwater Shale, Sandstone, *pp. 89, 92*

27 Coquina, *p. 94*

28   Limestone, *p. 94*

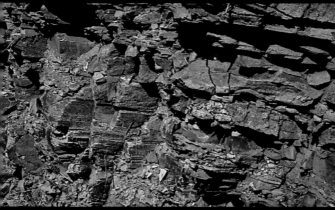

29   Lake Shale, *p. 89*

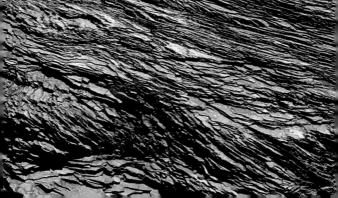

31 Coal, Shale, Sandstone, *pp. 89, 92*

32 Shale, *p. 89*

33 Shale, *p. 89*

Key to the Color Plates

The color plates on the following pages
are divided into 12 groups:

Scallop-shaped Fossils
Clam-shaped Fossils
Snail-shaped Fossils
Gastropods
Arthropods
Cystoids, Blastoids, and Crinoids
Cone-shaped Fossils
Disc-shaped Fossils
Sponges, Corals, and Bryozoans
Trace Fossils and Graptolites
Vertebrates
Plants

Thumb Tab
Guide:
To help you find the correct group, a
table of fossil silhouettes precedes the
color plates. On the left side of the
table, each group is represented by a
silhouette of a typical member of that
group. On the right, you will find the
silhouettes of fossils found within that
group. The representative silhouette for
each group is repeated as a thumb tab
at the left edge of each double page of
color plates, providing a quick and
convenient index to the color section.

Thumb Tab	Group	Plate Numbers
	Scallop-shaped Fossils	34–102
	Clam-shaped Fossils	103–165

Typical Shapes		Plate Numbers
	scallops and file shell	34–38
	cockles and other clams, rostroconch	39–45, 86, 100
	arks	46, 47
	brachiopods	48–81, 97, 98, 101, 102
	brachiopods	82–84
	brachiopods, clam	85, 87, 96
	brachiopods	88–93, 99
	brachiopods	94, 95
	brachiopods	103–109, 114, 119, 120
	brachiopods	110–113, 115, 118, 133, 134

Thumb Tab	Group	Plate Numbers
	Clam-shaped Fossils (continued)	103–165
	Snail-shaped Fossils	166–216

Typical Shapes		Plate Numbers
	arthropods	116, 117
	jingle shell, lucine, ark, mya, venus, rangia, and other clams	121–132, 135–138, 145, 146
	tellin and other clams	139, 143, 144
	geoduck and other clams	140–142
	ark, mussel, and other clams, brachiopod	147–149, 152–154
	oysters, mussel, pen shell, and other clams	150, 151, 155–159
	oysters, brachiopod, clam	160–165
	snails, worm tubes	166–180, 205, 206
	cephalopods	181–204, 207–209
	cephalopod, oysters, and snails	210–216

Thumb Tab	Group	Plate Numbers
	Gastropods	217–267
	Arthropods	268–315

Typical Shapes		Plate Numbers
	bellerophontids	217–219
	baby's ear, moon snail, and other snails	220–226
	top shell, carrier shell, and other snails	227–235, 238–240
	monoplacophore, limpet	236, 237
	olive and other snails	241–243, 252–258, 265
	vase shell, murex, and other snails, cephalopod	244–248
	cone, whelk, fig shell	249–251
	cerith, auger shell, turret shell, and other snails	259–264
	bubble shell, cowry	266, 267
	trilobites, cephalopod, horseshoe crab	268–305, 307

Thumb Tab	Group	Plate Numbers
	Arthropods (continued)	268–315
	Cystoids, Blastoids, and Crinoids	316–336
	Cone-shaped Fossils	337–357

Typical Shapes		Plate Numbers
	eurypterid	306
	shrimps	308, 309
	insects	310–315
	barnacles	316, 317
	crinoids, cephalopod	318–331
	cystoids, blastoids, crinoid	332–336
	cephalopods	337–339, 341, 342, 344
	tusk shell	340
	crinoids, tubes of uncertain affinities	343, 345–347
	corals	348–357

Thumb Tab	Group	Plate Numbers
	Disc-shaped Fossils	358–375
	Sponges, Corals, and Bryozoans	376–456

Typical Shapes		Plate Numbers
	corals, sponges	358–361, 370, 375
	sand dollars and other echinoids	362–369
	starfish, brittle star	371, 372
	echinoid	373, 374
	sponges	376, 443–445
	corals	377, 379, 389, 390, 401, 402
	bryozoans	378, 380–387
	bryozoans, sponge	388, 391–400, 423, 448–450, 453
	bryozoans	403–406, 418, 451
	sponges	407–410, 417, 421, 452

Thumb Tab	Group	Plate Numbers
	Sponges, Corals, and Bryozoans (continued)	376–456
	Trace Fossils and Graptolites	457–471
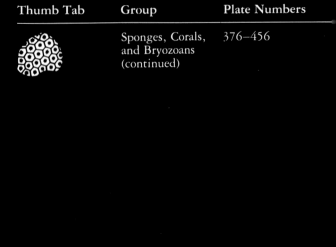	Vertebrates	472–486

Typical Shapes		Plate Numbers
	corals	411, 419, 420, 422, 424–442
	bryozoans	412–416
	sponge, coral	446, 447
	shark spine	454
	bryozoans	455, 456
	graptolites	457–463
	trace fossils, worm tubes	464–471
	dinosaur footprints	472–474
	vertebrate skulls and bone	475–477, 484
	vertebrate teeth	478–480

Thumb Tab	Group	Plate Numbers
	Vertebrates (continued)	472–486
	Plants	487–507

Typical Shapes		Plate Numbers
	bird eggs	481, 482
	coprolite	483
	turtle plate	485
	fish scales	486
	wood, alga, horsetail	487–489
	alga	490
	lycopods	491, 492
	compound leaves	493–503
	cordaite	504
	ginkgo, oak, and sycamore leaves	505–507

The color plates on the following pages correspond to the numbers preceding the text descriptions. The caption under each photograph gives the plate number, genus name, measurement, and page number of the text description. The measurement in inches indicates the maximum dimension of the specimen photographed, which may be length (*l.*), width (*w.*), height (*h.*), or for colonial corals, corallite width (*cor. w.*). The color plate number is repeated at the beginning of each text description.

Scallop-shaped Fossils

 Grouped here are shells that resemble
the familiar scallops both in shape and
in having radiating grooves and ridges
on the surface. Some, like the scallops,
are bivalves, while others are
brachiopods, whose two-part shells
resemble those of bivalves but are
oriented differently in life. Also
included is the rostroconch *Hippocardia,*
until recently thought to be a bivalve.

34   Aviculopecten, *l. 1″, p. 463*

35   Chlamys, *h. 2⅞″, p. 465*

36   Chesapecten, *l. 5½″, p. 465*

37 Pseudomonotis, *l.* 1⅛″, *p. 464*

38 Lima, *l.* 1⅛″, *p. 468*

39 Ambonychia, *h.* 1¼″, *p. 458*

40   Buchiola, *l.* ¼", *p. 451*

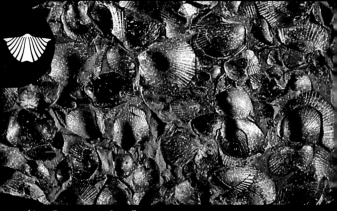

41   Oxytoma, *l.* ¼", *p. 460*

42   Cucullaea, *l.* 3⅜", *p. 454*

43   Hippocardia, *l.* 1¼″, *p. 444*

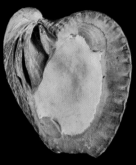

44   Venericardia, *h.* 4⅛″, *p. 480*

45   "Cerastoderma", *l.* 3½″, *p. 483*

46   Barbatia, *l.* 1⅜″, *p.* 452

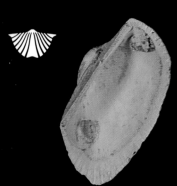 

47   Anadara, *l.* 2½″, *p.* 453

 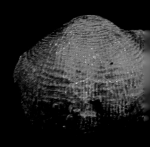

48   Reticulatia, *u.* 2⅜″, *p.* 662

49   Sowerbyella, *w.* ⅝″, *p. 649*

50   Mucrospirifer, *w.* 1⅝″, *p. 679*

51   Platystrophia, *w.* 1¼″, *p. 642*

52   Neospirifer, *w.* 2⅛″, *p.* 682

53   Spinocyrtia, *w.* 2⅜″, *p.* 680

54   Devonochonetes, *w.* ⅞″, *p.* 656

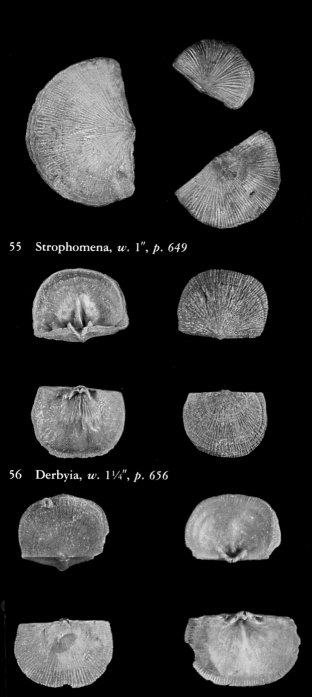

55 Strophomena, *w.* 1″, *p. 649*

56 Derbyia, *w.* 1¼″, *p. 656*

57 Schuchertella, *w.* ⅝″, *p. 655*

58   Mesolobus, *w.* ⅝″, *p.* 657

59   Linoproductus, *w.* 1¼″, *p.* 663

60   Juresania, *w.* ⅞″, *p.* 661

61   Leptaena, *w.* ⅞″, *p. 651*

62   Echinoconchus, *w.* 2″, *p. 660*

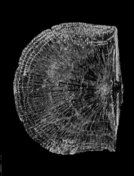

63   Strophodonta, *w.* 1¼″, *p. 652*

64   Douvillina, *w.* ¾″, *p. 653*

65   Rafinesquina, *w.* 1⅛″, *p. 650*

 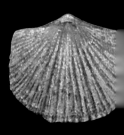

66   Glyptorthis, *w.* ¾″, *p. 639*

67   Dinorthis, *w.* 1⅜″, *p. 639*

68   Onniella, *w.* ½″, *p. 645*

69   Tropidoleptus, *w.* 1¼″, *p. 647*

70   Rhipidomella, *w.* 1⅜″, *p. 647*

71   Dolerorthis, *w.* 1⅛″, *p. 637*

72   Hesperorthis, *w.* ½″, *p. 638*

73 Dalmanella, *w.* ½″, *p. 644*

74 Desquamatia, *w.* 1⅜″, *p. 672*

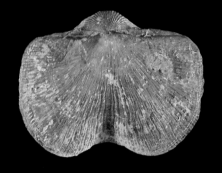

76   Hebertella, *w.* 1¾", *p. 642*

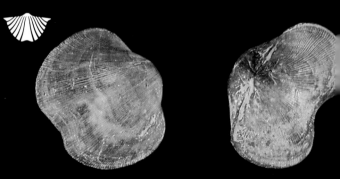

77   Schizophoria, *w.* 1", *p. 644*

78   Petrocrania, *l.* ½", *p. 632*

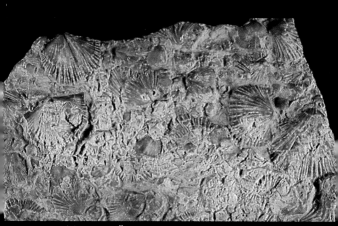

79   Eoorthis, *w.* ¾″, *p. 636*

80   Cupularostrum, *w.* ½″, *p. 668*

81   Zygospira, *w.* ¼″, *p. 671*

82   Conchidium, *l. 2", p. 664*

83   Rensselaeria, *l. 3⅜", p. 683*

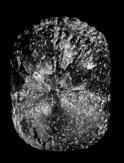

84   Uncinulus, *w. ⅝", p. 669*

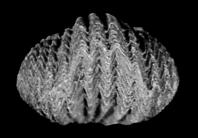

85 Rhynchotrema, *w.* ½″, *p.* 667

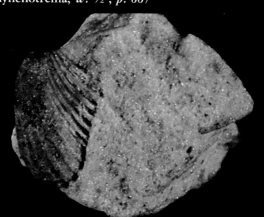

86 Hippocardia, *l.* 1⅛″, *p.* 444

87 Scabrotrigonia, *l.* 1⅛″, *p.* 476

88   Eatonia, *w.* 1″, *p.* 670

89   Rhynchotrema, *w.* 1½″, *p.* 667

90   Hustedia, *l.* ⅜″, *p.* 673

91 **Rhynchotreta**, *l. ½″, p. 666*

92 **Meekella**, *w. ¾″, p. 654*

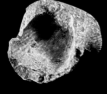

93 **Diaphragmus**, *l. ¾″, p. 659*

94   Cyrtina, *w.* ⅜", *p.* 678

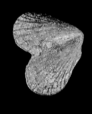 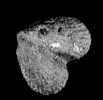

95   Dicoelosia, *w.* ⅜", *p.* 646

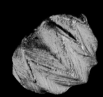

96   Enteletes, *w.* ⅞", *p.* 643

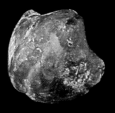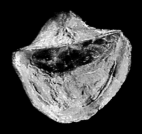

97   Ambocoelia, *w.* ⅜″, *p.* 678

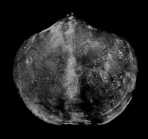

98   Eospirifer, *w.* 1⅜″, *p.* 677

99   Terebratulina, *l.* ¼″, *p.* 685

100   Protocardia, *l. 2¾″, p. 484*

101   Billingsella, *w. ⅜″, p. 636*

102   Finkelnburgia, *w. ⅜″, p. 640*

Clam-shaped Fossils

 These fossils include not only the
ancestors of the living clams, but also
some brachiopods and crustaceans
whose shells resemble those of clams in
shape. The shells in this group vary
from rounded to elongate, and their
surfaces often have concentric grooves.
Fossils of these shells often consist of
internal molds that clearly reproduce
fine details of the shell's original
structure.

103 Beecheria, *l.* ⅝″, *p. 684*

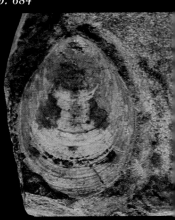

104 Lingulella, *l.* ⅝″, *p. 629*

105 Oleneothyris, *l.* 2″, *p. 684*

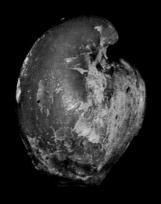

106 Gypidula, *u*. 1″, *p. 665*

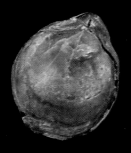 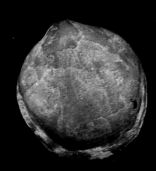

107 Waconella, *l.* ⅞″, *p. 686*

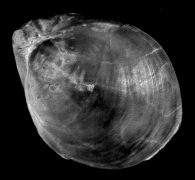

108 Cranaena, *l.* ⅝″, *p. 685*

109   Meristella, *w.* ⅝″, *p.* 674

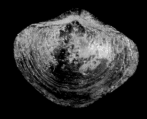

110   Athyris, *w.* 1¼″, *p.* 675

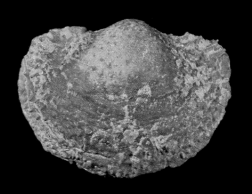

111   Productella, *w.* 1¼″, *p.* 658

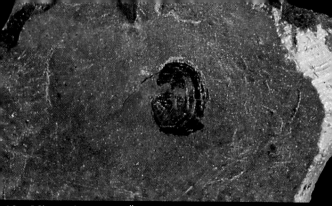

12 Micromitra, *w.* ⅛″, *p. 634*

13 Acrothele, *w.* ⅜″, *p. 630*

115   Obolella, *w.* ½″, *p. 633*

116   Leperditia, *l.* ½″, *p. 587*

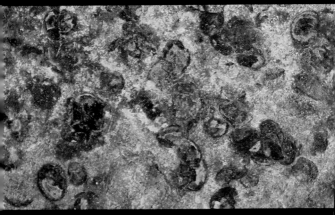

117   Cyzicus, *l.* ⅛″, *p. 588*

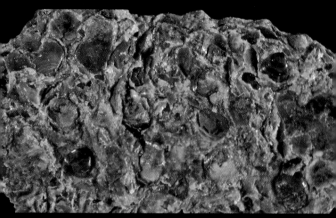

118 Dicellomus, *w.* ¹⁄₁₆″, *p. 628*

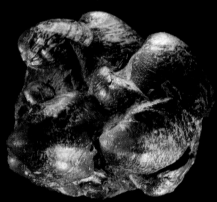

119 Pentamerus, *l.* 4″, *p. 664*

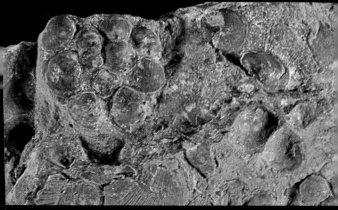

120 Schizocrania, *w.* ¹⁄₄″, *p. 631*

121   Pterinopecten, *l.* 1″, *p. 462*

122   Grammysia, *l.* 2″, *p. 498*

123   Edmondia, *l.* ⅞″, *p. 496*

124   Anomia, *l.* 1¼", *p.* 467

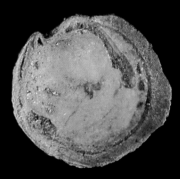

125   Codakia, *l.* ⅝", *p.* 478

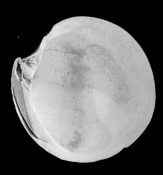

126   Dosinia, *l.* 2⅝", *p.* 489

127   Paracyclas, *l. ¾", p.* 479

128   Glycymeris, *l. 1⅝", p.* 455

129   Corbicula, *l. ¾", p.* 487

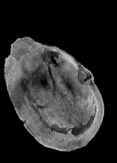

130   Mya, *l. 2⅞″, p. 493*

131   Pitar, *l. 1½″, p. 488*

132   Rangia, *h. 2⅜″, p. 485*

133   Cleiothyridina, *w.* ¾″, *p. 676*

134   Composita, *w.* 1″, *p. 677*

135   Mercenaria, *l.* 3¾″, *p. 492*

136   Lirophora, *l.* ¾", *p. 491*

137   Lirodiscus, *l.* 1", *p. 481*

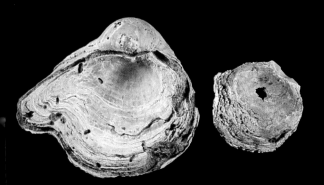

138   Pycnodonte, *l.* 4¾", *p. 469*

139   Corbula, *l.* 1″, *p. 494*

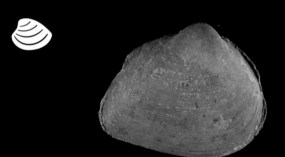

140   Schizodus, *l.* 1½″, *p. 476*

141   Mya, *l.* 1⅝″, *p. 493*

142   Panopea, *l. 6¼", p. 496*

143   Bathytormus, *l. 1½", p. 482*

144   Tellina, *l. 2¼", p. 486*

145   Nucula, *l.* 1⅛", *p. 448*

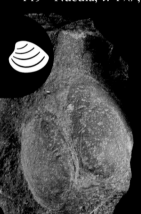

146   Palaeoneilo, *l.* ½", *p. 449*

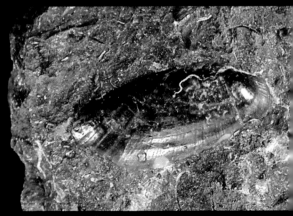

147   Solemya, *l.* ⅝", *p. 451*

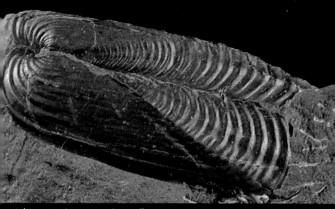

148 Orthonota, *l.* 1⅜″, *p. 498*

149 Parallelodon, *l.* ¾″, *p. 453*

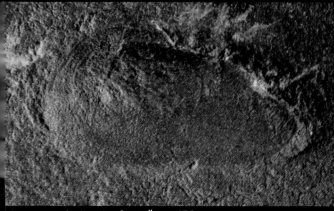

150 Modiolopsis, *l.* 1¼″, *p. 475*

151 Ctenodonta, *l. 1¼", p. 447*

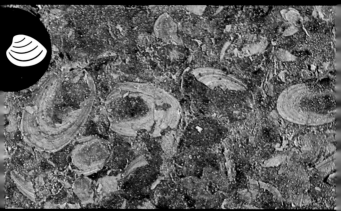

152 Lingula, *l. ⅝", p. 627*

153 Pholadomya, *l. 2½", p. 499*

154 Nuculana, *l.* 1¼″, *p. 450*

155 Promytilus, *h.* 1⅛″, *p. 456*

156 Leptodesma, *l.* 1⅜″, *p. 460*

157 Crassostrea, *h.* 2¼″, *p.* 472

158 Inoceramus, *h.* 6¼″, *p.* 461

159 Pinna, *h.* 5⅞″, *p.* 457

160   Ostrea, *h.* 5⅛″, *p. 473*

161   Spinatrypa, *w.* 1″, *p. 672*

162   Inoceramus, *h.* 3″, *p. 461*

163   Lopha, *h.* 3½″, *p.* 474

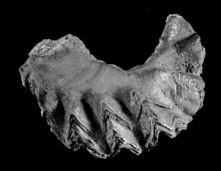

164   Agerostrea, *h.* ⅞″, *p.* 473

165   Myalina, *h.* 3⅛″, *p.* 459

Snail-shaped Fossils

The fossils in this group have distinctively coiled shells, and include many snails, as well as the cephalopods. Some are tightly coiled, while in others the shell is more loosely coiled, so that the whorls are not in contact. Many of these shells are sculptured and grooved, and some retain the pearly luster they had when the animal was alive.

166   Shansiella, *h.* ½", *p. 409*

167   Bembexia, *w.* ¾", *p. 405*

168   Maclurites, *w.* 1⅞", *p. 399*

169   Trepospira, *w.* 1⅛″, *p. 404*

170   Gyrodes, *w.* 1⅜″, *p. 429*

171   Straparollus, *w.* ½″, *p. 402*

172   Helicotoma, *w. 1⅝″, p. 401*

173   Planorbis, *w. ⅞″, p. 443*

174   Tropidodiscus, *w. 1″, p. 395*

175 Planorbis, *w.* 7/8″, *p.* 443

176 Bucanella, *w.* ½″, *p.* 394

177 Heliostoma, *w.* 3/4″, *p.* 491

178   Lecanospira, *w.* 1⅜″, *p.* 399

179   Maclurites, *w.* 4⅛″, *p.* 399

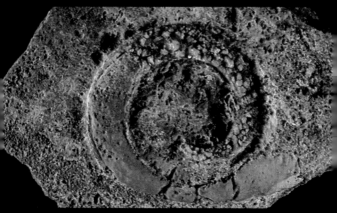

180   Ecculiomphalus, *w.* 1″, *p. 401*

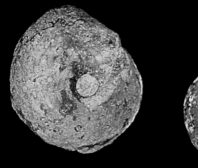

181　Waagenoceras, *l. 1¼", p. 522*

182　Desmoceras, *l. 10", p. 530*

83　Shumardites, *l. 3⅜", p. 521*

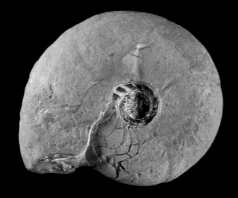

184   Muensteroceras, *l. 2⅛″, p. 523*

185   Meekoceras, *l. 3⅛″, p. 525*

186   Prionocyclus, *l. 2¾″, p. 533*

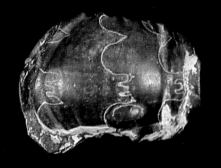

187   Gastrioceras, *l.* ⅝", *p. 524*

188   Gastrioceras, *l.* 1¼", *p. 524*

189   Eutrephoceras, *l.* 2⅛", *p. 516*

190 Muensteroceras, *l. 2⅛″, p. 523*

191 Imitoceras, *l. 3⅜″, p. 520*

192 Tornoceras, *l. 1¾″, p. 520*

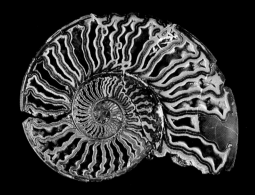

193   Placenticeras, *l. 9½″, p. 531*

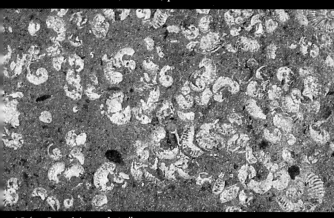

194   Scaphites, *l. ⅜″, p. 529*

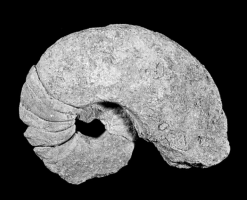

195   Centroceras, *l. 6¼″, p. 515*

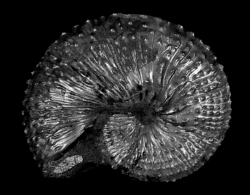

196   Hoploscaphites, *l. 4⅝", p. 529*

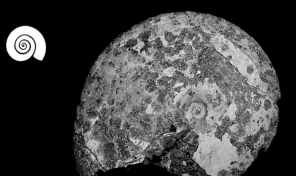

197   Placenticeras, *l. 9½", p. 531*

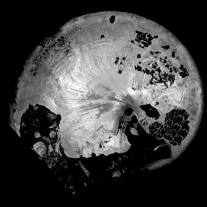

198   Sphenodiscus, *l. 3⅜", p. 535*

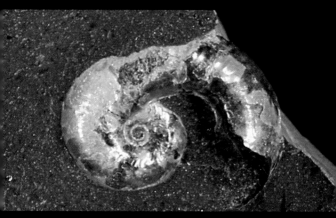

199   Scaphites, *l. ¾″*, *p. 529*

200   Temnocheilus, *l. 1¼″*, *p. 514*

201   Metoicoceras, *l. 2″*, *p. 533*

202   Texanites, *l.* 3⅞″, *p. 534*

203   Goodhallites, *l.* 3″, *p. 532*

204   Bickmorites, *l.* 8″, *p. 512*

205   Rotularia, *l.* ⅜", *p. 540*

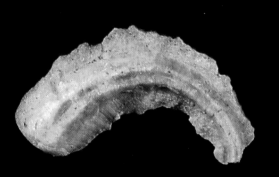

206   Hamulus, *l.* ¾", *p. 539*

207   Phragmoceras, *l.* 1¾", *p. 511*

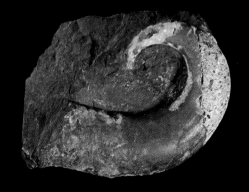

208   Oncoceras, *l.* 3⅛", *p. 510*

209   Oxybeloceras, *l.* 1⅜", *p. 526*

210   Turrilites, *l.* 1¼", *p. 528*

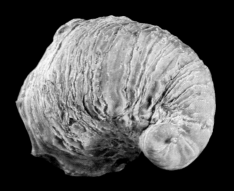

211   Exogyra, *h.* 4″, *p. 470*

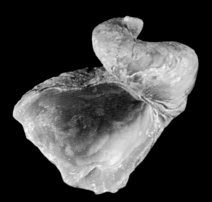

212   Ilymatogyra, *h.* 1¼″, *p. 471*

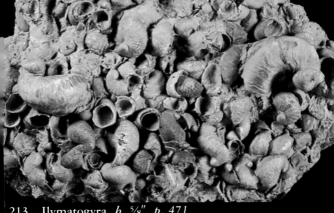

213   Ilymatogyra, *h.* ⅝″, *p. 471*

214   Crepidula, *w.* 2⅛″, *p.* 428

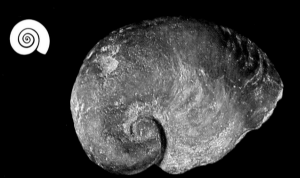

215   Platyceras, *w.* ¾″, *p.* 415

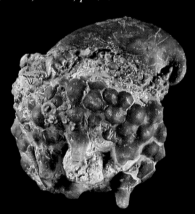

216   Platyceras, *h.* 1¼″, *p.* 415

Gastropods

The fossils in this group are shells that are coiled into a spire, or are elaborately ornamented with spines or with intricate grooves and ridges. Most of these fossils are snails, but *Cooperoceras*, a cephalopod whose spines resemble those of some snails, is also included in this group.

217 Knightites, *w.* 1⅛", *p.* 397

218 Bellerophon, *w.* 1", *p.* 396

219 Euphemites, *h.* ⅜", *p.* 394

220   Sinum, *h.* ⅝″, *p. 431*

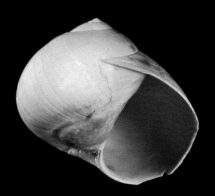

221   Polinices, *h.* 1½″, *p. 430*

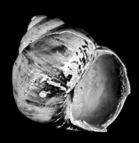

222   Viviparus, *h.* 1″, *p. 420*

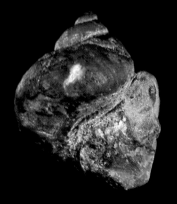

223   Holopea, *h.* 1″, *p. 413*

224   Naticopsis, *h.* 1⅝″, *p. 418*

225   Trochonema, *h.* 1″, *p. 411*

226   Cyclonema, *h.* 1″, *p. 414*

227   Trepospira, *w.* 1″, *p. 404*

228   Omphalotrochus, *w.* 1¾″, *p. 403*

229   Calliostoma, *w.* 1⅜″, *p. 416*

230   Calliostoma, *h.* ¾″, *p. 416*

231   Xenophora, *w.* 1¼″, *p. 427*

232 Glabrocingulum, *h.* ¾″, *p. 406*

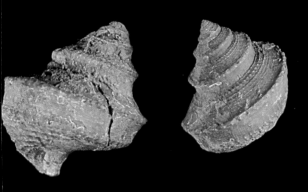

233 Worthenia, *h.* 1⅛″, *p. 408*

234 Loxoplocus, *h.* 1″, *p. 407*

235   Trochonema, *h. 2½″, p. 411*

236   Scenella, *h. ⅜″, p. 390*

237   Emarginula, *w. ⅝″, p. 412*

238 Ceratopea, *h.* 1⅛″, *p. 410*

239 Baylea, *h.* ⅜″, *p. 405*

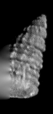 

240 Microdoma, *h.* ¼″, *p. 416*

241   Cancellaria, *h.* ½", *p.* 437

242   Fasciolaria, *h.* ¾", *p.* 436

243   Neptunea, *h.* 1½", *p.* 435

244   Ecphora, *h.* 2⅛*", p. 433*

245   Vasum, *h.* 3½*", p. 439*

246   Murex, *h.* 1⅜*", p. 433*

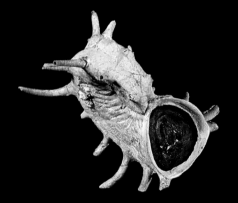

247   Cooperoceras, *l.* 5¼″, *p. 513*

248   Typhis, *h.* ⅝″, *p. 434*

249   Conus, *h.* 1½″, *p. 440*

50 Busycon, *h.* 5", *p.* 435

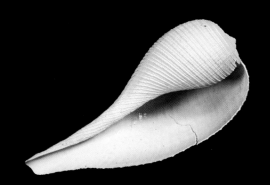

251 Ficus, *h.* 2¾", *p.* 432

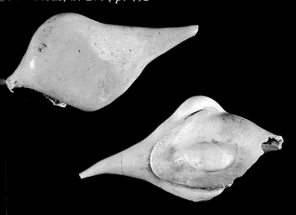

52 Calyptraphorus, *h.* 1¾", *p.* 426

253   Volutomorpha, *h.* 3⅜″, *p. 438*

254   Volutoderma, *h.* 6⅛″, *p. 439*

255   Anchura, *h.* 1¼″, *p. 427*

256   Ianthinopsis, *h.* 1⅛″, *p. 421*

257   Clathrodrillia, *h.* ¾″, *p. 441*

258   Goniobasis, *h.* ½″, *p. 425*

259   Loxonema, *h.* ⅝″, *p. 423*

260   Murchisonia, *h.* 1½″, *p. 419*

261   Meekospira, *h.* 1½″, *p. 422*

262   **Cerithium**, *h. 1⅛″, p. 424*

263   **Terebra**, *h. 1″, p. 441*

264   **Turritella**, *h. 2″, p. 424*

265   Olivella, *h.* ⅝″, *p.* 437

266   Bulla, *h.* ½″, *p.* 442

267   Cypraea, *h.* 1⅝″, *p.* 429

Arthropods

The arthropods are easily recognized by their clearly segmented bodies, which often bear leglike appendages. These appendages themselves are segmented, giving the group the name "arthropod," which means "jointed foot." Included here are the extinct trilobites and eurypterids, as well as the horseshoe crabs, true crabs and other crustaceans, and the insects, which still survive today.

268 Ptychagnostus, *l.* 1/4", *p. 552*

269 Homotelus, *l.* 5", *p. 570*

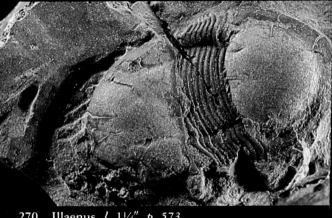

270 Illaenus, *l.* 1 1/4", *p. 573*

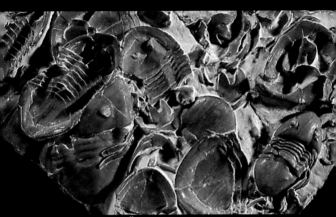

271   Homotelus, *l.* ¾″, *p.* 570

272   Bumastus, *l.* 1½″, *p.* 573

273   Dipleura, *l.* 1¾″, *p.* 582

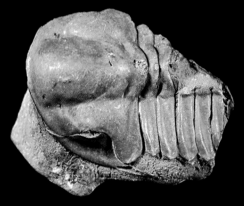

274 Olenoides, *l.* 1⅝″, *p. 558*

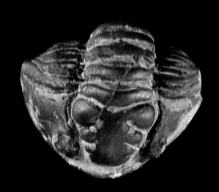

275 Flexicalymene, *l.* ¾″, *p. 581*

276 Encrinurus, *w.* 1″, *p. 579*

277 Griffithides, *l.* 1⅝", *p. 574*

278 Phacops, *l.* 1⅝", *p. 582*

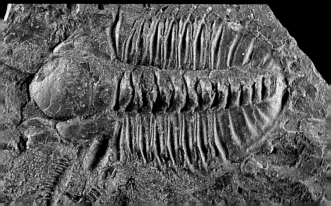

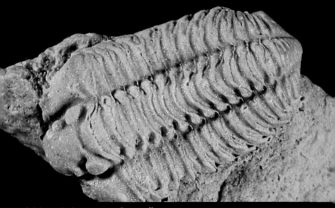

280　Calymene, *l.* 1⅛″, *p. 580*

281　Cedaria, *l.* ⅜″, *p. 569*

282　Bathyuriscus, *l.* ½″, *p. 561*

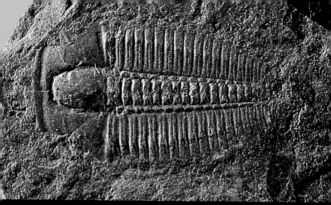

283 Elrathia, *l.* 1", *p.* 565

284 Ogygopsis, *l.* 3¼", *p.* 560

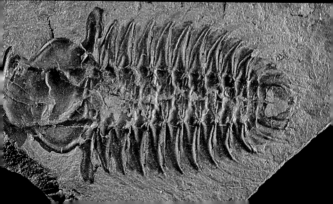

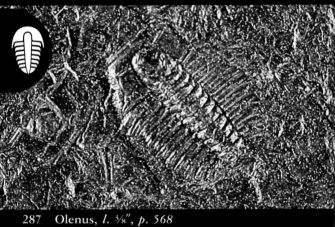

286 Triarthrus, *l.* ¾", *p. 568*

287 Olenus, *l.* ⅜", *p. 568*

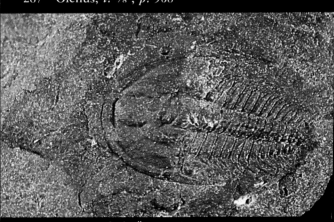

288 Holmia, *l.* 6¼", *p. 555*

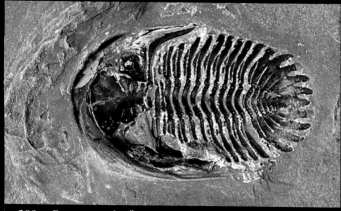

289   Greenops, *l.* 1″, *p. 585*

290   Dalmanites, *l.* 2¾″, *p. 583*

291   Albertella, *l.* ⅞″, *p. 563*

292   Olenellus, *l.* 3″, *p. 553*

293   Paedeumias, *l.* 1½″, *p. 554*

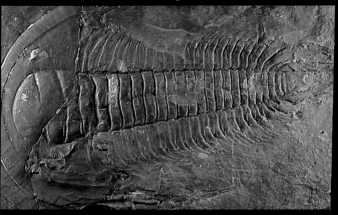

294   Paradoxides, *l.* 10½″, *p. 556*

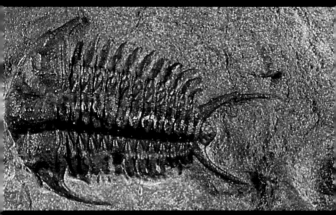

295 Ceraurus, *l.* 1¼", *p.* 578

296 Bathynotus, *l.* 2", *p.* 557

297 Tricrepicephalus, *l.* 2⅜", *p.* 566

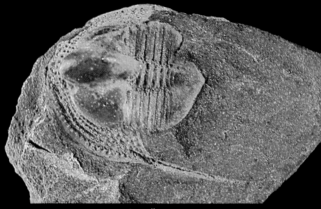

298   Cryptolithus, *l. ½", p. 575*

299   Gonioceras, *l. 2¾", p. 506*

300   Encrinurus, *l. ⅝", p. 579*

301   Pseudogygites, *w.* 1¼″, *p.* 572

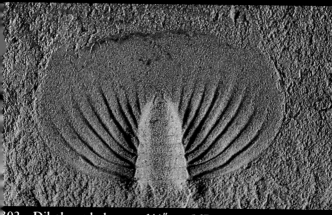

302   Dikelocephalus, *w.* 1¼″, *p.* 567

304   Euproops, *l.* 1″, *p. 593*

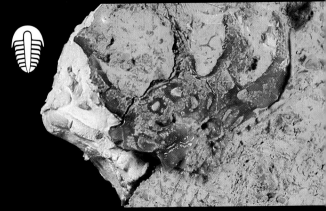

305   Cheirurus, *w.* 1¼″, *p. 577*

306   Eurypterus, *l.* 5½″, *p. 595*

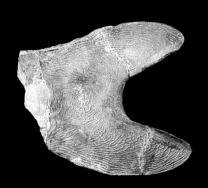

307 Homotelus, *l.* 2¾″, *p.* 570

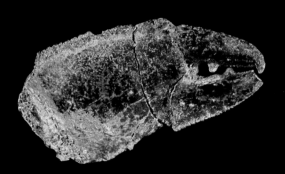

308 Callianassa, *l.* 1⅝″, *p.* 591

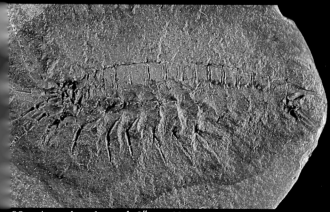

09 Acanthotelson, *l.* 2″, *p.* 591

310   Nematoceran, *l.* ⅛″, *p. 751*

311   Pronemobius, *l.* ⅜″, *p. 751*

312   Tipula, *l.* ⅜″, *p. 751*

313   **Eristalis,** *l.* ⅜″, *p. 751*

314   **Coleopteran,** *l.* ⅝″, *p. 751*

Cystoids, Blastoids, and Crinoids

These fossils have compact bodies whose surfaces are divided into geometric patterns by grooves or ridges. Many of them are borne on a stem made up of many small discs. Long, slender arms give many a flowerlike appearance, while those without arms often resemble a flower bud. Also included in this group are the barnacles, which look like clusters of flower buds, and a cephalopod whose long, segmented tube resembles a stem.

316   Balanus, *h.* 6¼″, *p. 589*

317   Balanus, *h.* ⅝″, *p. 589*

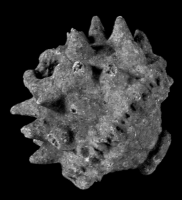

318   Batocrinus, *h.* 1⅛″, *p. 696*

519 Dorycrinus, *h. 1″, p. 695*

520 Michelinoceras, *l. 2″, p. 507*

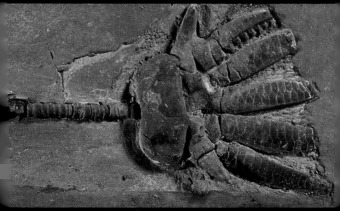

521 Delocrinus, *h. 2¾″, p. 702*

322 Taxocrinus, *h.* 4⅝″, *p.* 704

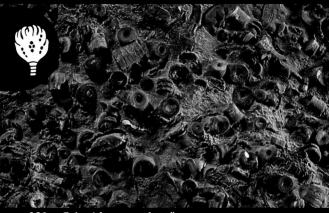

323 Crinoid stems, *l.* ¼″, *p.* 694

324 Onychocrinus, *h.* 2⅜″, *p.* 705

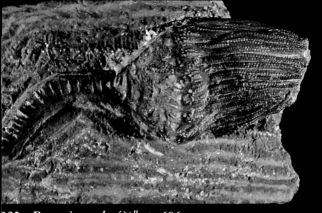

325 Batocrinus, *h.* 4⅜″, *p.* 696

326 Cyathocrinites, *h.* 3⅛″, *p.* 701

27 Platycrinites, *h.* 4″, *p.* 700

328 Uintacrinus, *h. 4¼", p. 706*

329 Eutrochocrinus, *h. 4", p. 697*

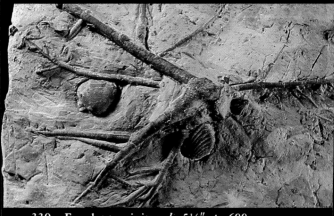

330 Eucalyptocrinites, *h. 5½", p. 699*

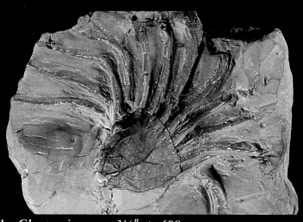

331 Glyptocrinus, *w.* 3⅛″, *p. 698*

332 Caryocrinites, *h.* 1⅜″, *p. 688*

33 Pentremites, *h.* 1¼″, *p. 692*

334   Troosticrinus, *h.* ⅞", *p. 690*

335   Holocystites, *h.* 2¾", *p. 689*

336   Eucalyptocrinites, *h.* 1⅜", *p. 699*

Cone-shaped Fossils

All of the fossils in this section taper to a point and form a cone that may be long and slender or short and thick. Some are straight, while others are gracefully curved. Included here are the shells of a scaphopod and some cephalopods, the tapering stem of a crinoid, the tubes of annelids and other burrowing animals, and the conical calices of some solitary corals.

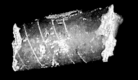
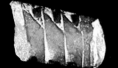
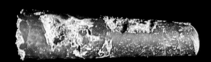

337   Bactrites, *l.* ½″, *p. 518*

338   Baculites, *l.* 3¾″, *p. 527*

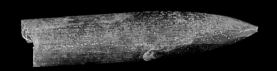

339   Belemnitella, *l.* 6¾″, *p. 536*

340   Dentalium, *l.* 1⅞", *p. 500*

341   Endoceras, *l.* 11¾", *p. 503*

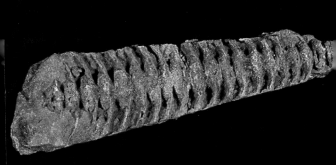

342   Actinoceras, *l.* 2¾", *p. 504*

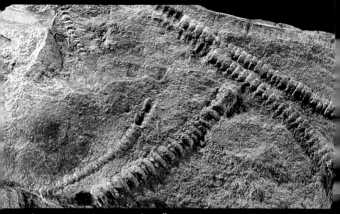

343 Crinoid stems, *l. ½″, p. 694*

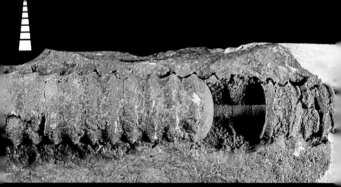

344 Dawsonoceras, *l. 6¼″, p. 508*

345 Cornulites, *l. ¾″, p. 545*

346 Tentaculites, *l. 2⅜″, p. 542*

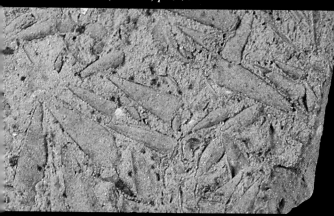

347 Hyolithes, *l. 1⅜″, p. 543*

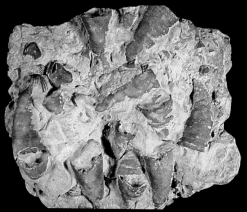

348 Grewingkia, *l. 2⅜″, p. 361*

349   Grewingkia, *l.* 3⅛″, *p. 361*

350   Heliophyllum, *l.* 1¼″, *p. 364*

351   Heterophrentis, *l.* 3½″, *p. 360*

352 Heterophrentis, *l. 2″, p. 360*

353 Parasmilia, *l. ⅞″, p. 375*

354 Flabellum, *l. ¾″, p. 376*

Cone-shaped Fossils

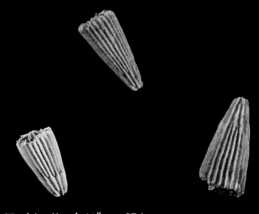

355 Turbinolia, *l.* ¼″, *p.* 374

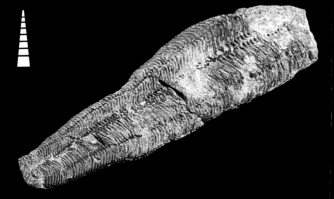

356 "Conularia," *l.* 4″, *p.* 355

357 "Conularia," *l.* 3⅛″, *p.* 355

Disc-shaped Fossils

Most of the disc-shaped fossils are round in outline and somewhat flattened. Their surfaces are often ornamented—with fine septa in corals, with petal-like rows of pores or sculptured knobs and ridges in echinoids, or with the openings of tubes in sponges. Also included here are a starfish and a brittle star, animals with disc-shaped bodies and long, tapering arms.

358   Trochocyathus, *w.* ⅜″, *p. 374*

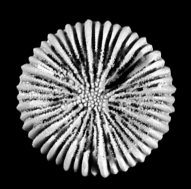

359   Trochocyathus, *w.* ⅜″, *p. 374*

360   Hadrophyllum, *w.* ½″, *p. 362*

361　Astylospongia, *l. 2″, p. 341*

362　Holectypus, *w. 2″, p. 714*

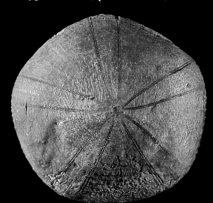

363　Holectypus, *w. 2″, p. 714*

364   Periarchus, *w.* 2¾″, *p. 715*

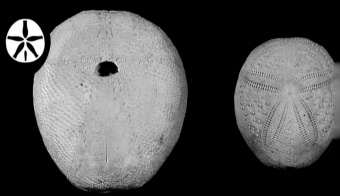

365   Eupatagus, *l.* 2⅛″, *p. 721*

366   Rhyncholampas, *l.* 1⅝″, *p. 719*

367 Hemiaster, *l.* 1½", *p.* 720

368 Dendraster, *w.* 1¾", *p.* 717

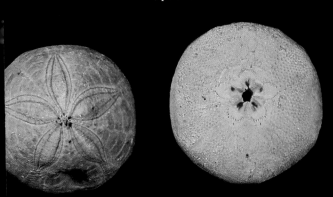

369 Hardouinia, *l.* 1⅞", *p.* 718

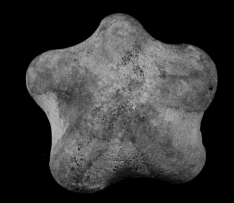

370   Brooksella, *w.* 1⅝″, *p. 354*

371   Hudsonaster, *w.* 1″, *p. 707*

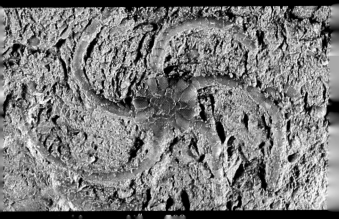

373 Stereocidaris, *w.* 3⅜″, *p. 713*

374 Stereocidaris, *w.* 3⅜″, *p. 713*

375 Brachiospongia, *l.* 4⅛″, *p. 345*

Sponges, Corals, and Bryozoans

 The fossils of sponges, corals, and bryozoans may consist of remains of whole animals or colonies, or of fragments. The surfaces of these fossils bear many small openings, pits, or cups. In sponges, the openings lead into tubes that penetrated the interior of the animal. In corals and bryozoans, the pits and cups each contained an individual animal when the colony was alive. Some of these fossils are rounded and massive, while others, especially among the corals and bryozoans, may be delicately branching. In still others, the colony was a thin sheet on the surface of a rock. Also included in this section is a slender spine of the shark *Orthacanthus*, whose finely serrated edge gives it a resemblance to a colonial bryozoan.

376 Parallelopora, *l. 2", p. 350*

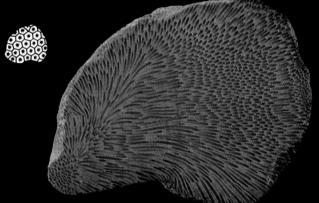

377 Favosites, *cor. w.* 1/16*", p. 381*

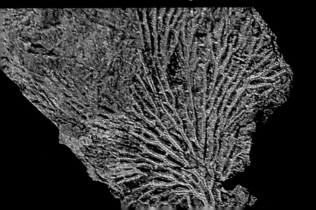

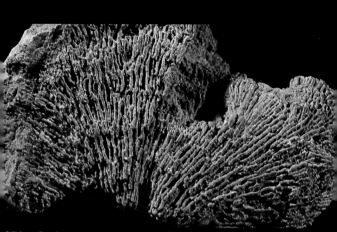

379 Syringopora, *cor. w.* 1/32", *p. 388*

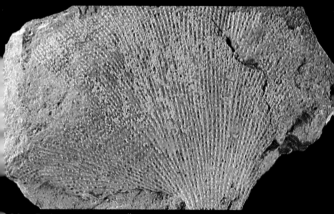

380 Polypora, *w.* 3¾", *p. 613*

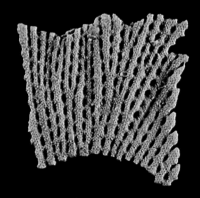

381 Polypora, *h.* ⅝", *p. 613*

382   Conopeum, *w.* 1″, *p. 620*

383   Schizoporella, *w.* 1⅜″, *p. 622*

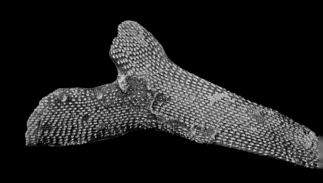

384   Rhinidictya, *l.* ½″, *p. 616*

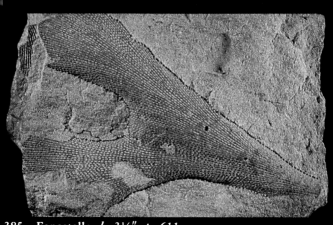

385   Fenestella, *h*. 3⅛", *p. 611*

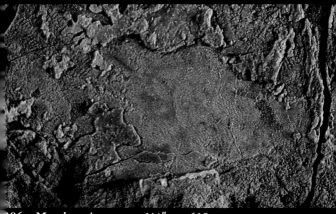

386   Membranipora, *w*. 1¼", *p. 619*

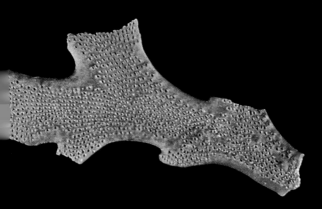

387   Sulcoretepora, *l*. ¾", *p. 617*

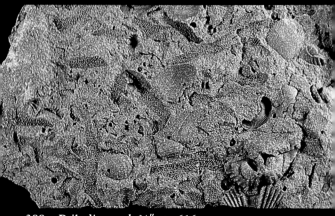

388   Ptilodictya, *l.* ¾″, *p. 616*

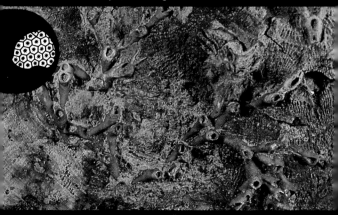

389   Aulopora, *cor. w.* ¼″, *p. 387*

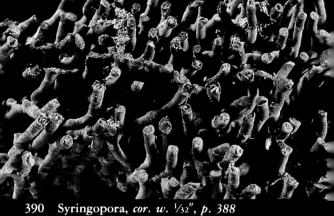

390   Syringopora, *cor. w.* ¹/₃₂″, *p. 388*

391   Dystactospongia, *l. 5⅞″*, *p. 342*

392   Monticulipora, *l. 4″*, *p. 606*

93   Parvohallopora, *l. 2⅛″*, *p. 609*

394   Acanthocladia, *l.* ⅝″, *p. 614*

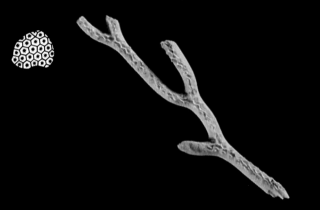

395   Idmidronea, *l.* ⅝″, *p. 599*

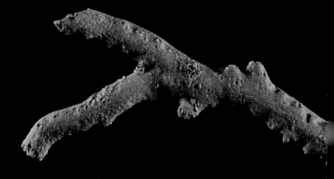

396   Hornera, *l.* ¾″, *p. 600*

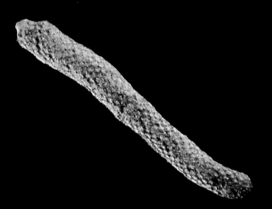

397   **Rhombopora,** *l.* ⅝*", p. 615*

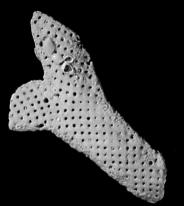

398   **Coscinopleura,** *l.* ¼*", p. 621*

99   **Tabulipora,** *l.* 1*", p. 608*

400   Rhombotrypa, *l.* 1⅝″, *p. 609*

401   Chaetetes, *cor. w.* ⅟₅₀″, *p. 378*

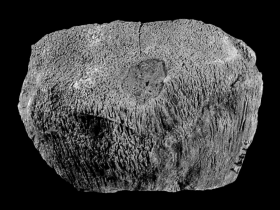

402   Tetradium, *cor. w.* ⅟₃₂″, *p. 379*

403　Ceriopora, *h.* ½″, *p. 600*

404　Celleporaria, *w.* 2⅜″, *p. 622*

05　Ropalonaria, *w.* 1″, *p. 618*

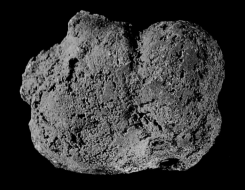

406   Ceramopora, *w.* 2⅛", *p. 603*

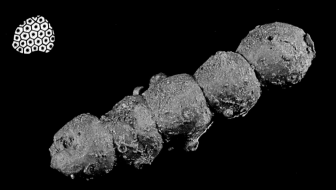

407   Girtyocoelia, *l.* 2¾", *p. 347*

408   Hindia, *l.* 2¾", *p. 341*

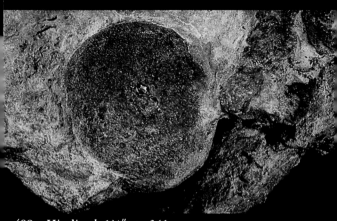

409 Hindia, *l.* 1⅜", *p. 341*

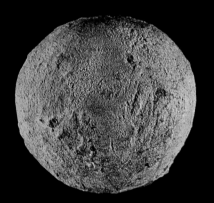

410 Hindia, *l.* 1¼", *p. 341*

411 Cladopora *cor. w.* 1/32", *p. 382*

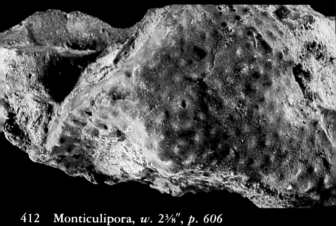

412   Monticulipora, *w.* 2⅜″, *p. 606*

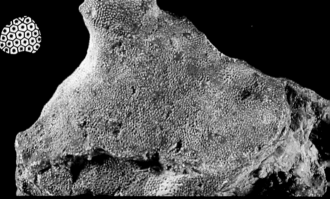

413   Fistulipora, *h.* 2⅜″, *p. 603*

414   Schizoporella, *w.* ¾″, *p. 622*

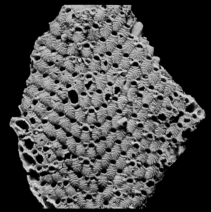

415   Cribrilaria, *h.* ⅝", *p. 621*

416   Fistulipora, *h.* ⅜", *p. 603*

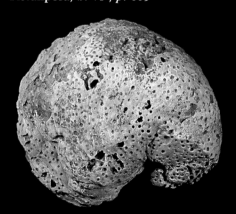

417   Cliona, *w.* 1/16", *p. 340*

418   Prasopora, *h.* ⅝″, *p.* 607

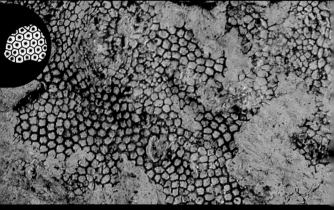

419   Tetradium, *cor. w.* 1/32″, *p.* 379

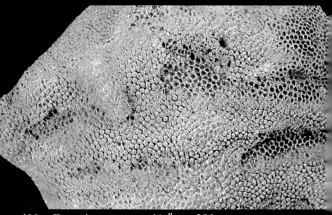

420   Favosites, *cor. w.* 1/16″, *p.* 381

421   **Stromatopora,** *w. 3½″, p. 349*

422   **Protaraea,** *cor. w. ⅟₃₂″, p. 385*

423   **Constellaria,** *l. 1¾″, p. 604*

424   Heliolites, *cor. w.* 1/16″, *p. 384*

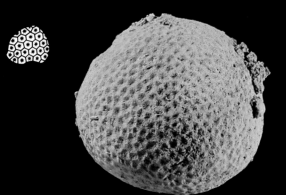

425   Siderastrea, *cor. w.* 1/8″, *p. 368*

426   Astrhelia, *cor. w.* 1/4″, *p. 372*

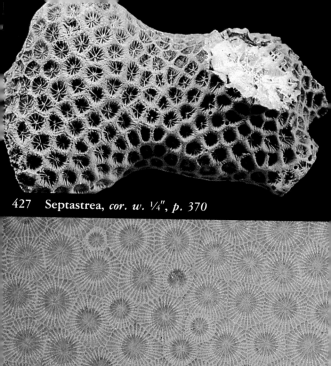

427 Septastrea, *cor. w.* ¼″, *p.* 370

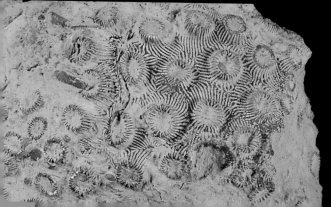

428 Prismatophyllum, *cor. w.* ⅜″, *p.* 363

430   Favistina, *cor. w.* ⅛″, *p. 359*

431   Acrocyathus, *cor. w.* ½″, *p. 366*

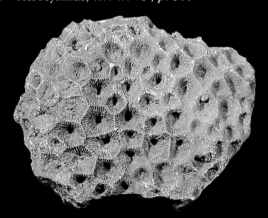

432   Prismatophyllum, *cor. w.* ½″, *p. 363*

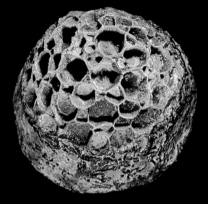

433 Pleurodictyum, *cor. w.* ⅜", *p. 382*

434 Foerstephyllum, *cor. w.* ⅛", *p. 380*

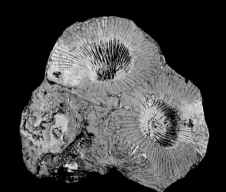

436   Oculina, *cor. w.* ⅛″, *p. 372*

437   Astrangia, *cor. w.* ¼″, *p. 371*

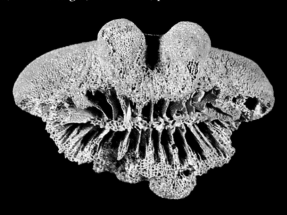

438   Endopachys, *l.* 1⅜″, *p. 377*

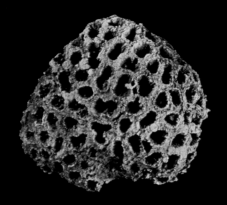

439   Favia, *cor. w.* ½″, *p. 369*

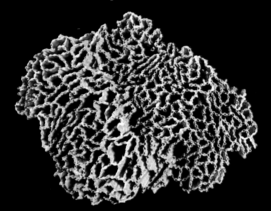

440   Halysites, *cor. w.* ¹⁄₁₆″, *p. 385*

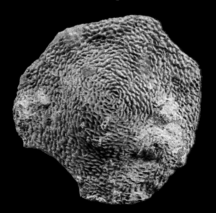

441   Alveolites, *cor. w.* ¹⁄₃₂″, *p. 383*

442   Meandrina, *cor. w.* ⅝″, *p. 373*

443   Actinostroma, *l.* 2⅛″, *p. 348*

444   Prismodictya, *l.* 3¾″, *p. 344*

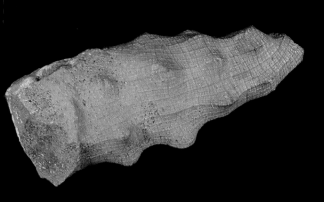

445 Hydnoceras, *l.* 8⅜″, *p. 344*

446 Aulacera, *l.* 4⅞″, *p. 351*

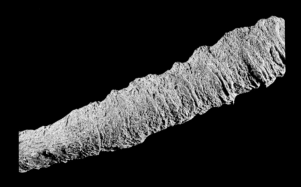

447 Cystiphylloides, *l.* 10½″, *p. 358*

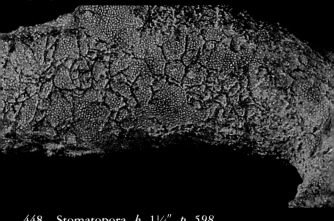

448   Stomatopora, *h.* 1¼″, *p. 598*

449   Hederella, *h.* 1¼″, *p. 602*

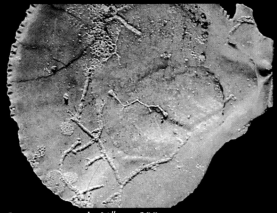

450   Stomatopora, *h.* ⅝″, *p. 598*

451　Lichenopora, *h.* ⅛", *p. 601*

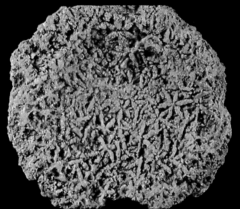

452　Astraeospongium, *l.* 2½", *p. 346*

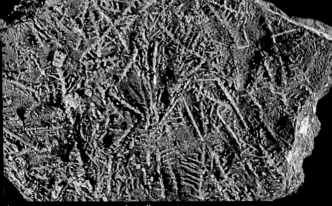

453　Penniretepora, *l.* ⁵⁄₄", *p. 614*

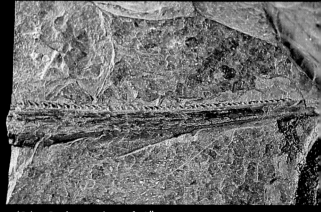

454   Orthacanthus, *l. 2″, p. 759*

455   Archimedes, *l. 4″, p. 612*

456   Archimedes, *l. 3⅛″, p. 612*

Trace Fossils and Graptolites

 Trace fossils are not the remains of animals themselves, but burrows, trails, and evidence of feeding. All of these kinds of trace fossils are shown in this section. Because the preserved tubes of graptolites often resemble the trails or burrows left by other animals, they are included in this section.

457 Orthograptus, *l.* ⅝″, *p.* 727

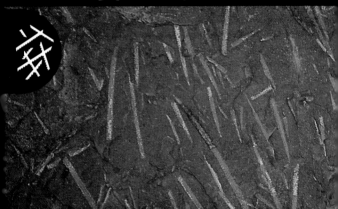

458 Climacograptus, *l.* 2⅛″, *p.* 728

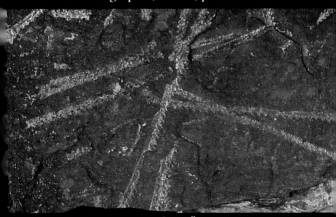

459 Didymograptus, *l.* 2⅜″, *p.* 725

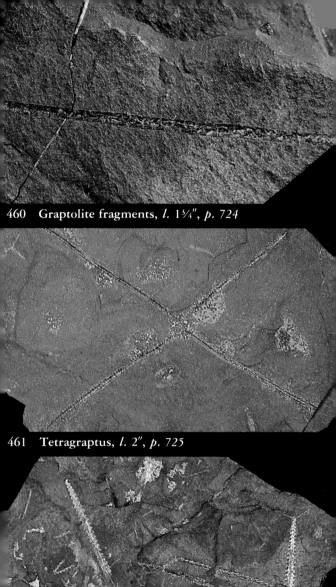

460   Graptolite fragments, *l.* 1¾″, *p.* 724

461   Tetragraptus, *l.* 2″, *p.* 725

462   Dicellograptus, *l.* ¾″, *p.* 727

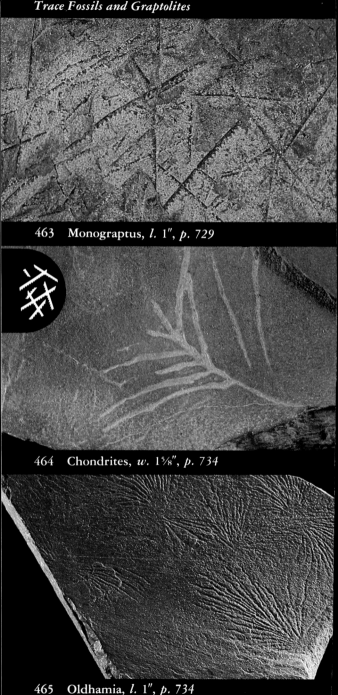

463 Monograptus, *l.* 1″, *p.* 729

464 Chondrites, *w.* 1⅝″, *p.* 734

465 Oldhamia, *l.* 1″, *p.* 734

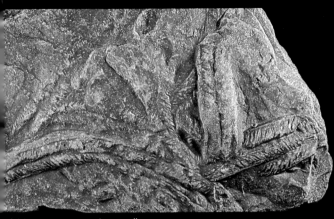

466   Cruziana, *l.* 11¼″, *p.* 732

467   Planolites, *l.* ¾″, *p.* 735

468   Planolites, *l.* 2¾″, *p.* 735

469   Serpula, *l.* 1¼", *p. 538*

470   Skolithos, *l.* 2¾", *p. 732*

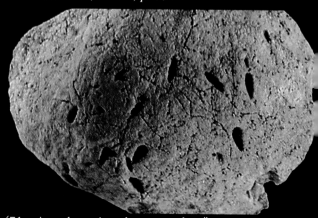

471   Acrothoracican burrows, *l.* ⅛", *p. 590*

Vertebrates

Ancient vertebrates have left a varied
fossil record—skulls and other bones,
teeth, eggs, preserved pieces of skin
with scales, and fossilized feces, or
coprolites. All of these are included in
this section, along with the tracks of
dinosaurs.

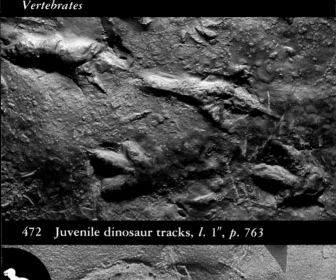

472   Juvenile dinosaur tracks, *l.* 1″, *p.* 763

473   Eubrontes, *l.* 12″, *p.* 763

474   Anchisauripus, *l.* 4¾″, *p.* 763

475　Mesohippus, *l.* 8″, *p.* 766

476　Subhyracodon, *l.* 14″, *p.* 766

477　Merycoidodon, *l.* 6″, *p.* 766

478   Carcharodon tooth, *h.* 5″, *p.* 759

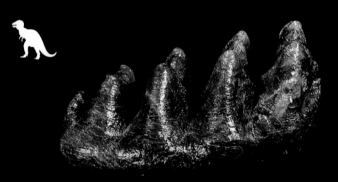

479   Mammut tooth, *l.* 6″, *p.* 766

480   Ptychodus tooth, *w.* 1½″, *p.* 759

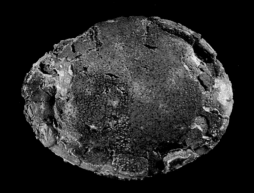

481 Bird egg, *l. 2″*, *p. 765*

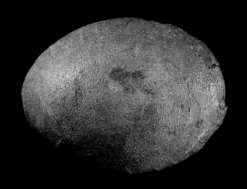

482 Bird egg, *l. 2″*, *p. 765*

483 Shark coprolite, *l. 1½″*, *p. 759*

484   Dinosaur bone, *l.* 3", *p.* 763

485   Trionyx, *l.* 4½", *p.* 763

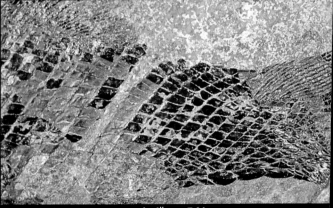

486   Semionotus, *l.* 6", *p.* 761

Plants

The plant fossils shown in this section are typical of those most commonly found—impressions of leaves and fronds, sections of stems, pieces of mineralized wood that show growth rings, and slabs of bark. Fossils of receptaculitid algae are masses of stone with a delicately lacy surface.

487   Celtis, *w. 4″, p. 749*

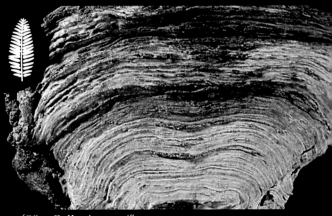

488   Collenia, *w. 4″, p. 741*

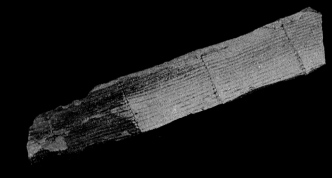

489   Calamites, *l. 8¼″, p. 745*

490   Receptaculitid, *w. 7⅛″, p. 742*

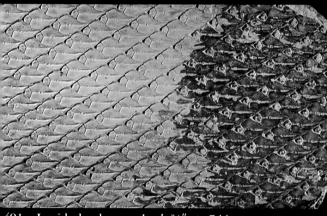

491   Lepidodendron scale, *l. ⅞″, p. 744*

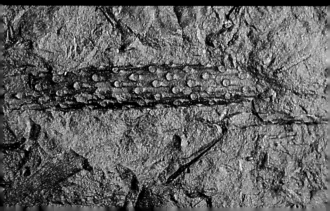

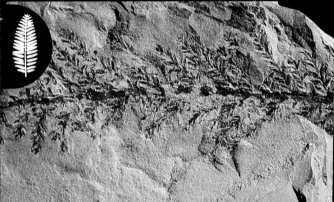

493   Annularia, *l.* 4″, *p. 745*

494   Pagiophyllum, *l.* 6″, *p. 748*

495   Pagiophyllum, *l.* 2¾″, *p. 748*

496 Metasequoia, *l. 3″*, *p. 748*

497 Sequoia, *l. 6″*, *p. 748*

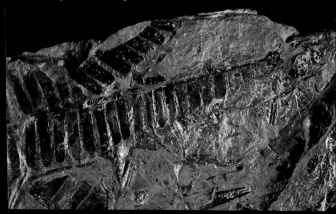

498 Nilssonia, *l. 4″*, *p. 747*

499   Pecopteris, *l. 4⅞″, p. 745*

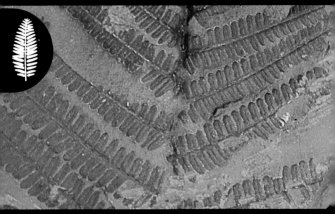

500   Pecopteris, *w. 8″, p. 745*

501   Alethopteris, *l. 9″, p. 746*

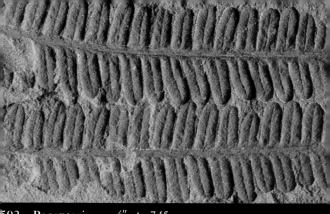

502   Pecopteris, *w. 4"*, *p. 745*

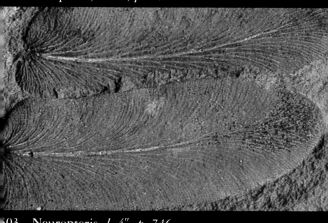

503   Neuropteris, *l. 4"*, *p. 746*

Plants

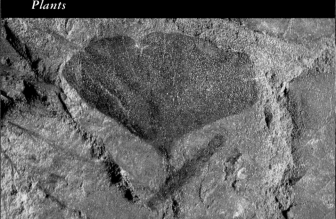

505 Ginkgo, *w.* 1¾″, *p.* 747

506 Quercus, *l.* 5″, *p.* 749

507 Platanophyllum, *w.* 6″, *p.* 749

The numbers preceding the genus
descriptions on the following pages
correspond to the plate numbers in the
color section. When descriptions have
no plate number, drawings illustrate
the text.

Phylum Porifera
(Precambrian to Recent)

Sponges are simple animals that vary in size and shape. Although sponges were probably common marine animals as early as the Cambrian, their fossil record is poor. We do know that fossil sponges are often vaselike or spherical, whereas living ones are more apt to be massive or branching.

Sponges lack true tissues: they have no nerves, muscles, or glands. They have an internal skeleton usually composed of needlelike, mineralized *spicules* and *spongin,* a fibrous, horny substance. The skeletons of some sponges consist solely of *calcareous* or *siliceous* spicules, whereas others consist of spongin alone. In most the spicules are not joined, and when the sponge dies, they scatter. Since spongin decays, the only sponges likely to be preserved are those whose spicules are fused into a rigid framework. A few of these forms are fairly common in Paleozoic rocks.

The simplest sponges are vase-shaped, with small pores in the walls through which water passes to the interior. Chambers inside are lined with collared cells bearing small whiplike hairs called *flagella.* As these hairs beat, small food particles are strained out by the collars and are passed in and engulfed. There are only 2 layers of cells in a sponge, with a jellylike material between them that holds the skeleton. In more advanced sponges, the vase wall is folded, creating many chambers lined with collared cells.

Sponges are almost entirely marine. Lacking the power of movement, they generally attach to a firm object, but are also found in sand or mud. They vary from less than 1″ (25 mm) to more than 2′ (61 cm) across.

CLASS DEMOSPONGEA
(Precambrian to Recent)

Most living sponges belong to this class. The skeleton consists of spongin alone or of spongin and siliceous spicules. The spicules are sometimes straight rods, but may have 4 rays diverging at angles of 60° or 120°. The walls of these sponges are commonly thick and folded.

417 Cliona

Description: Commonly preserved only as borings in shell; borings appear as shallow, branching tunnels opening to shell surface through small holes ¹⁄₁₆″ (2 mm) long and wide. Usually found in fossil brachiopod and mollusk shells.

Age: Devonian to Recent.

Distribution: Widespread in North America.

Comments: Since the tissues of *Cliona* decay quickly after death, the sponge itself is not preserved in fossil form. *Cliona* is still found very commonly today, however, living in dead shells on the continental shelves. It is known as the boring sponge because of its hole-excavating habit. When alive, the sponge is usually bright yellow, and the shell that it inhabits appears to be covered with yellow polka dots: clusters of tiny pores through which water enters, and larger single pores from which water leaves the sponge. The shell offers *Cliona* protection from predators and water agitation, and allows the sponge to exist in a favorable environment without being washed away. At the same time, *Cliona* is an important agent in shell destruction, since its borings weaken a shell until it finally crumbles. *Clionolithes* is another boring sponge, found in the Devonian through the Pennsylvanian in the United States. The tunnels and holes in the outer surface of its host shell are smaller than those of *Cliona,* averaging only 1/50" (0.5 mm) long and wide. *Clionoides,* found in the Devonian of Iowa, is distinctive because of the irregular branching and wide spacing of its tunnels.

361 Astylospongia

Description: 2" (51 mm) long and wide. Almost spherical, but slightly flattened top to bottom, with shallow depression at top containing many large, regularly arranged pores. Surface covered with scattered, fainter pores and many irregular vertical grooves. Base evenly rounded; evidence of attachment lacking.

Age: Ordovician through Silurian.

Distribution: New York, Ontario, Ohio, Indiana, Kentucky, Tennessee, Illinois, Iowa.

Comments: *Astylospongia* is a Silurian fossil sponge, particularly common in Tennessee. The pores on the sides of the sponge may have functioned to take in water, and the large pores in the depression at the top probably served for water removal. The irregular vertical grooves were most likely canals through which the water travelled, and would have been covered over by tissue when the sponge was alive.

The marginal drawing shows a side view of *Astylospongia.*
Another common fossil sponge, found in the Silurian of Tennessee, is *Caryomanon,* which is shaped like *Astylospongia* but lacks the grooves and has a lumpier surface.

408, 409, 410 Hindia

Description: 2¾" (70 mm) long and wide (pl. 408); 1⅜" (35 mm) long and wide (pl. 409);

1¼" (32 mm) long and wide (pl. 410). Spherical; surface smooth, covered with very tiny pores. Stalk and depression absent.

Age: Ordovician through Permian.
Distribution: Eastern Canada, New York, New Jersey, Maryland, Virginia, Georgia, Ohio, Kentucky, Tennessee, Wisconsin, Illinois, Missouri, Minnesota, Iowa, Nevada.
Comments: *Hindia* fossils are illustrated both embedded in rock (pl. 409) and split, revealing the dense, narrow canals that radiate from the center of the sphere (pl. 408). Since there is no evidence of a top or bottom in the sponge, and no attachment, *Hindia* may have been adapted to roll about freely on the sea floor.

391 Dystactospongia

Description: 5⅞" (149 mm) long, 3½" (89 mm) wide. Massive; irregularly shaped but somewhat cylindrical, with top usually wider than base. Shallow depression at top. Surface finely pitted. Skeleton obscure.
Age: Ordovician through Silurian.
Distribution: Virginia, Ohio, Tennessee, Illinois.
Comments: Tiny radiating canals are visible if the surface of *Dystactospongia* is eroded. These canals probably channeled water from tiny pores in the outer surface to larger openings through which the water left the sponge. In the process, tiny bits of food would be strained out of the water by special cells, called collar cells, that lined the canals. The sponge is abundant in parts of Ohio, and in reefs in western Virginia and Tennessee, but is not common elsewhere. This is not an unusual pattern for fossil sponges: they tend to be locally abundant, but very narrowly distributed.
Zittelella, from the Middle Ordovician

of Canada, is superficially similar to *Dystactospongia,* but is much more regularly conical, and its straight, radiating canals, arranged horizontally, are clearly visible on the surface. *Zittelella* is often so regular that it looks like a conical coral.

CLASS HEXACTINELLIDA
(Lower Cambrian to Recent)

The glass sponges have skeletons consisting only of siliceous spicules that have 6 or fewer rays that diverge at 90° angles. These spicules are generally cross-shaped, and are often fused into a rigid framework. This class is sometimes called Hyalospongea.

Protospongia

Description: ¾" (19 mm) long and wide; spicules ¼" (6 mm) long. Spicules usually found separated, simple, cross-shaped; may form reticulated meshwork. Sponge funnel-shaped, thin-walled, with slender root tuft and opening at top.

Age: Lower and Middle Cambrian.

Distribution: Widespread in North America.

Comments: *Protospongia* is one of the earliest sponges, and one of the simplest. It was an ascon, a single vaselike form, hollow in the middle with an opening at the top. Water entered through pores in the walls that were supported by the reticulated spicules. While the water passed through to the hollow interior, flagella in small collar cells beat and forced water toward the cell and through the sievelike collar that surrounded the flagellar base. Here food particles were collected, to be engulfed by the cell at the base of the collar. Water then passed into the wide center cavity, and up and out the opening on top. The spicules are commonly preserved as fossils, although the whole sponge is not.

The marginal drawing shows the reticulated mesh formed by the spicules.

445 Hydnoceras

Description: 8⅜″ (21 cm) long, 3½″ (9 cm) wide. Large; roughly conical, but usually somewhat crushed laterally, with 8 pronounced, sharp, vertical ridges crossed by horizontal ridges, producing conical swellings at intersections. Entire surface covered by coarse and fine rectangular lattice pattern of narrow, threadlike ridges.

Age: Upper Devonian through Pennsylvanian.

Distribution: New York, Pennsylvania.

Comments: *Hydnoceras* is a glass sponge, very distantly related to living glass sponges. Glass sponges are now mostly restricted to deep water, but in the Paleozoic they were very common at all depths. In glass sponges, siliceous spicules form a strong, tubular skeleton. A *Hydnoceras* fossil is preserved not as the tubular, conical skeleton itself, but as an internal filling of that skeleton. The distinctive rectangular, latticelike pattern is an impression of the inside of the skeleton.

444 Prismodictya

Description: 3¾″ (95 mm) long, 1⅝″ (41 mm) wide. Tall, narrow, vaselike, with 8 prismlike sides. Swellings usually absent. Rectangular, latticelike pattern of fine threads covers surface.

Age: Upper Devonian through Mississippian.

Distribution: New York, Indiana.

Comments: Glass sponges have a skeleton of siliceous spicules, which are often fused together to form a rigid framework that does not disintegrate with the death of the sponge. These sponges contain 1 large canal through which water is carried out of the sponge, and many small pores along the sides through which it enters. Glass sponges are usually cup- or vase-shaped. The fossils commonly occur either as internal fillings or as impressions of the animal. The fossil of a glass sponge formed when the sponge was uprooted and covered by mud, which also filled the sponge's interior. When the sponge's tissues rotted away, and after the mud surrounding and filling the sponge hardened to shale, the animal's siliceous skeleton dissolved, and the external and internal impressions of the sponge remained in the mud. The shale then split in the area of weakness caused by the cavity previously filled with the sponge's tissues, revealing the characteristic pattern, depending on where the split occurred, either of the outside or of the inside of the sponge. The specimen illustrated shows an internal impression. A glass sponge with a shape and lattice pattern similar to that of *Prismodictya* is *Dictyospongia,* from the Devonian of New York. *Dictyospongia,* however, is cylindrical, and lacks the prism faces of *Prismodictya*. Plate 444 shows a specimen fragment turned on its side. A complete specimen is shown in the margin.

---

375 Brachiospongia

Description: 4⅛" (105 mm) long, 3¾" (95 mm) wide. Radiating projections surrounding a central opening. Periphery divided into 8–12 (usually 10) fingerlike lobes that curve down, probably forming base. Upper surface has fairly large, round, elevated opening connecting to hollow lobes. Walls thick. Root tuft absent.

Age: Ordovician.

Distribution: Ontario, Ohio, Kentucky.

Comments: The distribution of fossil sponges often seems puzzling. *Brachiospongia* is found abundantly in Kentucky and rarely in Ohio and Ontario, but nowhere else in the world except in Scotland. The form of this sponge is unusual. Water most likely entered it through tiny pores in the sides of the lobes, circulated through the hollow interior, and then left the sponge through the large central opening. The specimen illustrated is shown with the tips of the fingerlike lobes broken, revealing their hollow interiors.

CLASS CALCAREA
(Ordovician to Recent)

The Calcarea are the most primitive sponges, with skeletons consisting only of calcareous spicules, usually fused into a rigid framework, or of beadlike spheres or flakes. This class is sometimes called Calcispongia.

452 Astraeospongium

Description: 2½" (64 mm) long and wide. Saucer-shaped, with no trace of attachment. Skeleton is feltlike mass of 8-rayed spicules, 6 rays in one plane and 2 perpendicular, inconspicuous rays, well developed in interior but not on exterior. Spicules on surface resemble tiny 6-petaled flowers.

Age: Silurian through Devonian.

Distribution: New York, Ohio, Kentucky, Tennessee, Illinois, Iowa.

Comments: Both the saucer shape and the

flowerlike spicules, which are easily
visible, make this a very distinctive
genus. *Astraeospongium* is spelled
Astraeospongia in older books.

407 Girtyocoelia

Description: 2¾" (70 mm) long, ¾" (19 mm) wide.
Small. Linear series of distinct hollow
spheres pierced by large axial tube.
Outer wall covered with few large,
spoutlike pores. Branches may be
present.
Age: Pennsylvanian.
Distribution: Kansas, Oklahoma, Texas.
Comments: The marginal drawings show a cross
section and an external view of
Girtyocoelia. This is a genus distinct
from *Girtycoelia,* which is also
Pennsylvanian and is found in Kansas
and Texas. To add to the confusion,
Girtycoelia is very similar to *Girtyocoelia,*
but the spheres of the former overlap
more, it has no axial tube, and the
openings in its side walls are larger.
Amblysiphonella is another similar
sponge, found in the Pennsylvanian
through the Permian of Nebraska.
Amblysiphonella has no large pores—
only tiny pores covering the outer
surface. *Girtyocoelia* is called *Heterocoelia*
in older books.

CLASS STROMATOPOROIDEA
(Cambrian to Cretaceous)

Until recently these important reef-
builders from the Silurian and
Devonian were classed as cnidarians.
Their classification is still debated, and
their inclusion among the sponges is
based on their striking similarity to
modern sclerosponges, which have
calcareous skeletons and siliceous
spicules. Stromatoporoids are massive,

encrusting, or cylindrical; the surface of the massive or encrusting genera may be covered with small grooves that form a radiating rootlike pattern. Most stromatoporoids have a layered structure. The skeleton consists of separate or joined horizontal and vertical elements. The average diameter of skeletal masses is only about 4" (102 mm), but they may grow to several meters wide.

We have included 4 genera of stromatoporoids here to show the features of this class. Identifying stromatoporoids to genus requires specialized techniques beyond the scope of most collectors.

443 Actinostroma

Description: 2⅛" (54 mm) long, 1¼" (32 mm) wide. Massive, dense skeletal deposit. Surface may show small, shallow grooves radiating from tiny pores. In longitudinal section, thin skeletal elements are vertical and horizontal. Vertical elements fairly continuous, crossed at regular intervals by discontinuous horizontal elements. In some specimens horizontal elements periodically form dense layers, giving section a banded appearance, with bands a few millimeters thick.

Age: Cambrian through Mississippian.

Distribution: Widespread in North America.

Comments: The specimen illustrated has been sliced open longitudinally—that is, cut perpendicular to the surface—and polished. The dark brown skeletal elements are clearly visible against the beige background. The beige rock represents a filling of spaces previously probably occupied by soft tissue. The horizontal banding in the specimen is a common characteristic of stromatoporoids, the bands often becoming distinct when the fossil weathers. Paleontologists have wondered if the bands are periodic, that is, deposited in synchrony with some natural cycle like the day, lunar month, or year. The dark layers may represent a slowing or cessation of growth caused by adverse conditions related to tides or seasons. But, so far, periodicity has not been demonstrated. *Actinostromaria* is the Mesozoic form of *Actinostroma,* and is found from the Jurassic through the Cretaceous widespread in North America. In longitudinal section, the vertical elements of *Actinostromaria* are heavier than the horizontal elements, and both are less regularly arranged than those of *Actinostroma,* with the vertical elements often discontinuous. In cross section, *Actinostromaria* shows an irregular branching and fusing of the skeleton.

421 Stromatopora

Description: 3″ (76 mm) long, 3½″ (89 mm) wide. Massive or sheetlike calcareous skeletal deposit; sometimes hemispherical. Surface usually has bumps which may be low or high, and which may have prominent pits in centers from which shallow grooves radiate. Longitudinal section may show fine lattice pattern, with stout horizontal layers a fraction of a millimeter thick, transected by numerous strong, vertical pillars, or a less regular pattern with loss of clear layering. Horizontal bands several millimeters thick usually present, set off by concentration of skeletal material in horizontal plane, or by change in arrangement of skeletal elements. Pattern may become less regular toward surface of skeleton.

Age: Silurian through Devonian.

Distribution: Widespread in North America.

Comments: It is not certainly known if stromatoporoids were colonial animals, like many corals, or, more likely, single animals perhaps related to the sponges. *Stromatopora* means "layer-pores," and refers to the layered structure visible in cross section. The photograph shows the surface of *Stromatopora*. The small bumps once bore starlike systems of canals that led into a central pore. These pores were probably water outlet pores and were fed by the canals. The intake pores may have been scattered over the surface and are too small to be recognized. If *Stromatopora* was a sponge, it must have fed by filtering tiny organisms from the water that flowed through the canals. A longitudinal view of *Stromatopora* is

shown in the marginal drawing. *Clathrodictyon* is a similar stromatoporoid from the Cambrian through the Devonian, found throughout eastern Canada and the eastern and central United States. The vertical pillars in *Clathrodictyon* are confined between 2 horizontal sheets, and the structure in cross section looks like a network of cells. The genus *Stromatopora* is called *Coenostroma* in some older books.

376 Parallelopora

Description: 2″ (51 mm) long, 1¼″ (32 mm) wide. Massive, sheetlike, or encrusting calcareous skeletal deposit with irregular surface, sometimes covered with distinct bumps. Some species covered with small, shallow grooves radiating from tiny pits. Longitudinal section shows very fine mesh of vertical and horizontal elements, with chambers usually elongated vertically and crossed by thin, horizontal platforms. Horizontal layers prominent and closely spaced; may be grouped to give thicker appearance in weathered specimens.

Age: Silurian through Devonian.

Distribution: Widespread in North America.
Comments: The photograph shows a natural longitudinal section of *Parallelopora* that has been weathered. The domed bands, which are groupings of tiny horizontal layers, are clearly visible. Like other stromatoporoids, *Parallelopora* may have harbored symbiotic algae in its tissues, as reef corals do today. These algae would have provided nourishment and, perhaps more importantly, would have aided in the precipitation of calcium carbonate for the skeleton of *Parallelopora* by removing carbon dioxide from its tissues, thus preventing the formation of an acidic environment in which the calcium carbonate could not be produced.

446 Aulacera

Description: 4⅞" (124 mm) long, 1⅜" (35 mm) wide. Solid cylindrical mass; surface covered with wrinkles or bumps. Cross section shows large vertical canal, divided horizontally by strongly arched, sheetlike platforms and surrounded by irregular layers of cystlike plates.
Age: Ordovician.
Distribution: Widespread in North America.
Comments: A similar genus, related to *Aulacera,* is *Labechia,* also found in the Ordovician in North America. *Labechia* is massive or sheetlike, contains many layers of cystlike plates, and may have strong bumps on its upper surface. Illustrated is the exterior of *Aulacera.*

Phylum Cnidaria
(Precambrian to Recent)

Cnidarians include the hydroids, jellyfishes, sea anemones, and corals. They are characterized by a simple, radially symmetrical, saclike body enclosing a digestive cavity with only 1 opening, the mouth, which is usually ringed by tentacles. The body wall contains 2 cell layers. Nerve cells, skin tissue, and connective tissue are all present. The tentacles or the body wall bear stinging cells. Skeletons are present in corals and are made of calcite or aragonite. Asexual reproduction by budding is common, and many groups are colonial.

With few exceptions, cnidarians are marine. Most are sessile, or sedentary, but others, like the jellyfish, are mobile; in some, generations alternate between a sessile polyp stage and a mobile jellyfish stage. The stinging cells, relatively large gut cavity, and expandable mouth allow cnidarians to eat a wide range of food, and most are predators. A few are suspension feeders. Because of the durability of their skeletons, corals have left a very clear fossil record. Almost all forms are corals of the class Anthozoa and belong to 3 subclasses: Rugosa, Scleractinia, and Tabulata.

CLASS PROTOMEDUSAE
(Latest Precambrian through
Ordovician)

The class Protomedusae consists of only
1 genus, *Brooksella,* whose members are
jellyfishlike animals with bodies
divided into lobes by deep grooves. On
top of these lobes may be 4 or more
smaller lobes. A protomedusan has no
tentacles and no central mouth.

370 Brooksella

Description: ¾″ (19 mm) high, 1⅝″ (41 mm) wide.
Disc- or blob-shaped, consisting of 4–
15 or more radially arranged, swollen,
pie-shaped segments with smooth
surfaces. 2 stacked layers of segments
may be visible. 4–5 smaller, narrower
lobes sometimes preserved on top of
large segments.

Age: Middle Cambrian through Ordovician
(possibly also in the Late Precambrian).

Distribution: Alabama, Wyoming, Arizona.

Comments: *Brooksella* fossils are interpreted as casts
of a kind of jellyfish. The animals may
have been washed up on a beach, where
they died and decayed. Their
impressions in the beach were later
filled in with mud or fine sand. The
sand over each jellyfish impression then
hardened to form the fossil. These
fossils are called star cobbles, and are
abundant in the Middle Cambrian
Conasauga Shale of the Coosa Valley in
Alabama.

CLASS SCYPHOZOA
(Latest Precambrian to Recent)

Scyphozoans are true jellyfishes whose
body consists of a gelatinous bell, with
4 sheets of tissue, or *septa,* that divide
the gastric cavity into a central stomach

and 4 pouches. Only 1 order, Conulariida, questionably assigned to the class, is known to have had a skeleton. All other jellyfishes are very rare as fossils.

ORDER CONULARIIDA
(Middle Cambrian through Triassic)

Conulariids are cone-shaped to pyramidal shells composed of chitinophosphate. The top of the shell projects as triangular flaps that could fold down to form a cover. The shell is usually marked with pronounced transverse ribs and faint longitudinal ribs. Some genera have a midline, or interruption in the transverse ribs, on each of the 4 sides of the pyramid, perhaps marking the point of attachment of internal septa like those of scyphozoans. But it is not clear that these animals are scyphozoans, or even cnidarians.

356, 357 "Conularia"

Description: 4″ (102 mm) long, 1″ (25 mm) wide (pl. 356); 3⅛″ (79 mm) long, 1⅜″ (35 mm) wide (pl. 357). 4-sided, elongate pyramid, square to rectangular in cross section. Corners not thickened; indented by longitudinal furrows. Sculpture consists of well-defined, closely spaced horizontal ribs, with tiny vertical ribs in spaces between them. Sculpture not interrupted at corners.

Age: Upper Cambrian through Permian.

Distribution: Widespread in North America.

Comments: The genus "Conularia" has been broken up into several new genera, but these are often difficult to distinguish. We are calling the specimens illustrated "Conularia" to indicate that we are using the genus name in the old,

unrestricted sense. Plate 356 shows a
specimen removed almost whole from
the matrix, and only slightly deformed.
Plate 357 shows a specimen that has
been crushed on a bedding plane, and 2
of the 4 sides are visible.

Each well-preserved, small *"Conularia"*
has a little disc at the apex of the
pyramid that served to attach the
animal to some firm surface. Larger
animals were probably free-living, and
possibly could even swim. They may
have been jellyfishlike, with 4 large
tentacles, 1 projecting from each corner
of the tube. They have been variously
interpreted as hydrozoans, scyphozoans,
or members of a separate, extinct
phylum. Most paleontologists now
accept them as scyphozoans.

CLASS ANTHOZOA
(Latest Precambrian to Recent)

Anthozoans are all marine, and include
the sea anemones, corals, and soft corals
like sea fans, sea pens, and sea feathers.
Anthozoans may be solitary or colonial.
They have only a polyp form, in which
the central cavity is divided by
longitudinal folds. Hollow tentacles
surround the mouth. Sea anemones
have no skeletons and are very rare as
fossils. Many soft corals have an
internal spicular skeleton that rarely
fossilizes, and are therefore not treated
here.

A fossil solitary coral consists of a
single *corallite,* the skeleton secreted by
an individual polyp; if the coral is
colonial, the fossil consists of a group of
corallites fused together. A corallite is
usually a conical or cylindrical structure
with a cuplike opening at the upper
end that held the polyp, and is called
the *calice.* In many corals, the inside
walls of the calice are lined with small
plates, or septa, that converge toward

the center. As the coral grows upward, the polyps secrete new bottoms in the corallite calices, usually in the form of thin horizontal or convex platforms. When a corallite is sectioned longitudinally, these platforms can be seen beneath the calice. The polyp also secretes new outer walls that appear either solid or cellular when the coral is sectioned.

SUBCLASS RUGOSA
(Ordovician through Permian)

Rugosans are also called tetracorals, horn corals, or cup corals. They are usually large, and may be solitary or colonial, with a solid external skeleton that is readily fossilized. Their basic shape is that of an inverted cone, but some of the colonial forms developed slender and tubular or prismatic corallites.

Most horn corals have well-developed radial septa in the calice, and less well-developed transverse plates, or platforms, under the calice. Long septa usually alternate with short ones. Rugose corals are sometimes known as tetracorals because the septa are inserted in groups of 4. A new polyp first secretes 4 septa, dividing the calice into quadrants. Then it secretes 2 more septa very close to 1 of the original septa. After this, new septa are secreted in groups of 4, 1 septum in each of the original quadrants. This pattern is usually difficult to see in the mature calice, but by making a series of transverse cuts across the corallite, from the apex to the calice floor, the pattern of introduction of new septa can be seen. The walls of the corallites are composed either of the thickened bases of the septa or of small, arched plates inclined toward the center of the cone. The platforms across the middle of the

cone are flat or arched. There may be a
raised structure in the center of the
calice, or a deep or shallow depression
on 1 side of the calice floor.
Solitary rugose corals are curved or
erect, and can be short and wide or tall
and cylindrical. Some are even disc-
shaped. Corallites in some colonial
rugose corals are closely packed; in
others they do not touch, but are joined
by horizontal tubes or sheets.

ORDER CYSTIPHYLLIDA
(Middle Ordovician through Middle
Devonian)

This is a small group of corals that are
mostly solitary and have inconspicuous
septa that may appear as narrow radial
rows of small spines on the walls of the
calice, or may be absent altogether.
The walls of the corallites are composed
of small, arched plates that continue
under the calice to form the entire
interior structure of the corallite.

447   Cystiphylloides

Description: 10½″ (27 cm) long, 2″ (5 cm) wide.
Conical to cylindrical, solitary or
weakly aggregated. Outer wall strongly
wrinkled. Floor of calice concave; septa
absent or represented by discontinuous
spines or ridges inside calice. Corallite
composed of small, domed plates,
forming cystlike enclosures that fill
corallite interior, giving vesicular
structure. Plates may cause inside of
calice to appear bumpy.

Age: Lower and Middle Devonian.

Distribution: Widespread in North America.

Comments: *Cystiphylloides* and related Devonian
genera mark the culmination of the
development of the vesicular structure
in the cone and the reduction of the

septa. Rugose corals after the Devonian
did not show these features. This fine
structure resulted in a skeleton that was
strong but relatively light.
Cystiphylloides and related genera were
originally included in the genus
Cystiphyllum, which ranges from the
Upper Ordovician to the Middle
Devonian. *Cystiphyllum* is very similar
to *Cystiphylloides,* and can only be
distinguished by detailed laboratory
examination of the septal development.

ORDER STAURIIDA
(Middle Ordovician through Upper
Permian)

Most rugose corals are included in this
order, which is characterized by the
presence of well-developed septa.
Stauriids have horizontal, arching or
sagging platforms that fill the center of
the corallite. These partitions are
distinct from the wall structure, which
is composed either of small plates, as in
the cystiphyllids, or of dense shell.

430 Favistina

Description: Corallite ⅛" (3 mm) long and wide,
bounded by distinct walls. Colonies
massive and compound, hemispherical
or globular; long, prismatic, polygonal
in cross section. 12–15 large septa
reach ⅓–½ distance from rim of calice
to axis; very short septa between large
septa. Longitudinal section shows
regularly spaced platforms, flat except
for downturned edges, that reach
completely across corals.
Age: Middle and Upper Ordovician.
Distribution: Widespread in North America.
Comments: Compound, or colonial, rugose corals
evolved independently many times in
all 3 major groups of fossil corals,

Rugosa, Tabulata, and Scleractinia.
Some rugose coral genera, like
Favistina, look very similar to modern
colonial corals. They probably lived in
a similar manner also, their cups
occupied by small tentacled polyps.
They built reefs as modern corals do,
and they may even have harbored
symbiotic algae in their tissues as reef
corals do today. *Favistina* is called
Favistella in most older references.

351, 352   Heterophrentis

Description: 3½" (89 mm) long, 2⅛" (54 mm) wide
(pl. 351); 2" (51 mm) long, 1⅜" (35
mm) wide (pl. 352). Curved, conical;
always solitary. Exterior covered with
longitudinal ridges and encircling
growth lines. Distinct outer wall
present. Calice large, with flat or
irregular bottom containing radiating
ridges and slightly raised center. Septa
extend to, or almost to, center of base.
Longitudinal section shows domed,
irregular platforms, depressed in
center, under calice.

Age: Lower and Middle Devonian.

Distribution: Widespread in eastern North America.

Comments: The platforms under the calice, visible
when the cone of *Heterophrentis* (and
that of most other rugose corals) is cut
open, are former floors of the calice.
Instead of depositing shell continually
over a single base, coral animals
periodically lifted up their "bottoms"
by secreting a new base, leaving an
open space underneath. This process
was advantageous to the coral because it
kept the cup reasonably light and less
top-heavy than it would have been if
shell secretion had been continuous
over the floor of the cup.

The marginal drawing shows a view of
Heterophrentis, looking into the calice.

348, 349 Grewingkia

Description:

2⅜" (60 mm) long, 1" (25 mm) wide (pl. 348); 3⅛" (79 mm) long, 1⅛" (28 mm) wide (pl. 349). Solitary, conical, usually with curved sides. Covered with longitudinal grooves and horizontal growth lines and wrinkles. Outer wall thick, formed of dense material. Calice deep, with narrow, straight walls. Septa short, ending before center of calice bottom; fingerlike septal lobes extend to axis, forming irregular network that projects as low mound on calice floor. Longitudinal section shows domed platforms below base of cup.

Age: Upper Ordovician.

Distribution: Widespread in North America.

Comments: *Grewingkia* had no method of attachment. It is usually preserved in shales, which were mud when the animal was living. Its heavy cup probably sank into this mud, and new mud then settled around the cup— with luck, at the same rate at which the cup grew upward. If the mud settled too fast, the animal would have been smothered; if too slowly, the cup would have become top-heavy and fallen over. *Streptelasma,* found in the Middle Ordovician through the Middle Silurian in New York and Michigan, is a common genus in the same family. It is like *Grewingkia* except that its septa are long, extending to the base of the calice. *Enterolasma,* found in Silurian and Lower Devonian strata in New York, Tennessee, and Oklahoma, is usually smaller, and has fewer but thicker septa that bear knoblike projections on the sides. In the specimen of *Grewingkia* illustrated, the outer surface was worn smooth before fossilization occurred.

360　Hadrophyllum

Description:　⅛" (3 mm) high, ½" (13 mm) wide. Disc-shaped, with 1 side of disc sometimes drawn down into very short, wide cone with apex slightly curved. Other side of disc convex, covered with many heavy septa. Narrow, slotlike depression, divided by narrow ridge, extends from center to margin. Septa either terminate on walls of depression or are radially arranged. Some septa branch.

Age:　Lower and Middle Devonian.

Distribution:　Widespread in eastern North America.

Comments:　Small, disclike corals very similar to *Hadrophyllum* are still living today, but they belong to the subclass Scleractinia, and are not directly related. These living corals lie unattached on sandy bottoms, and *Hadrophyllum* probably behaved the same way.

 The marginal drawing shows a side view of an entire *Hadrophyllum* specimen.

Other disc-shaped rugose corals found throughout North America are *Microcyclus,* from the Middle Devonian of Ontario, Illinois, and Missouri, which is thinner, has a flat base, and septa more radially arranged; and *Baryphyllum,* from the Mississippian of Kentucky, Tennessee, and Alabama, which has an upper surface like *Hadrophyllum,* but septa extending out over the sides and down across the base.

429　Pachyphyllum

Description:　Corallite ⅝" (16 mm) long and wide. Massive, compound, with individual corals not bounded by walls. Septa extend out of calices and join with septa from adjacent calices. Septa smooth, lacking small ridges on sides. Platforms under calices domed; shell at

sides of platforms cellular in
appearance.

Age: Upper Devonian.

Distribution: Widespread in North America.

Comments: The calices of *Pachyphyllum* held small
animals called polyps, which were
probably very similar to modern coral
animals. If so, they consisted of little
more than a saclike gut below a mouth
ringed with tentacles. The tentacles
may have borne stinging cells used to
poison small animals that touched
them. The tentacles would then draw
the prey down into the gut, where it
would be digested. A very similar
compound coral is *Asterobillingsa,* which
is found in Lower and Middle Devonian
regions in North America. Small,
transverse ridges are present on the
septa of *Asterobillingsa.*

428, 432 Prismatophyllum

Description: Corallite ⅜" (10 mm) long and wide
(pl. 428); ½" (13 mm) long and wide
(pl. 432). Massive, colonial; corallites
separated by thin walls but closely
packed, with prismatic shape, often
with 6 sides. Calice covers entire top of
corallite, and has broad, sloping sides
and depressed center. Septa thin, with
many tiny cross-bars; longest septa may
meet or intertwine in center of calice.
Longitudinal section shows area under
center of calice with horizontal
platforms; area under sides of calice has
dense cellular structure. Lower side of
colony covered by dense, wrinkled
layers.

Age: Lower and Middle Devonian.

Distribution: Widespread in eastern North America.

Comments: *Prismatophyllum* is the "Petoskey stone"
of northern Michigan, and the official
stone of that state. Petoskey stones are
pieces of *Prismatophyllum* colonies that
have been broken off and subsequently
rounded by streams. They are often

brilliantly colored and can be polished to show the structure of the corallites in 3 dimensions. *Hexagonaria* is also called "Petoskey stone" in some references. *Hexagonaria* is a similar genus that is found in the Devonian of western North America, but does occur as far east as Michigan. *Hexagonaria* tends to have thicker walls and septa. The septa appear beaded in transverse section. Plate 428 shows a specimen that has been cut in a plane parallel to the top of the colony. This section has been polished. Plate 432 shows a whole colony.

350, 435   Heliophyllum

Description: 1¼" (32 mm) long and wide (pl. 350). Corallite 1⅝" (41 mm) long and wide (pl. 435). Large; conical or cylindrical; solitary or weakly aggregate. Exterior usually covered with pronounced horizontal wrinkles. Calice deep, but usually somewhat restricted in width. Septa extend to axis of coral, and have small cross-bars that give latticed appearance to sides of calice. Longitudinal sections show flat platforms in center, and cellular zone near periphery.

Age: Lower and Middle Devonian.

Distribution: Widespread in eastern North America; rare in West.

Comments: *Heliophyllum* is sometimes preserved with a new, small cone growing out of the calice of the old cone. The process that produces the new cone is called rejuvenescence, and occurred again and again in some rugose corals. The new corals produced by rejuvenescence are genetically identical to the parent. The marginal drawing shows a cross section of a *Heliophyllum* calice.

Lophophyllidium

Description: 1¼" (32 mm) long, ⅝" (16 mm) wide. Solitary, conical, with straight or curved sides. Exterior has longitudinal grooves that reflect placement of septa. Calice moderately deep, with 30–50 alternating long and short septa. Base of calice has solid, conical projection that may be higher than calice rim if rim is broken. Cross section shows thin platforms, arching up below calice, that extend across from wall to wall. Walls lack linings of cellular structure.

Age: Pennsylvanian through Permian.

Distribution: Widespread in North America.

Comments: *Lophophyllidium* is one of the most common rugose corals in the Pennsylvanian and Permian periods. It lived with the calice upright. The polyp that lived in the calice probably caught smaller animals, like larvae and tiny fish, with its tentacles.

The marginal drawing shows a side view of an entire specimen of *Lophophyllidium.*
Lophophyllidium is called *Lophophyllum* in some older books.

Caninia

Description: 4¾" (121 mm) long, 1⅜" (35 mm) wide. Large, solitary, cylindrical or conical; exterior covered with longitudinal ridges and many concentric wrinkles and growth lines. Calice relatively large, with evenly distributed septa extending partway to axis across bottom. Conspicuous, slotlike depression to 1 side of calice. Longitudinal section shows flat platforms with downturned margins, extending almost entirely across cone. Walls of cone lined by narrow zone of cellular structure.

Age: Mississippian.

Distribution: Widespread in North America.

Comments: *Caninia* is very common in the
Mississippian. It often has an irregular
shape, with sharp bends in the cone
that resulted from its life on a soft
substrate. When the larva settled, it
probably attached to a small, hard
object that soon no longer supported it,
and it toppled over. The coral then
resumed upward growth, but with a
90° angle in the shell. This toppling
apparently sometimes occurred more
than once, and created repeated bends
in the shell. Other long, cylindrical
rugose corals such as *Heliophyllum* and
Siphonophrentis, from the Lower and
Middle Devonian of New York,
Ontario, Ohio, Indiana, and Kentucky,
show similar changes in angle.
The marginal drawings show a
complete specimen of *Caninia* at the
top, and a cross section at the bottom.

431 Acrocyathus

Description: Corallite ½″ (13 mm) long and wide.
Composed of closely jointed, prismatic
corallites with shared walls, highly
variable in size within a single colony.
Short mounds rise from base of calices,
and septa continue almost to apex of
mounds. Longitudinal section shows
arched platforms under calice with
cellular structure near well-defined
walls.

Age: Mississippian.

Distribution: Widespread in North America.

Comments: *Acrocyathus* is one of the most
distinctive of colonial rugose corals
because of the prominent cones rising
out of its calices. It is also one of the
largest, with colonies over 1′ (30 cm)
wide not uncommon. The colony
increased in size when new polyps
formed by budding off the edges of the
old ones. As the colony grew upward,
the new polyps secreted their own
walls. As a result, there is great

variation in the sizes of the corallites in an *Acrocyathus* colony. *Acrocyathus* is sometimes known as *Lithostrotionella*, and has also been called *Lithostrotion*.

SUBCLASS SCLERACTINIA
(Middle Triassic to Recent)

All living corals with stony skeletons are scleractinians. Their ancestors were probably sea-anemonelike creatures without skeletons, perhaps related to the rugose corals. Scleractinians may be solitary or colonial. The colonial forms may be massive, encrusting, branching, or sheetlike. They have a light, porous skeleton consisting of radially arranged septa, usually with some kind of external sheathing that forms a cup. The development of scleractinian septa is different from that of other corals. First 6 septa form, then 6 more septa are secreted between the first, and then another 12, then 24, and so on. Scleractinians were fairly rare in North America, except along the Pacific coast, until the Cretaceous, when they built reefs in Texas and Mexico. By the late Cretaceous, non-reef corals were common in many faunas of North America. In the Oligocene, reefs appeared near shore in the southeastern United States, West Indies, and Mexico. During the later Tertiary, non-reef builders were more abundant than reef corals in North America. In the Pleistocene, reefs flourished approximately where they do today, south of about 35° N latitude.

ORDER FUNGIIDA
(Middle Triassic to Recent)

This order contains corals that are solitary or colonial. Most are reef builders. The septa within the corallites are perforated and linked to adjacent septa by tiny calcareous rods, and the edges of the septa are toothed. The corallites are relatively large, usually more than $\frac{1}{16}''$ (2 mm) long and wide. Some, like the living solitary mushroom corals of the genus *Fungia,* grow to several inches across. These solitary corals can often live on soft substrates rather than on rock.

425 Siderastrea

Description: Corallite $\frac{1}{8}''$ (3 mm) long and wide. Massive, branching, or encrusting; colonial. Corallites and calices closely packed and irregular in outline. Calices funnel-shaped, lined by numerous septa that appear beaded on edges and are joined to adjacent septa by tiny cross-bars. Septa extend as ridges to edges of corallites, but do not join septa from adjacent corallites.

Age: Cretaceous to Recent.

Distribution: Atlantic and Gulf coastal plains.

Comments: *Siderastrea* has probably always been restricted to warm, shallow, clear water, and today is still a common reef coral on the southern coast of Florida and in the Bahama Islands, where it is known as the starlet coral. It is a hermatypic coral, which means that when it is alive, its tissues contain tiny algae that supply the coral with oxygen and nutrients and remove waste products.

ORDER FAVIIDA
(Middle Triassic to Recent)

This is the largest of the scleractinian orders and contains many fossil genera. Faviid corals may be solitary or colonial.

Their septa are composed of vertical pillars of radiating fibers arranged in 1 or more fan-shaped structures. Each pillar may be the height of the septum; or a fan-shaped structure may be composed of several transverse rows of short pillars. The tops of the septa are toothed. In some genera, the septa may be perforated with tiny holes, or reduced to small spines. The teeth can be seen with the naked eye on well-preserved specimens, but a microscope is needed to see the arrangement of the pillars. The walls of the corallite are not perforated.

439 Favia

Description: Corallite ½″ (13 mm) long and wide. Colonial; massive, encrusting, or sheet-like. Corallites round, triangular, or oval. Calices have raised rims and septa that continue over rim as ridges to join septa from adjacent corallites. Septa have small ridges, giving beaded appearance. Longitudinal section shows area under corallites with numerous close, horizontal or concave platforms.

Age: Cretaceous to Recent.

Distribution: Atlantic and Gulf coastal plains.

Comments: *Favia* is now, and has been in the past, an important component of reefs. For growth, it requires water that is warm, shallow, and clear. A reef coral must grow rapidly in order to counter all the destructive forces on a reef, such as storms and animals that bore into and chew off coral. Without rapid growth, a coral is quickly overgrown or shaded by other organisms. Rapid growth requires the help of zooxanthellae, the symbiotic algae that live in coral tissues, supplying the coral with nutrients and removing wastes, especially carbon dioxide which inhibits the precipitation of aragonite, the mineral from which the coral builds its skeleton. The zooxanthellae, in turn, must have light, so the coral must live where the water is shallow and clear. The water must be warm because calcium carbonate (which becomes aragonite in the coral skeleton) is less soluble in warm water, and therefore easier for the coral to remove and precipitate as aragonite.

Montastrea

Description: Corallite ⅜″ (10 mm) long and wide. Colonial; massive, encrusting, or sheet-like. Calices round, their walls formed by thickening, and sometimes union, of septa. Septa numerous, bladelike; 12 septa reach to center of calice. Edges of septa have regular transverse ridges. Some septa extend beyond calice and may join with septa from adjacent corallites.

Age: Upper Jurassic to Recent.

Distribution: Atlantic and Gulf coastal plains.

Comments: *Montastrea* is one of the most common corals in the reefs at the southern tip of Florida, where it is known as the star coral, as well as in fossil reefs along the coast. When alive, it is hermatypic, containing symbiotic algae called zooxanthellae that help nourish it, provide it with oxygen, and dispatch its wastes.

The marginal drawings show a complete colony of *Montastrea* (top), and a cross section illustrating the calices (bottom).

427 Septastrea

Description: Corallite ¼″ (6 mm) long and wide. Colony large, branching. Corallites prismatic from being closely packed. Calices relatively large, covering entire surface of corallites, commonly with 6

long and 6 short septa; longer septa
unite in center of calice to form solid
axial structure. Septa have minute
transverse ridges on edges.

Age: Miocene through Pliocene.

Distribution: Atlantic coastal plain.

Comments: *Septastrea* was not a reef-forming coral,
and did not harbor symbiotic algae in
its tissues; nor did it have the
dependence on warm, shallow water
that reef corals have. It could live in
deeper, colder water where reefs do not
grow.

 The marginal drawing shows a portion
of the surface of a *Septastrea* colony.
Actinastrea (known as *Astrocoenia* in
older books) is a coral that resembles
Septastrea. It ranges from Upper Triassic
to Recent strata in North America. It is
usually massive, but may be encrusting
or branching. It has corallites very
similar to those of *Septastrea,* with the
calices also covering the entire surface
of the corallite. The walls between the
corallites are lower, however, and less
distinctive in *Actinastrea,* and there are
more long septa, with 8–12 reaching
the center of the calice. *Actinastrea* is a
reef-building coral.

437 Astrangia

Description: Corallite ¼″ (6 mm) long and wide.
Colony encrusting. Corallites small,
low, pressed closely together or united
basally by thin, calcified sheet. Septa
have fine-toothed edges. New corallites
form basal expansions at edges of
colony. Longitudinal section shows
vesicular material uniting corallite
walls; centers of corallites composed of
radiating fibers.

Age: Middle Cretaceous to Recent.

Distribution: Atlantic, Pacific, and Gulf coastal
plains.

Comments: *Astrangia* does not form coral reefs, and
is one of the few scleractinian corals to

live in cold, shallow water. One species, the Northern Stony Coral (*Astrangia danae*), is the only coral found north of Cape Hatteras on the East Coast. It is often found encrusting breakwaters north to Cape Cod.

436 Oculina

Description: Corallite ⅛″ (3 mm) long and wide. Colony treelike, branching, with corallites widely spaced, extending out, tubelike, from surface of branches. Corallites tend to spiral around branches. Tops of corallites ribbed by septa extending out from calices. Most septa end in small pillarlike structures, giving centers of calices grainy appearance. Calcified tissue between corallites dense.

Age: Cretaceous to Recent.

Distribution: Widespread in North America.

Comments: Some living *Oculina* species, known as bush corals, are hermatypic, harboring symbiotic algae and building reefs; others are not. There are no structures in the skeleton to distinguish these 2 kinds of *Oculina,* so it is impossible to tell which fossil *Oculina* were reef-builders and which were not, unless the fossils are actually found as components of reefs. *Archohelia* is a closely related genus, found from the Middle Cretaceous through the Pliocene on all 3 coasts of North America. *Archohelia* differs from *Oculina* only in that there is a large corallite on the tip of each of its branches, from which all the other corallites bud.

426 Astrhelia

Description: Corallite ¼″ (6 mm) long and wide. Colony of stout branches, with small, round calices evenly separated by

dense, smooth skeleton. Septa very
finely toothed. Rims of calices slightly
raised.

Age: Miocene.
Distribution: Atlantic coastal plain.
Comments: *Astrhelia* was probably not a hermatypic
or reef-building coral, and had a
broader tolerance for water temperature
and light levels than did reef-building
corals. A reef-building coral that looks
very similar to *Astrhelia* is
Haimesastraea, which occurs in Eocene
rocks of the Gulf coastal plain.
Haimesastraea differs from *Astrhelia* in
that its calices are closer together, and
their rims are not raised. The calices
are separated by thickened septa.

442 Meandrina

Description: Corallite ⅝" (16 mm) long and wide.
Colonial, massive; covered with
continuous or discontinuous V-shaped
grooves lined with septa that end in
bottoms of grooves, where there is
discontinuous sheetlike ridge. Septa
from each groove meet those from
adjacent grooves at tops of crests
separating grooves.

Age: Recent (possibly also in the Eocene).
Distribution: Southern coast of North America.
Comments: The polyps of *Meandrina,* a genus
known today as brain coral, line the
grooves of the colony, and are not
separated as in most corals. Instead,
they share a common gut, which grows
longer as the colony grows outward.

ORDER CARYOPHYLLIIDA
(Jurassic to Recent)

Caryophylliids may be colonial, but
most are solitary and do not build reefs.
Their septa are always smooth on the
margins and composed of a single row

of contiguous vertical pillars of
radiating crystal fibers. Each pillar is
the height of the septum.

358, 359 Trochocyathus

Description: ⅜″ (10 mm) long and wide. Small,
solitary, disc-shaped or conical. Calice
round; septa extend over rim and part
way down sides of corallite as
pronounced ridges. Center of calice has
spongy structure of short columns.

Age: Middle Jurassic to Recent.

Distribution: Widespread in North America.

Comments: *Trochocyathus* lives today at depths of
100–5000′ (30–1524 m). It does not
form reefs at this depth, and cannot
harbor symbiotic algae. The disc-
shaped species live free on the bottom;
the conical species attach. *Trochocyathus*
probably fed by catching small animals
with its tentacles, and may have
collected food particles that settled on
its surface. Illustrated are both sides of
a disc-shaped specimen. *Micrabacia*,
found from the Cretaceous to the
Recent in North America, is also
a small, disc-shaped coral, but its
septa appear beaded on the sides of the
disc and extend to the center of the
slightly concave base. The septa form a
starlike pattern on the calice.

355 Turbinolia

Description: ¼″ (6 mm) long, ⅛″ (3 mm) wide.
Small, solitary, conical, with circular
calice. Calice has 24 or more septa, 6 of
which fuse in bottom to form raised
cone. Septa prolonged into ribs on
outside of corallite. Corallite wall
between ribs is pitted.

Age: Eocene through Oligocene.

Distribution: Atlantic, Pacific, and Gulf coastal
plains.

Comments: When *Turbinolia* was alive, the polyp extended out of the calice to cover the entire corallite down to the base. *Turbinolia* shows no evidence of attachment, so it must have rested in soft sediment with the tip of the cone buried. Although *Turbinolia* still lives off the coast of Florida at depths of 600–1860′ (183–567 m), it is found as a fossil only in the Eocene and Oligocene.

353 Parasmilia

Description: ⅞″ (22 mm) long, ¼″ (6 mm) wide. Small, solitary, cylindrical or conical, with area of attachment at base. Calice circular; septa numerous, granular, extending out of calice and down outer sides. Alternate septa reach to base of corallite.

Age: Lower Cretaceous to Recent.

Distribution: Gulf coastal plain.

Comments: Living *Parasmilia* attach by cementing the base to a firm substrate. Today they are found at depths of 1030–1200′ (314–366 m), and are not components of coral reefs. In the fossil record of scleractinian corals in North America, small solitary forms like *Parasmilia, Flabellum,* and *Turbinolia* are more common than the massive reef-building corals like *Montastrea* and *Meandrina.* Reef-building corals are more sensitive than the small solitary forms to temperature and light and are more easily killed by too much sediment in the water. In the Mesozoic and Cenozoic, conditions in North America were seldom favorable to reefs. In Europe, however, reefs flourished during much of the Mesozoic, and at times during the Cenozoic.

The marginal drawing gives a top view of a *Parasmilia* calice.
Tiarasmilia, found in the Lower Cretaceous of Texas, has the same

conical to cylindrical shape as *Parasmilia,* but has fewer septa, and these are very thick and unequal in length. Only the 6 primary septa extend to the center of the calice.

354 Flabellum

Description: ¾″ (19 mm) long, ⅝″ (16 mm) wide. Solitary, wedge-shaped or compressed cone. Calice oval, containing numerous septa with smooth margins and smooth or granular sides. 12 septa thicker and longer than others. Center of calice has small, oval, raised area. Septa do not extend outside calice. Exterior of corallite has faint or pronounced ridges and concentric growth lines.

Age: Eocene to Recent.

Distribution: Atlantic, Pacific, and Gulf coastal plains.

Comments: *Flabellum* alive today do not attach, but live free at depths from 9 to 9500′ (3 to 2896 m), with their bases partially buried in sediment. *Endopachys* is a coral with the same wedge shape as *Flabellum,* but its septa extend partway down the outside of the cup, and it has 2 large, short, knoblike ridges on both flat sides of the cup.

ORDER DENDROPHYLLIIDA
(Upper Cretaceous to Recent)

The crystal fiber pillars in the septa of dendrophylliids are arranged vertically, and are restricted to 1 row, as in the caryophylliids, but the dendrophylliid septa are perforated. The walls of the corallites are formed from small rods that bridge the outer edges of the septa and are also perforated. The margins of the septa are either smooth or slightly toothed. These corals are solitary or colonial, and most do not build reefs.

438 Endopachys

Description: 1⅜" (35 mm) long, 1" (25 mm) wide.
Solitary, wedge-shaped, not attached,
oval in cross section. Many septa
present, fusing toward center; center
has raised, elongated, spongy structure,
to which longest septa are joined.
Outer wall of coral porous, usually with
ribs on outside that are extensions of
septa. 2 extended or knoblike ribs on
each flat side of coral. Longer axis may
be extended out into "wings."

Age: Eocene to Recent.

Distribution: Atlantic, Pacific, and Gulf coastal
plains.

Comments: Today, *Endopachys* does not live in coral
reefs, but is found at greater depths of
121–1982' (37–604 m). It does not
attach, but rests in soft sediment with
the calice up. The polyp extends out of
the calice and down the sides, where it
continues to secrete septa. The
knoblike ribs on the sides of the coral
and the "wings" stabilize the cup as it
rests on the bottom.
Balanophyllia, also found from the
Eocene to the Recent on all coasts of
North America, is another small,
solitary coral that is usually more oval
than round in cross section, but may
also be horn-shaped. *Balanophyllia*
cemented its base to something firm, so
the fossils show an attachment scar.

SUBCLASS TABULATA
(Ordovician through Permian)

Like the rugose corals, tabulates are a
strictly Paleozoic group. Tabulates are
characterized by inconspicuous septa
and well-developed horizontal
platforms under the calices. The
platforms are usually straight and
complete. All tabulates are colonial.
Their corallites are typically long,
straight, slender tubes, and may be

elliptical, circular, or polygonal in cross section. The walls of the tubes may have pores. The septa, if present at all, are always short and may be spinelike.

There are about 300 described genera of tabulates. They are abundant and conspicuous fossils in most Middle Paleozoic rocks.

ORDER CHAETETIDA
(Ordovician through Upper Permian)

The corallites of chaetetids are very small, slender, and closely packed. Septa and wall pores are absent. New corallites form when radial elements grow from the opposite sides of a calice to meet in the center, dividing the corallite in two. Chaetetids may not be corals, and it has been variously suggested that they be included among the algae, the sponges, or the bryozoans.

401 Chaetetes

Description: Corallite $\frac{1}{50}''$ (0.5 mm) long and wide. Massive or encrusting, thin-walled; colony of long, slender tubes, polygonal to rounded in cross section. Septa absent in calices. Thin walls shared between adjacent corallites; pores absent. Longitudinal section shows complete, numerous or widely spaced platforms under calices. Division of corallites often incomplete, producing oval corallites.

Age: Middle Devonian through Permian.

Distribution: Widespread in North America.

Comments: Some paleontologists have proposed that the group of tabulates to which *Chaetetes* belongs is related to a class of sponges, the schlerosponges. Both groups have tubular units of the same size and arrangement and a similar method of reproduction by fission across the individual units. In both the chaetetids and the sponges, walls are shared rather than individually secreted by each unit. Recently some chaetetids have been found with spicules that resemble those of sponges.

The marginal drawings show a cross section (top) and a longitudinal section (bottom) of *Chaetetes.*

ORDER TETRADIIDA
(Middle and Upper Ordovician)

The corallites of this group are very slender and usually have 4 sides. There are no pores in the walls and 4 septa, and there are only a few platforms under the calices. New corallites form when 4 radial elements (septa), 1 in the middle of each wall, grow out to meet in the middle of the calice and divide the corallite into 4 corallites.

402, 419 **Tetradium**

Description: Corallite $\frac{1}{32}''$ (1 mm) long and wide (pls. 402, 419). Colonies massive, sheetlike, or branched. Corallites tiny, long, narrow, tightly packed, round or almost square in cross section. Most calices have 4 distinct septa, 1 in middle of each wall. Septa do not reach center of calice. Some species have tiny septa between larger septa. Pores in walls absent. Longitudinal section shows numerous or rare platforms under calices.

Age: Middle and Upper Ordovician.
Distribution: Widespread in North America.
Comments: When a corallite of *Tetradium* reproduced asexually, the 4 large septa grew inward across the calice until they met in the center. The calice was thus

divided into 4 parts, and each part became a new corallite. A colony usually shows several individual corallites in the process of division.

ORDER SARCINULIDA
(Middle Ordovician through Devonian)

Sarcinulid corallites are slender and closely packed. Their calices contain septa that are short, stout at their bases, and either equal or alternating in size. The platforms under the calices are horizontal. Corallites may lack walls and be embedded in a common calcareous tissue.

434   Foerstephyllum

Description: Corallite ⅛″ (3 mm) long and wide. Colonies massive or encrusting. Corallites polygonal, sometimes with more than 16 spinelike, wedge-shaped septa of equal length radiating into calices about ¹⁄₁₀ diameter of calice. Longitudinal section shows well-developed corallite walls, sometimes with sparse pores; regularly spaced horizontal platforms cross tubes.

Age: Middle and Upper Ordovician.

Distribution: Widespread in North America.

Comments: *Billingsaria,* from the Middle Ordovician of Quebec, New York, and Tennessee, is in the same family as *Foerstephyllum. Billingsaria* corallites look flowerlike because of 16 thick, short septa that radiate into the center.

ORDER FAVOSITIDA
(Middle Ordovician through Permian)

Favositid corallites are slender, with pores in their walls. Septal spines or small plates project like eaves from the calice walls. Platforms under the calices are complete and horizontal. Some genera have a vaulted upper wall and reduced lower wall in the calices, giving the openings an oblique orientation.

377, 420 Favosites

Description: Corallite ⅟₁₆″ (2 mm) long and wide (pls. 377, 420). Colony massive or branched; corallites long, narrow, closely packed tubes, polygonal in cross section. Septa absent or reduced to short, irregular spines. Walls between corallites thin but distinct, with coral colonies often splitting vertically along them; walls have small, rounded pores, usually located near middle. Platforms under calices well developed, complete, numerous.

Age: Upper Ordovician through Middle Devonian.

Distribution: Widespread in North America.

Comments: *Favosites* is commonly known as the honeycomb coral. This genus and other members of the family Favositidae are the most common tabulates in the Paleozoic. The pores in the walls of the corallites may have served to transfer nutrients from one corallite to another. The marginal drawings show (top) a surface view with a longitudinal section and (bottom) a cross section of *Favosites*. *Emmonsia,* from the Lower and Middle Devonian of New York, Ontario, Ohio, Michigan, Indiana, Kentucky, Texas, and New Mexico, is very similar to *Favosites,* and is in the same family. *Emmonsia,* however, has discrete, flattened projections on the corallite walls that look like spines in longitudinal section. *Paleofavosites,* from the Upper Ordovician through the Upper Silurian, and found at widely scattered localities in the Arctic,

Anticosti Island (Quebec), Ontario, Manitoba, Texas, Wyoming, New Mexico, and Alaska, has pores in the corallite walls that are located predominantly at the angles of the walls.

---

433 Pleurodictyum

Description: Corallite ⅜″ (10 mm) long and wide. Colonies disc-shaped or hemispherical, with flat, concentrically wrinkled base. Corallites large, thick-walled, round or polygonal in cross section. Septa on calices faint or absent. Walls thick, with large, irregularly distributed pores. Longitudinal section shows few very thin, complete platforms under calices.

Age: Upper Silurian through Middle Devonian.

Distribution: Widespread in North America.

Comments: Some specimens of *Pleurodictyum* are found growing around sipunculid worm tubes. Apparently the calcareous tube was a favorable settling place for the *Pleurodictyum* larva. As the coral grew, budding off new corallites, the worm tube also grew, maintaining its opening above the colony. *Michelinia,* from the Upper Devonian through the Permian and widespread in North America, has colonies that look very similar to those of *Pleurodictyum.* A longitudinal section of *Michelinia* shows many more platforms, however, and those present are incomplete.

---

411 Cladopora

Description: Corallite ⅟₃₂″ (1 mm) long and wide. Colonies branching, massive, or flattened. Corallites polygonal, with circular calices oblique to surface. Septa absent. Pores in walls common.

Longitudinal sections show platforms under calices to be common or rare.

Age: Silurian through Devonian.

Distribution: Widespread in North America.

Comments: *Striatopora,* found from the Silurian through the Devonian in New York and Alaska, is a tabulate coral very similar to the branching form of *Cladopora,* but *Striatopora* has faint ridges on the sides of the calice, and the calice openings are not oblique to, but in the same plane as the colony surface. Illustrated are both cross sections and longitudinal sections of several branches of *Cladopora* enclosed in a matrix of black rock.

441 Alveolites

Description: Corallite $\frac{1}{32}$" (1 mm) long and wide. Colonial, massive, branching, or encrusting; corallites compressed, small, irregularly shaped. Each calice opens at oblique angle to surface of colony, with raised lip on side of corallite away from direction of colony growth. Opposite lip low or flush with base of calice. Septal spine commonly in base of calice; other septa inconspicuous. Large pores at base of upper walls of calices.

Age: Upper Silurian through Devonian.

Distribution: Widespread in North America.

Comments: This is a very distinctive genus, which does not at first look like a coral. It is often mistaken for a sponge or a bryozoan.

ORDER HELIOLITIDA
(Middle Ordovician through Middle Devonian)

The corallites in this group are embedded in tubular calcareous tissue and form massive colonies. Each corallite has 12 septa or 12 rows of septal spines. The platforms are commonly complete and horizontal.

424 Heliolites

Description: Corallite ¹⁄₁₆″ (2 mm) long and wide. Colonial, globular to fan-shaped, rarely branching. Corallites cylindrical, bounded by distinct, unperforated walls. Calices rounded, widely spaced, set in common mass of complex calcareous tissue. Each calice commonly has 12 spiny septa, which may reach almost to center of calice; but septa may be lacking. Tiny prismatic tubes without septa surround corallites. Platforms more closely spaced under tiny tubes than under calices.

Age: Lower Silurian through Middle Devonian.

Distribution: Widespread in North America (in Middle Devonian, in western North America only).

Comments: *Propora,* from the Middle Ordovician through the Upper Silurian, and widespread in North America, has similar widely spaced corallites, but the calcified tissue between the corallites is not tubular. It is more amorphous than in *Heliolites,* with many irregular, domed platforms of various sizes showing in longitudinal section. Another genus with large corallites surrounded by tiny prismatic tubes, and in the same family as *Heliolites,* is *Plasmopora,* found in the Silurian through the Middle Devonian of Indiana and Kentucky. *Plasmopora* differs from *Heliolites* mainly in that each of its corallites is surrounded by 12 tiny tubes whose radial walls are continuous with the corallite septa. *Plasmopora* lacks distinct platforms under the tiny tubes, and has vesicular material present instead.

422 Protaraea

Description: Corallite $\frac{1}{32}$" (1 mm) long and wide.
Colonies encrusting or disc-shaped,
sometimes nodular or branching. Each
tiny, rounded calice has 12 equal,
spiny septa extending about $\frac{1}{3}$ distance
across calice to form star-shaped
opening. Longitudinal section shows
oblique platforms under calices
between many vertical tubes.

Age: Middle and Upper Ordovician.

Distribution: Eastern and central North America.

Comments: *Protaraea* is often found as a thin crust
on the upper valves of brachiopods.
Protaraea probably colonized living
brachiopods and benefited from the
feeding currents that the brachiopods
created. A coral made the upper valve
of a brachiopod heavier and therefore
more difficult to open, but this
disadvantage may have been offset if
Protaraea had stinging cells (like all
cnidarians living today), and thus could
provide the brachiopod with extra
protection from enemies.

ORDER HALYSITIDA
(Middle Ordovician through Upper
Silurian)

The corallites of halysitids are slender,
thick-walled, circular or elliptical in
cross section, and arranged in chains
that branch and join with other chains.
The calices may have 12 longitudinal
rows of septal spines.

440 Halysites

Description: Corallite $\frac{1}{16}$" (2 mm) long and wide.
Large, meshlike colonies formed of
chains of corallites joined together.
Corallite tubes oval in cross section,
with long axis coincident with trend of

chain. Small, rounded or angular tubes between larger oval corallites. 12 spinelike septa may be present in corallites. Longitudinal section shows well-developed, horizontal or gently arched platforms under large corallites. Platforms in intercorallite tubes more closely spaced.

Age: Silurian.

Distribution: Widespread in North America.

Comments: *Halysites* and similar forms are called chain corals. Other common chain corals are: *Catenipora,* found in the Upper Ordovician through the Silurian in North America, which lacks small tubes between the larger corallites; and *Cystihalysites,* from the Silurian of Canada, Tennessee, and Utah, in which the corallites are separated by cellular walls. The small corallites of *Halysites* are usually obscured, however, and the only way to tell these genera apart may be to examine a polished cross section. The corallites are more widely spaced in *Halysites* than in many other tabulate corals, and the open areas between the chains saved the colony from having to secrete extra calcium carbonate without the benefit of adding polyps. At the same time, the corallites were linked end-to-end into chains that increased the strength of each. This colony form, then, maximized distance between corallites to lessen competition while they fed, and minimized calcium carbonate secretion to save energy. The marginal drawings show cross sections of *Halysites* corallites. The enlarged view at bottom shows the septa and the tubes connecting the calices.

ORDER AULOPORIDA
(Lower Ordovician through Upper Permian)

Colonies of auloporids may be shrubby looking, or flat and encrusting. Their corallites may be short, prostrate tubes, with or without other corallites rising above them, or long, slender, vertical tubes connected by short tubes or pores in their walls. The septa consist of longitudinal rows of fine spines. Platforms in the corallites may be numerous, sparse, or absent. If present, they are usually funnel-shaped.

389 Aulopora

Description: Corallite ¼" (6 mm) long and wide. Colonial; forms flat chains, networks, or crusts. Corallites short, trumpet-shaped, with each growing outward from point along rim of another, forming chain that may branch and join other parts of colony to form network. Colonies spread out over foreign surface, with each individual cemented along side away from calice, which is circular and faces up. Calices have faint septal ridges in some species. Few, if any, platforms inside corallites.

Age: Upper Ordovician through Pennsylvanian.

Distribution: Widespread in North America.

Comments: *Aulopora* is often found growing on shells of brachiopods and gastropods and the outer walls of other corals. Each corallite cemented itself to its host shell, making a much stronger colony than would have been possible if the colony grew upright or lay on the sediment. The movements of the brachiopod or snail host kept sediment from collecting over the colony, and also aided in feeding. *Aulopora* may have offered some protection to its host, since almost all corals have stinging cells on their tentacles. It is difficult to imagine how *Aulopora* could have benefited a coral host, but perhaps there was no way for a coral to prevent *Aulopora* from settling on it in the first place. 2 other corals with similar habits and appearance to *Aulopora* are: *Aulocystis,* found in the Middle Devonian in North America, which can be distinguished by the coarse, steeply inclined platforms inside the corallites; and *Aulocaulis,* found in the Upper Devonian in North America, whose calices open vertically into expanded apertures 1½–4 times the tube width.

379, 390   Syringopora

Description:
Corallite ⅟₃₂″ (1 mm) long and wide (pls. 379, 390). Massive, colonial; composed of long, cylindrical but somewhat irregular corallites, usually separate but sometimes closely set. Corallites grow upward in approximately parallel positions, connected by small cross-tubes at regular or irregular levels. Septa absent or consisting of 12 vertical rows of small spines on walls of calices. Platforms inside corallites funnel-shaped, with funnels pointing away from surface of colony.

Age: Silurian through Lower Permian.

Distribution: Widespread in North America.

Comments: *Syringopora* is very similar to the living Indo-Pacific genus *Tubipora,* the organ pipe coral. *Tubipora* is an octocoral, which means that the polyps always have 8 tentacles, and it is not at all closely related to *Syringopora.* But *Syringopora* and *Tubipora* are closely convergent—that is, both corals evolved a particular form, probably to fill identical or nearly identical niches in the ocean. *Syringopora* colonies are common in many formations, attaining heights of 2′ (61 cm) or more. A similar genus is *Chonostegites,* also found in the Lower and Middle Devonian in eastern North America. *Chonostegites* ha corallites that are connected at regular intervals by horizontal plates instead of tubes.

Phylum Mollusca
(Lower Cambrian to Recent)

Mollusks are the most familiar of all marine invertebrates. Almost all have shells—the common seashells that wash up on beaches. These shells are readily fossilized, and mollusks have the most complete fossil record of any phylum. In the Mesozoic and Cenozoic they dominate the marine fossil record almost to the exclusion of all other groups. In addition to the gastropods and bivalves—the best known of the mollusks—the phylum also includes the cephalopods: living squids, nautiloids, cuttlefish, and octopods, as well as ammonoids, which are known exclusively as fossils. A number of smaller groups of mollusks have poor fossil records: the Monoplacophora, caplike shells that may have been the ancestors of gastropods; the Polyplacophora, or chitons; the Rostroconchia, bivalvelike shells; and the Scaphopoda, or tusk shells.

The typical mollusk body is divided into 4 regions: a *head* with eyes and tentacles, well developed in most classes but reduced in some; a ventral, muscular *foot;* a *visceral mass* on the dorsal side, containing the internal organs; and a *mantle,* a sheet of tissue that covers the body and secretes a calcareous shell with an organic matrix. The first mollusks had a one-piece shell, and monoplacophorans, gastropods, scaphopods, and cephalopods are still univalves. Bivalves have a shell composed of 2 valves, hinged at the top, and enclosing the soft body. A chiton's shell is divided into 8 pieces. In many advanced mollusks, the shell has been reduced or lost altogether.

Mollusks evolved in the ocean and most classes are still strictly marine, but gastropods and bivalves have colonized fresh water, and gastropods have also moved onto land.

CLASS MONOPLACOPHORA
(Cambrian to Recent)

Most monoplacophoran genera are
known only as fossils from the early
Paleozoic. This group of mollusks with
cap-shaped shells had been thought
extinct until recently, when several
were dredged up from the deep sea.
Monoplacophoran fossils are known to
have a series of paired *muscle scars* inside
the shell. The living specimens also
have serially repeated parts like gills
and excretory organs. No other
mollusks show segmentation, and some
biologists think that monoplacophorans
demonstrate a link between mollusks
and annelids.

236 Scenella

Description: ⅜" (10 mm) high, ¼" (6 mm) wide.
Small; cap-shaped; apex tilted, usually
toward front. Base of cap oval.
Ornamented with moderately strong,
concentric wrinkles crossed by fine
radial ridges. Interior has 6–7 pairs of
muscle scars.

Age: Cambrian.

Distribution: Widespread in North America.

Comments: Until 1957, all mollusks similar to
Scenella had been thought to have
become extinct in early Devonian
times. In that year, however, a living
monoplacophoran, called *Neopilina,* a
close relative of *Scenella,* was dredged
up from the ocean depths off the
western coast of Mexico. The soft
anatomy showed clearly that
monoplacophorans do not belong with
the gastropods, where they had been
placed, but in a class of their own. A
number of the structures and organic
systems, like the gills and muscles,
were present in multiple pairs. Many
biologists think that the living
monoplacophorans clearly link the

mollusks with the annelids, which also have multiple paired organs. The specimen shown in Plate 236 is an internal mold. The marginal drawing shows a side view of the outside of the shell.

Helcionella, found throughout North America in the Cambrian, has the same caplike shape as *Scenella,* but its concentric sculpture is more pronounced, and the radial ribs are absent. *Helcionella* is classed with the gastropods, although muscle scars are unknown in that genus. *Proplina,* from the Upper Cambrian through the Lower Ordovician in North America, has an uncoiled apex that is shifted to overhang the anterior end. Its shell is smooth, or covered with faint concentric lines. There are 6 pairs of muscle scars inside the shell, identifying it clearly as a monoplacophoran.

CLASS GASTROPODA
(Cambrian to Recent)

Gastropods, living and fossil, are the most diverse of the mollusk classes. They include the snails, as well as the shell-less nudibranchs, sea hares, pteropods (planktonic oceanic snails), and garden slugs. Gastropods probably evolved from monoplacophorans, possibly very early in the Cambrian. The shells of the first gastropods were very much like those of the monoplacophorans, but the gastropod body had undergone *torsion:* the gut had been twisted around into a figure-8 so that the anus opened just over the head, with gills on either side of the head. The shell had also twisted around, so that the rear of the shell became the front.

Gastropods have a well-developed head, with eyes and tentacles. The mouth has a rasping jaw, called a *radula,* consisting of many tiny teeth arranged on a movable plate.

Most gastropods have a heavy, solid shell not easily broken or destroyed. The shell material is almost always aragonite, a form of calcium carbonate that is more easily dissolved than calcite, the shell material of many bivalves and brachiopods. Consequently, many gastropod shells, especially from the Paleozoic, are preserved as internal or external molds, or as casts. The typical gastropod shell is a long, narrow cone coiled into a disc or, more usually, a spire with an opening at the bottom. Each complete turn of the coil is called a *whorl.* The opening is the *aperture,* through which the gastropod withdraws its head and foot when threatened. Many gastropods bear an *operculum,* a horny or calcareous disc on the side of the foot used to seal off the aperture when the foot is withdrawn.

Most shells are coiled so that the aperture is on the observer's right when the spire is upright. The shell is carried so that the coiling axis is almost horizontal, with the spire extending out on the gastropod's right—the observer's left as he faces the snail. There may be a channel in the aperture on the gastropod's left for water intake, and a *sinus* or *slit,* or both, in the aperture on the right for water disposal. Gastropods are divided into 3 subclasses: the Prosobranchia, the Opisthobranchia, and the Pulmonata.

SUBCLASS PROSOBRANCHIA
(Lower Cambrian to Recent)

The Prosobranchia form the largest gastropod subclass, and include most marine gastropods and most fossil gastropods. Some prosobranchs live in fresh water and a few are terrestrial. They almost always have shells, which can be variously shaped. The soft parts show complete torsion—that is, anus, gills, and mantle cavity are located in the anterior.

ORDER ARCHAEOGASTROPODA
(Lower Cambrian to Recent)

These are the most primitive of the gastropods. They have gills with *filaments,* wedge-shaped or fingerlike sheets of tissue arranged on both sides of a central axis. There are usually 2 of these gills, but in some more advanced archaeogastropods the right gill is reduced in size or lost altogether. Archaeogastropods lack a *proboscis,* an extension of the mouth in later gastropods, used for feeding. Moreover, the archaeogastropod mantle is not protruded into a *siphon,* the tube contained in the *siphonal canal* through which later gastropods take in water. The earliest shells of archaeogastropods are coiled and are bilaterally symmetrical; they must have been held upright over the head. Shells soon appeared in which the coils were drawn out into spires, and these shells must have been carried to the right side of the body like most gastropod shells today. There is usually either a slit or a sinus, or both, through which the animal expels water, in the outer side of the aperture.

SUPERFAMILY BELLEROPHONTACEA
(Upper Cambrian through Lower Triassic)

The shells of this group are distinctive because their coils have not been drawn

out into spires, and the shells are
bilaterally symmetrical. There is always
a sinus or slit, or both, on the outer
edge of the aperture. In many genera,
the slit or sinus generates a *slit-band,* a
narrow, raised or depressed area with
crescent-shaped growth lines on the
periphery of the whorl.

176 Bucanella

Description: ⅜" (10 mm) high, ½" (13 mm) wide.
Small; loose, flat, disc-shaped coil,
with right and left sides symmetrical.
Whorls have 3 lobes: 1 central lobe,
visible only on outermost whorl, and 1
lobe to each side. Growth lines present
in all species; later species may have
fine revolving ridges. At wide aperture,
central lobe is cut back into broad
sinus. Slit absent; umbilicus totally
open, showing all whorls.

Age: Ordovician through Devonian.

Distribution: Widespread in North America.

Comments: The *Bucanella* specimen illustrated is
half buried in matrix, and part of the
last whorl is broken away, but the
central lobe and 1 lateral lobe show
clearly. Complete *Bucanella* shells, free
from the matrix, are extremely rare.
The marginal drawing is a
reconstruction of the complete shell.
Bucanella is a member of an important
Paleozoic suborder called
Bellerophontina. Did these gastropods
carry their shells with the sinus over
the head or over the back end?
Scientists do not agree, but the shell
would probably have balanced better if
the sinus were over the front.

219 Euphemites

Description: ⅜" (10 mm) high and wide. Globular,
tightly coiled; right and left sides

symmetrical, only outer whorl visible. Parietal area marked with about 10 sharp, revolving ridges beginning far within aperture and continuing out halfway to edge of aperture. Aperture broad, kidney-shaped. Outer lip of aperture has small U-shaped sinus in center. Umbilicus absent.

Age: Mississippian through Permian (possibly also in the Devonian).

Distribution: Widespread in North America.

Comments: The earliest gastropods, the bellerophontids like *Euphemites,* had bilaterally symmetrical shells. Most living gastropods have asymmetrical shells—that is, the coils are drawn out into a spire.

Sinuites, widespread in North America in the Ordovician, has the same tight coiling as *Euphemites* and, like *Euphemites,* lacks a slit and umbilicus, but *Sinuites* lacks spiral ribbing and has a larger sinus.

Owenella, found widespread in the Upper Cambrian, has a larger sinus like *Sinuites,* but is even more globular than either *Sinuites* or *Euphemites,* and does have an umbilicus.

174 Tropidodiscus

Description: ½″ (13 mm) high, 1″ (25 mm) wide. Small; disc-shaped, bilaterally symmetrical coil, compressed laterally, with periphery bent into sharp angle. Outer lip of aperture drawn out and point bisected by long, narrow slit. Slit-trace on outside of coil appears as band of crescent-shaped growth lines on crest of outer whorl. Umbilicus wide, showing earlier whorls.

Age: Lower Ordovician through Devonian.

Distribution: Widespread in North America.

Comments: During development, the gut in all gastropods twists around so that the anus is located in the front. This change of position creates a problem, since wastes are discharged over the animal's head. In many early gastropods the slit probably evolved to help move wastes to the rear. Water would have entered at the sides of the aperture, flowed up and over the gills, and carried wastes from the anus out the slit, and to the rear.

Bucanopsis

Description: ⅜″ (10 mm) high and wide. Small; tight coil with wide, flaring aperture and right and left sides symmetrical. Outer lip of aperture has sharp sinus. Sinus traces revolving band of crescent growth lines on ridge running from sinus along middle of outer surface of whorl. Surface ornamented with fine spiral threads. Large parietal area continues around sides of aperture as flange. Umbilicus small.

Age: Middle Ordovician through Silurian.

Distribution: Widespread in North America.

Comments: Today gastropods can be classified into 17 different feeding types. *Bucanopsis* probably belonged to the most primitive of these. The marginal drawing shows a front view of *Bucanopsis*.

Bucania, widespread in North America in the Middle Ordovician through the Silurian, has a large, open umbilicus and a flange smaller than that of *Bucanopsis* around the aperture. *Carinaropsis,* also widespread from the Middle and Upper Ordovician, has a very small coil and a greatly expanded aperture lip. The inside of the aperture has a well-developed keel on the floor.

218 Bellerophon

Description: ¾″ (19 mm) high, 1″ (25 mm) wide. Almost globular, with right and left

sides symmetrical, inner coils almost
completely covered by outer coils. Coils
rounded, usually with center crest.
Ornamentation variable, from fine
growth lines to broad concentric ribs or
regularly arranged small bumps.
Aperture flares out from broad parietal
deposit; slit at middle of outer lip of
aperture. Slit-band of crescent-shaped
growth lines on periphery of whorl.

Age: Silurian through Lower Triassic.
Distribution: Widespread in North America.
Comments: *Bellerophon* was the very first fossil
gastropod to be described—by
Montfort, in 1808.
Bellerophon occurs worldwide, and there
are many described species with a wide
variation in surface ornamentation. The
marginal drawing shown here is a
reconstruction of the whole shell of the
subgenus *Pharkidonotus,* which is
common in North America. A
reconstruction is often used to illustrate
a fossil genus. It is a drawing not of an
actual specimen, but of an idealized
one, based on many fossils.
Bellerophon and other bellerophontid
gastropods are sometimes confused with
coiled cephalopods, since the right and
left sides are symmetrical. Generally,
however, bellerophontids are much
smaller than cephalopods.

217 Knightites

Description: ¾″ (19 mm) high, 1⅛″ (28 mm) wide.
Coiled, with right and left sides
symmetrical. Sculpture of concentric
lines or threads present; some species
have spiral ribs that form lattice pattern
over surface. Wide, flaring lip around
aperture usually present, and
sometimes drawn out to sides in
hornlike projections. Lip notched with
deep, broad slit that traces depressed
slit-band around shell. Spouts or
bumps to sides of slit-band mark

inhalant channels in some species; other species lack conspicuous channels.

Age: Devonian through Middle Permian.
Distribution: Widespread in North America.
Comments: In the illustrated specimen of *Knightites,* the aperture is turned away and the rear of the shell is visible. The animal is oriented so that the slit, which is usually shallow, is down. This is how the snail probably carried its shell in life, with the slit just over its head. The hornlike projections probably served to take in water, which may have travelled up under the knoblike ridges to the exit at the slit. Other species of *Knightites* have spouts instead of bumps to the sides of the slit-band that have been interpreted as vehicles for taking in water.

SUPERFAMILY MACLURITACEA
(Upper Cambrian to Triassic; possibly also in the Upper Cretaceous)

Macluritaceans are an early group of coiling gastropods that have lost the symmetry of the bellerophontaceans. The coil of their shells is either pushed in to form a depressed top or pulled out into a spire. Those with spires appear to be left-coiling because, when the spire is pointed up, the aperture is to the observer's left. However, macluritaceans always have a sinus in the side of the aperture away from the spire, and this is assumed to be a channel for water disposal, or *exhalant channel.* Since such a channel would function better if located over the animal's head, it was probably so located, and the shells must be right-coiling; the snail must have carried the spire down rather than up.

178 Lecanospira

Description: ⅝" (16 mm) high, 1⅜" (35 mm) wide. Disc-shaped, all whorls visible. Top of coil depressed, bottom of coil flat. Whorls narrow and numerous, flat or round on base, sharply rounded above. Ornamentation of growth lines only. Aperture almost triangular, with sharp apex toward spire. Deep V-shaped sinus in top of aperture. Umbilicus very wide.

Age: Lower Ordovician.

Distribution: Widespread in North America.

Comments: *Lecanospira* is one of the earliest gastropods to develop an asymmetrical shell, which, judging from the flat base, was dragged by the animal over the sea floor.

The specimen illustrated shows the base of what is probably an internal mold. Gastropods are commonly preserved as internal molds, which are often difficult to identify because they show no surface ornamentation or details of the aperture. Paleozoic gastropods in particular are often preserved as molds because their shells are made of aragonite, a form of calcium carbonate that dissolves away relatively easily from the mud that fills the shell to create the mold, and 225 million years or more is usually ample time for such dissolution to occur.

The marginal drawing shows a reconstruction of a *Lecanospira* specimen.

168, 179 Maclurites

Description: ¾" (19 mm) high, 1⅞" (48 mm) wide (pl. 168); 1¾" (44 mm) high, 4⅛" (105 mm) wide (pl. 179). Large; modified disc-shape, with flat base. Outer whorl large, upper surface strongly convex. Growth lines present; pronounced revolving ridges also present in some species. Aperture high, broad, with slight sinus at top. Umbilicus deep, steep-walled.

Age: Ordovician.

Distribution: Widespread in North America.

Comments: The specimen illustrated in Plate 168 is an almost complete shell seen from the side, with the aperture forward. The other specimen is the flat base, with most of the shell material gone and the internal mold showing underneath (pl. 179).

Maclurites was named in honor of William Maclure (1763–1840), who has been called "the father of American geology." Maclure was born in Scotland, made a fortune early in life, and came to the United States in 1796. He was president of the Academy of Natural Sciences in Philadelphia for 22 years.

Maclurites (called *Maclurea* in older books) is very abundant in Ordovician rocks, but is usually poorly preserved. Many *Maclurites* fossils are found when limestones or shales are cut open with a saw. This procedure cuts the shell also, but the whorls can be seen as a spiral on the surface of the cut rock.

SUPERFAMILY EUOMPHALACEA
(Lower Ordovician through Upper Cretaceous)

Most euomphalaceans have right-coiling shells and are disc-shaped; some genera have fairly high spires. The area at which the angle of the shell changes on the outer rim of the aperture is believed to have contained an exhalant channel. In some genera, this angle change bears a slit that generates a slit-band. The whorls are usually not tightly coiled—in some genera the whorls are not in contact—and there is usually a large, open *umbilicus*. If the original shell material is preserved, the outer layer is calcite with a prismatic crystal structure, and the inner layer is aragonite with a sheetlike crystal structure.

172, 177 Helicotoma

Description: ¾" (19 mm) high, 1⅝" (41 mm) wide (pl. 172); ⅝" (16 mm) high, ⅞" (22 mm) wide (pl. 177). Disc-shaped, but with slightly raised spire and convex base. Sharp crest at edge of shoulder. Ornamented with faint revolving grooves and growth lines. Aperture has V-shaped sinus, with notchlike slit at crest of shoulder ridge. Slit leaves trace of crescent growth lines on top of ridge. Umbilicus wide.

Age: Lower and Middle Ordovician.

Distribution: Widespread in North America.

Comments: Plate 172 shows an internal mold of *Helicotoma,* with the top up. The specimen illustrated in Plate 177 is a *Helicotoma* shell embedded in limestone with the top exposed. Almost ½ the top of the last whorl is broken away. The fossil is composed of the original shell material. Calcareous shells are less apt to dissolve in limestone than in shale or sandstone.

Polhemia, from the Lower Ordovician in North America, is in the same family as *Helicotoma* and very similar to it. *Polhemia* has a deep groove at the upper sutures, a slightly concave shoulder, and a low ridge at the periphery of the shoulder. Its base has fine revolving ridges.

180 Ecculiomphalus

Description: ⅜" (10 mm) high, 1" (25 mm) wide. Disc-shaped, but openly coiled; later whorls out of contact with earlier ones. Coil round, but with high, thin frill-like crest at upper outer edge. Early part of coil closed off by internal partitions. Sharp, fine growth ridges follow shape of aperture. Slit and slit-band absent.

Age: Lower Ordovician through Silurian.

Distribution: Widespread in North America.

Comments: *Lytospira* is another open-coiled gastropod, from the Lower Ordovician through the Middle Silurian of North America. *Lytospira* has no crest, but does have a broadly angular sinus in the outer lip that makes a low ridge near the upper surface of the whorls. *Macluritella,* widespread in North America in the Lower Ordovician, also has whorls that do not touch, but the whorls lack a crest and have instead a slight angle change in the same place, with a shallow sinus on the aperture where the angle changes.

The specimen of *Ecculiomphalus* illustrated is an internal mold. The shell partially filled with mud after the death of the animal, but the mud did not reach all the way to the inner part, which filled instead with crystals of calcite that grew in the space. Interiors of shells and empty spaces in rocks will quite often fill with calcite, silica, or clay minerals.

The marginal drawing shows the top view of a reconstruction of *Ecculiomphalus.*

171 Straparollus

Description: ⅜″ (10 mm) high, ½″ (13 mm) wide. Shape variable, from conical with moderately high spire to disc-shaped with concave top. Base flat or concave. Early whorls partitioned off internally. Growth lines and sometimes faint revolving ridges present. Slight angle at outer, upper margin of aperture makes faint to pronounced ridge on upper surface of coil. Umbilicus wide.

403

Age: Silurian through Middle Permian.
Distribution: Widespread in North America.
Comments: *Straparollus* species, of which there are
many, especially in the Mississippian
and Pennsylvanian, are useful in dating
Paleozoic strata. Because of a printing
error in the original description,
Straparollus has been spelled *Straparolus*
until recently.

The specimen illustrated has a low
spire and rounded whorls. The
marginal drawings are reconstructions
of *Straparollus* that show the side and
the base of the shell.

228 Omphalotrochus

Description: 1⅜" (35 mm) high, 1¾" (44 mm)

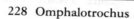

wide. Conical. Whorl profile shows
shoulder sloping down to abrupt angle
change, then sloping inward to become
almost vertical. Revolving ridge may
be present just above suture. Narrow
channel inside lower part of whorl.
Growth lines but no revolving
ornamentation present. Aperture
almost circular, but with broad sinus in
outer, upper edge. Bottom of aperture
protrudes forward. Wide umbilicus
open.
Age: Pennsylvanian through Middle
Permian.
Distribution: Widespread in North America.
Comments: Early archaeogastropods have 2 gills;
more advanced gastropods only 1. In
Omphalotrochus the broad sinus may not
have functioned to channel water.
Instead it may have served as a window
out of which *Omphalotrochus* could see
without having to extend its full body.
Babylonites, widespread throughout
North America in the Middle Permian,
has a spire that is almost a smooth
cone. The whorls are gently concave on
their upper ⁷⁄₁₀, then have a slight
ridge, with the band below either
concave or convex.

The specimen illustrated is unusually well preserved for a Paleozoic gastropod. The aperture is broken slightly so the sinus cannot be seen.

SUPERFAMILY PLEUROTOMARIACEA
(Upper Cambrian to Recent)

Many of the most common Paleozoic genera of gastropods belong to this superfamily. Their shells are mostly conical, but in a few genera are disc- or ear-shaped. The inner shell layer, made of aragonite, is pearly. There is almost always a sinus or a slit in the outer edge of the aperture, and usually a slit-band. Many living pleurotomariaceans have a deep slit in the aperture and are sometimes called slit shells. Others, like the abalone, have slits modified as a series of holes in the shell for the exhalant channel.

169, 227   Trepospira

Description: ¾″ (19 mm) high, 1⅛″ (28 mm) wide (pl. 169); ¾″ (19 mm) high, 1″ (25 mm) wide (pl. 227). Low cone with convex base. Ornamentation of very fine growth lines and revolving row of bumps just below upper suture. Aperture broad. V-shaped sinus on outer lip, with short slit in middle. Slit traces a slit-band just above outermost edge of body whorl; slit-band covered by upper edge of coil on earlier whorls. Flat slit-band has crescent-shaped growth lines. Umbilicus covered by small parietal region.

Age: Devonian through Middle Permian.

Distribution: Eastern United States; Mississippi Valley.

Comments: Paleozoic archaeogastropods tend to have little ornamentation other than fine growth lines; *Trepospira* is a distinctive exception.

Plate 227 shows a side view of a well-preserved specimen. The very top of the spire is broken and the slit does not show because the aperture is filled with rock. Plate 169 shows the top of another specimen. The drawing is a reconstruction based on many specimens.

239 Baylea

Description: ⅜″ (10 mm) high, ¼″ (6 mm) wide. Small, with broad, conical spire and deep, convex base. Whorls have sloping shoulders and periphery almost parallel to long axis of shell. Fine revolving ridges on whorls. Aperture deep, with narrow, shallow sinus near upper edge ending in shallow, notchlike slit that forms slit-band on outer edge of shoulder just above change in slope. Slit-band bordered by fine, narrow revolving ridges. Umbilicus small or absent.

Age: Mississippian through Middle Permian.
Distribution: Widespread in North America.
Comments: *Baylea* is sometimes preserved with its color pattern still visible; fossil color patterns are best seen with ultraviolet light.

Plate 239 shows a species of *Baylea* that is uncharacteristically high-spired and on which the shell is abraded so that not all features show clearly. Most species are more similar to the drawing in the margin, which shows a reconstruction of the side view of *Baylea*.

167 Bembexia

Description: ⅝″ (16 mm) high, ¾″ (19 mm) wide. Conical, with convex base. Whorls rounded. Slit-trace concave, in middle of whorls, bordered by fine revolving ribs. Another revolving ridge just above slit-band. Aperture extended outward, with moderately deep slit on upper, outer edge. Sutures moderately deep. Umbilicus open or absent.

Age: Lower Devonian through Mississippian.

Distribution: Widespread in North America.

Comments: The marginal drawing shows a side view of a reconstructed specimen of *Bembexia*.

Bembexia can be distinguished from *Baylea* by its more rounded whorls and its aperture, which is extended outward more than in *Baylea*. In the same family with *Bembexia* are 2 similar and common gastropods, both also found throughout North America: *Mourlonia*, from the Middle Ordovician through the Lower Permian, has a smoother spire, with only revolving ridges bordering the slit-band, and a deeper slit; and *Eotomaria*, from the Middle Ordovician through the Silurian, is squatter and wider than *Bembexia* or *Mourlonia* and has a much shallower slit.

232 Glabrocingulum

Description: ¾" (19 mm) high, ⅝" (16 mm) wide. Shaped like a top with rounded base. Aperture has shallow sinus and deep slit that leaves concave slit-band bordered by threads and situated on periphery of whorls. Sutures either at lower edge of slit-band, resulting in smooth spire, or well below, with whorl face vertical between slit-band and suture, resulting in stepped spire. Ornament of small, concentric ribs crossed by spiral ribs just below suture and on base, producing pattern of nodes on these areas.

Age: Mississippian through Middle Permian.

Distribution: Widespread in North America.

Comments: *Glabrocingulum* is one of the more
common of the many genera of top-
shaped Paleozoic gastropods with slits
in the aperture and slit-bands on the
whorls. The differences between the
genera may be quite inconspicuous and
especially difficult to see on fossil
specimens, since the whole gastropod is
seldom recovered.

Phymatopleura is a fossil very similar to
Glabrocingulum, and is found
throughout North America in the
Pennsylvanian. The shape of the whorls
is somewhat different in the 2 genera:
in both *Glabrocingulum* and
Phymatopleura the slit-band marks the
periphery, but in *Glabrocingulum* the
slit-band is at a 45° angle, while in
Phymatopleura it is vertical.

The ornamentation on *Phymatopleura* is
so similar to that of *Glabrocingulum* that
that element alone is insufficient to
distinguish the 2 genera, except that in
Phymatopleura there is always a thread
or a line in the middle of the slit-band.

Clathrospira, widespread in North
America from the Middle Ordovician
through the Silurian, has a body form
identical to that of *Phymatopleura* and
the smooth-spired species of
Glabrocingulum. But the slit is shorter
than in the other genera, and the only
ornament is the slit-band, which is
placed slightly above the periphery, as
in *Glabrocingulum.*

On the specimen illustrated in Plate
232, the spiral ribs are visible, but the
nodes are delicate and not very evident.
In the marginal drawing, which shows
the side view of a reconstructed
specimen, the ornamentation is more
apparent.

234 Loxoplocus

Description: 1" (25 mm) high, ⅝" (16 mm) wide.
Conical to very high-spired; later

whorls may be out of contact with earlier ones. Whorls with 1–3 ridges where direction of slope changes. Numerous fine growth lines present; some species have faint revolving ridges. Ridge on periphery of whorls shows trace of slit that is set into deep sinus on outer lip. Small umbilicus may be present on base.

Age: Ordovician through Silurian.

Distribution: Widespread in North America.

Comments: *Loxoplocus* has been found in fossil beds with brachiopods that have had small holes bored through their shells. Some paleontologists think that *Loxoplocus* may have been responsible, although no archaeogastropods living today bore into shell.

233 Worthenia

Description: 1⅛″ (28 mm) high, 1″ (25 mm) wide. Fairly high-spired, with convex base. Whorl profile slopes outward from suture to ridge, then downward, becoming concave. Ornamentation of fine growth lines and narrow, delicate revolving ridges. Aperture wide, almost circular, with deep sinus ending in slit. Slit-band forms knobbed ridge that turns around periphery of whorls at angle change. Area below slit-band on last whorl covered by spiral rows of bumps in some species. Umbilicus tiny or absent.

Age: Mississippian through Middle Triassic.

Distribution: Widespread in North America.

Comments: *Worthenia* was named for the paleontologist Amos Henry Worthen (1813–1888). Worthen was born in Illinois and worked in the Midwest for a good part of his life, mostly on the fossils of Mississippian and Pennsylvanian rocks. He discovered and described dozens of new fossil genera and species.
Worthenia may be confused with those

species of *Glabrocingulum* that have a stepped spire. On *Glabrocingulum*, however, the slit band is concave and smooth, while that on *Worthenia* is convex and knobbed.

166 Shansiella

Description: ½" (13 mm) high and wide. Spire low, whorls rounded. Lip has short slit that generates concave slit-band just above suture line just below mid-whorl. Ornament of spiral threads or cords dominant.

Age: Mississippian through Lower Permian.

Distribution: Widespread in North America.

Comments: The marginal drawing shows the side view of a reconstructed specimen of *Shansiella.*

Yunnania, another widespread gastropod found in the Devonian through the Middle Permian, has the same shape and dominant spiral ornament as *Shansiella* but does not have a slit or slit-band. *Euryzone,* found in North America in the Upper Ordovician through the Middle Devonian, also has the same shape, as well as a slit-band, but the slit-band is placed relatively high, above the middle of the whorls. Spiral ornamentation is faint on *Euryzone,* and the concentric ornamentation dominates.

Pleurotomaria

Description: 3" (76 mm) high, 2¾" (70 mm) wide. Conical, with round base; spire moderately high to depressed. Whorls steplike, with shoulder slightly sloping to angle change, which carries revolving row of small bumps. Whorl below angle fairly flat, sloping gently outward to lower suture. Ornamented with narrow revolving ridges and fine growth lines. Aperture large, round, with slit near middle of outer edge. Broad slit-band of crescent-shaped growth lines near middle of whorls. Umbilicus may be present.

Age: Lower Jurassic through Lower Cretaceous.

Distribution: Widespread in North America.

Comments: Some members of the family of *Pleurotomaria* are still living, although not in North America. Much about the biology of the extinct species can be deduced from these living members. *Pleurotomaria* carried its shell with the spire pointing down and to the right rear and the slit over the head and to the left, with paired gills at either side of the slit. Water entered the shell from the sides of the aperture, passed up and over the gills, and flowed out through the slit, travelling over the shell and to the rear. The slit served to keep wastes from being excreted on the animal's head.

The marginal drawing shows the side view of a complete *Pleurotomaria* specimen.

238 Ceratopea

Description: 1⅛″ (28 mm) high, ¾″ (19 mm) wide. Relatively large. Operculum horn-shaped, somewhat flattened. Deep, oval depression in base for muscle insertion. Prominent concentric growth lines.

Age: Lower Ordovician.

Distribution: Newfoundland to west Texas.

Comments: Many snails have an operculum that seals off the aperture after the body has been pulled into the shell. The calcareous operculum of this snail is commonly preserved, and was found and named long before the associated shell was discovered. The shell itself, which is conical, with round, smooth whorls, is rare. The operculum of

Ceratopea is scarcely ever found in association with its shell. *Ceratopea* is one of the most useful guides to the upper part of the Lower Ordovician since it is widespread, highly distinctive and ornamented, and found only in rocks of this age.

SUPERFAMILY TROCHONEMATACEA
(Middle Ordovician through Middle Permian)

These shells are turban-shaped. There is a channel or, sometimes, a sinus on the outer edge of the aperture where the angle of the whorl changes, about midway between the sutures or higher on the whorl. The shells have a pearly lining and are ornamented with spiral threads or cords or with rows of nodes, crossed by growth lines or heavier concentric threads.

225, 235 Trochonema

Description: 1″ (25 mm) high and wide (pl. 225); 2½″ (64 mm) high, 2¼″ (57 mm) wide (pl. 235). Low- to high-spired; base deep, round. Final whorls in old shells may not contact earlier whorls. Whorl profile shows abrupt angle change near upper suture that is trace of inner channel, and usually a 2nd near lower suture; whorl between angulations flat or concave. Aperture large, flared, with angle just above middle of outer lip containing channel within. Some species have ridge circling wide to minute umbilicus.

Age: Middle Ordovician through Lower Devonian.

Distribution: Widespread in North America.

Comments: *Trochonema* is very abundant and has many species in the Ordovician. It may have lost its right gill in the process of evolution, but since its entire superfamily is extinct, we can only hypothesize this loss. The marginal drawing shows the side view of a complete specimen of *Trochonema.* *Cyclobathmus,* found in the Middle Permian throughout North America, is in the same family as *Trochonema.* *Cyclobathmus* has a similarly shaped spire, but lacks the 2nd angle change, and is ornamented with spiral cords. Plate 225 depicts a species of *Trochonema* that has only 1 angle change on the whorls. Most of the specimen is an internal mold, but there is a bit of the shell partially covering 1 whorl. Plate 235 shows a specimen in which the upper whorls are well preserved, but the body whorl and aperture are broken.

SUPERFAMILY FISSURELLACEA
(Triassic to Recent)

The shells in this group are conical or cap-shaped, but not coiled (except at the tip in juveniles); the tip may be curved over in adults. There is a slit in the edge of the aperture, or a hole on the side or at the apex of the cone for the passage of water out of the shell. Inside the shell there is a horseshoe-shaped muscle scar, with the prongs pointing to the anterior.

237 Emarginula

Description: ½″ (13 mm) high, ⅝″ (16 mm) wide. Small; caplike, with apex bent toward rear and down. Slit on opposite, front margin. Muscle scar horseshoe-shaped, with inside toward rear, and opening toward front. Slight groove from slit continuing partway to apex, in some species bordered by 2 ribs. Ornamented with radial ribs crossed by faint to strong concentric ridges.

Age: Lower Jurassic to Recent (possibly also in the Middle Triassic).

Distribution: Widespread in North America.

Comments: Gills of living *Emarginula* cannot screen out much sediment from the water, and, in consequence, *Emarginula* is usually found only on hard substrates. It is found today on rocks, from Rhode Island south on the Atlantic coast and in the Gulf of California. Many species live in the ocean's depths.

Diodora, the Keyhole Limpet, found throughout North America in the Upper Cretaceous to the Recent, is another cap-shaped shell ornamented with ribs and ridges. However, *Diodora* lacks the marginal slit and has a hole in the apex instead.

SUPERFAMILY PLATYCERATACEA
(Lower Ordovician through Middle Permian)

These are mostly turban-shaped shells, although some are horn- or cap-shaped. There is no sinus or slit in the aperture, but there is a notch where the outer lip joins the body whorl. Most genera have a pearly inner shell layer.

223 Holopea

Description: 1″ (25 mm) high, ¾″ (19 mm) wide. Small; conical, moderately high-spired. Whorls round and smooth; final whorl much larger than earlier ones. Later whorls may not be in contact with earlier ones. Sutures deep. Final whorl has narrow umbilicus, round base, sometimes with coarse, revolving ribs. Aperture large, round, smooth, with no reflected lip or sinus.

Age: Middle Ordovician through Devonian.

Distribution: Eastern North America; Mississippi Valley.

Comments: A genus similar to *Holopea* is *Isonema*, which is widespread in the Lower Devonian. *Isonema* has a lower spire, whorls that are more sharply rounded, an aperture that is more oval, and a thickened parietal deposit covering the umbilical region. *Holopea* is sometimes preserved with a color pattern.

226   Cyclonema

Description: 1″ (25 mm) high, ¾″ (19 mm) wide. Most species have medium-high conical spire with slightly rounded whorls and large, rounded aperture with thickened columellar lip, lacking notch or slit. Other species have flattened spire with last whorl not in contact with earlier whorls. Ornamentation characteristic, with sharp spiral threads crossed by finer concentric threads.

Age: Middle Ordovician through Lower Devonian.

Distribution: Widespread in North America.

Comments: Like certain other gastropods, *Cyclonema* subsisted on the feces of crinoids. Some of the early species, like the *C. bilix* illustrated, were not noticeably modified for this mode of life; but some later genera like *Platyceras* became specialized: the shape of the aperture would change as the animals grew, depending on the shape of the host. Thus, the same species produced individuals with very different shell forms, making the systematics of this group very difficult.

Elasmonema, from the lower Devonian and also widespread, has a shape very similar to that of the primitive *Cyclonema*. *Elasmonema* has no spiral ornament, but does have a small umbilicus.

215, 216 Platyceras

Description: ½" (13 mm) high, ¾" (19 mm) wide
(pl. 215); 1¼" (32 mm) high and wide
(pl. 216). Variable in shape, from
tightly coiled cones to completely
separate whorls. Most common genera
have 1–2 early whorls coiled atop deep,
caplike cone. Some species horn-
shaped. All thin-shelled, with fine
growth lines and sometimes revolving
lines; some species have hollow spines.
Shape of aperture highly variable.
Umbilicus may be present.

Age: Silurian through Middle Permian.

Distribution: Widespread in North America.

Comments: *Platyceras* is one of the most abundant
gastropod genera in the Paleozoic,
possibly, in part, because the shell has
an outer layer of calcite that is more
resistant to dissolution than the usual
aragonite outer layer. *Platyceras* has a
most interesting life history. These
animals lived by attaching themselves
over the anus of crinoids and eating
their feces. *Platyceras* is occasionally
found preserved still attached to its
crinoid host. The crinoids probably
tolerated *Platyceras,* or even encouraged
it as a sanitary convenience.
Plate 216 shows *Platyceras* at the top
left still attached to its crinoid host.
Plate 215 shows a side view of the most
common form of *Platyceras.*

**SUPERFAMILY
MICRODOMATACEA**
(Middle Ordovician through Middle
Permian)

These shells are turban- or awl-shaped,
with simple, round apertures. The lip
at the *columella* is turned back slightly,
and the inner shell layer is pearly.

240 **Microdoma**

Description: ¼" (6 mm) high; ⅛" (3 mm) wide. Small; spire high with many whorls. Sutures shallow, sides of whorls not rounded. Beginning on 4th whorl, spire ornamented with broad concentric ribs, with fine threads between. On later whorls, 2 spiral grooves develop that break concentric ribs into 1 or more rows of bumps. Aperture rounded, simple, with slightly reflexed columellar lip. Umbilicus small.

Age: Lower Devonian through Lower Permian

Distribution: Widespread in North America.

Comments: Another small gastropod that looks similar to *Microdoma* is *Pseudozygopleura* (sometimes considered a subgenus of *Palaeostylus*), found throughout North America in the Mississippian through the Middle Permian. *Pseudozygopleura* has a very high, pointed spire, usually with relatively large, rounded, concentric ribs, except on its base, which is smooth. Both spiral ornamentation and umbilicus are lacking.

SUPERFAMILY TROCHACEA
(Triassic to Recent)

These shells are turban-shaped, globular, or smoothly conical, with no slit or sinus. The inner shell layer is pearly.

229, 230 **Calliostoma**

Description: 1" (25 mm) high, 1⅜" (35 mm) wide (pl. 229); ¾" (19 mm) high and wide (pl. 230). Conical, with slightly rounded base. Whorls not rounded, but have small shoulder below sutures. Ornamented with spiral threads that may be smooth or beaded. Outer lip extended forward; aperture discontinuous: outer lip overlaps columellar lip so that aperture is not all in same plane. Umbilicus absent.

Age: Oligocene to Recent.

Distribution: Atlantic, Pacific, and Gulf coastal plains.

Comments: Living *Calliostoma,* known as top shells, range along both coasts of North America from boreal to tropical regions, and are also found in the Gulf of Mexico. They prefer a rocky habitat near shore, and are often found between the tide lines.

Some species of *Margarites* are very similar to *Calliostoma.* Widespread in the Upper Cretaceous to the Recent, *Margarites* is a common fossil genus that still lives on the Atlantic and Pacific coasts and is placed in the Trochidae family with *Calliostoma.* While most *Margarites* species have well-rounded whorls, some have almost straight whorls and look like *Calliostoma.* However, *Margarites* has an umbilicus, and, although the lip is interrupted as in *Calliostoma,* it lies more nearly in a single plane, with the outer lip not inclined as far forward.

SUPERFAMILY NERITACEA
(Middle Devonian to Recent)

Neritacean shells are commonly globular, with flat or slightly protruding spires. There are few whorls, and the *body whorl* is usually much expanded. A few genera are limpet-shaped. The outer shell layer is calcite, which is more likely to be preserved in fossils than aragonite, and often retains the color pattern. The inner shell layers are thick and aragonitic, and are not pearly. The right gill has been lost in living forms.

224 Naticopsis

Description: 1⅝″ (41 mm) high, 1¼″ (32 mm)
wide. Spire small; final whorl large,
globular. Growth lines may be
thickened into short ridges just below
suture. Aperture very large, round to
almost square. Sinus absent. Parietal
area thickened. Umbilicus and
revolving ornamentation absent.

Age: Middle Devonian through Triassic.
Distribution: Widespread in North America.
Comments: The marginal drawing shows a side
view of a complete specimen of
Naticopsis.

Trachydomia, found throughout North
America from the Pennsylvanian
through the Middle Permian, is closely
related to *Naticopsis,* and similar to it,
except that *Trachydomia* is covered with
revolving rows of bumps of different
sizes.

Naticopsis is believed to have lived in
shallow water, perhaps even above the
low-tide line. It probably fed like
Littorina, the periwinkle, a common
gastropod living today along all North
American coasts. *Littorina* uses its
radula, or set of tiny teeth, to scrape
algae off rocks and to eat virtually
anything else lying on the shore.

SUPERFAMILY MURCHISONIACEA
(Lower Ordovician through Upper
Triassic; possibly also in the Upper
Cambrian)

These shells have high spires with
many whorls. The outer lip has a sinus
or a slit that generates a slit-band. In
some, the lower edge of the aperture is
drawn down into a U-shape.

260 Murchisonia

Description: 1½" (38 mm) high, ⅝" (16 mm) wide. Spire very high, tapered; whorls numerous, round. Aperture elongate, oval, with broad sinus in outer lip terminating in V- or U-shaped notch. All whorls bear pronounced slit-band just below mid-whorl; slit-band is concave and may have narrow ridges as borders. Some species have 3rd ridge in middle of slit-band.

Age: Ordovician through Triassic.

Distribution: Widespread in North America.

Comments: *Murchisonia* was named for Roderick Murchison (1792–1871), who was an English army officer and gentleman farmer before his wife's interest in sketching rock strata and fossils persuaded him to study geology at Oxford and Cambridge. Later, Murchison studied the rocks and fossils of southern Wales, and in 1839 published his description of a new system of rocks, the Silurian System, which still forms the basis of our knowledge of the Silurian Period. *Hormotoma,* considered a separate genus until recently, is now included as a subgenus of *Murchisonia.* *Ectomaria,* widespread throughout North America in the Middle and Upper Ordovician, looks very much like *Murchisonia,* but has no slit—only a deep sinus in the outer lip. The slit-band is always bordered by strong spiral ridges. *Stegocoelia,* found widespread throughout North America in the Mississippian through the Middle Permian, has a high spire like *Murchisonia,* but its slit-band is above mid-whorl, bordered by spiral threads, and there are 2 or more additional spiral threads. The whorls may be rounded or have an angulation at the whorl periphery. *Michelia* (formerly known as *Coelocaulus*), widespread in the

Ordovician through the Devonian, has the same form as *Murchisonia*, but its slit is short, without parallel sides, and consequently the slit-band is not sharply defined on the whorls.
In the Middle Devonian, *Murchisonia* produced many new species, some of which had elaborate and bizarre forms.

ORDER MESOGASTROPODA
(Ordovician to Recent)

Almost all mesogastropods are characterized by a single gill with only 1 row of filaments, in contrast to the archaeogastropod gill, which has 2 rows. Fossil shells, in which the gills are not preserved, are classified on the basis of their similarity to living mesogastropods. The group, in general, has porcelaneous shells and simple, unnotched apertures.

SUPERFAMILY CYCLOPHORACEA
(Mississippian to Recent)

In this group of freshwater and terrestrial mesogastropods, the gills have been lost and the mantle cavity transformed into a lung. The shell is usually conical, with smooth, rounded whorls.

222 Viviparus

Description: 1″ (25 mm) high, ¾″ (19 mm) wide. Conical, with a moderately high spire. Whorls smooth, rounded or flattened; sutures depressed. Aperture circular. Sinus and channels absent. Umbilicus minute or absent.

Age: Jurassic to Recent.

Distribution: Widespread in North America.
Comments: *Viviparus* is a freshwater gastropod. As
its name indicates, it bears its young
alive—that is, the young leave the
parent as tiny shelled gastropods, not as
eggs or larvae. This adaptation is useful
for a freshwater species because fresh
water would diffuse by osmosis into the
cells of tiny, unprotected eggs or
larvae, which would then swell up and
rupture. This problem of osmosis does
not occur in salt water because the
concentration of salts in sea water is
roughly the same as that in the cell
fluids of an animal. The black marks on
the specimens in Plate 222 are stains,
perhaps from the enclosing rock, and
not a color pattern.

SUPERFAMILY SUBULITACEA
(Middle Ordovician through Permian)

Subulitaceans have spindle- or awl-
shaped shells, commonly slender and
pointed at the apex of the spire. The
surface of the whorls is smooth, and
rarely has faint spiral or concentric ribs.
Some genera have spiral ridges on the
columella, and some have a slight
notch in the lowest part of the
aperture.

256 Ianthinopsis

Description: 1⅛" (28 mm) high, ⅝" (16 mm) wide.
Spindle-shaped to almost spherical.
Surface smooth. Whorls almost flat to
slightly round, except for more convex
body whorl; width greatest at middle of
last whorl. Base tapering or round.
Aperture may extend down to form
broad siphonal channel at base of
columella. Outer lip almost straight.
Parietal fold present. Umbilicus absent.
Age: Devonian through Permian.

Distribution: Widespread in North America.
Comments: A fossil found very commonly in Pennsylvanian and Permian faunas, *Ianthinopsis* is usually well preserved in shales. *Subulites,* widespread throughout North America from the Middle Ordovician through the Lower Devonian, has some species with smooth spires like *Ianthinopsis,* but *Subulites* has no parietal fold. *Bulimorpha,* found in the Mississippian throughout North America, resembles the more globular forms of *Ianthinopsis,* but has a columella that is arched, and no parietal fold.

261 Meekospira

Description: 1½" (38 mm) high, ⅝" (16 mm) wide. Small; thin-shelled, high-spired cone with apex pointed; width greatest just above convex base. Whorls between sutures almost flat; sutures shallow. Surface smooth and polished, with only fine growth lines. Aperture teardrop-shaped, round at bottom. Outer lip nearly straight. Sinus, siphonal channel, and columella folds absent.
Age: Ordovician through Permian.
Distribution: Widespread in North America.
Comments: Common in Pennsylvanian and Permian faunas, *Meekospira* was named for Fielding Bradford Meek (1816–1876), a paleontologist who worked with F.V. Hayden in the West, and participated in the first geological surveys of Iowa, Wisconsin, and Minnesota.

SUPERFAMILY LOXONEMATACEA
(Ordovician through Cretaceous)

These shells are high-spired and slender, with many whorls. The outer

lip of the aperture commonly has a
deep to shallow sinus. The inner shell
layer is not pearly.

259 Loxonema

Description: ⅝″ (16 mm) high, ¼″ (6 mm) wide.
High-spired cone with many round
whorls. Base convex. Sutures
moderately deep. Outer lip with deep,
U-shaped sinus; notch and slit absent.
Growth lines or thin transverse ribs
form crescent pattern on whorls, with
crescents concave toward aperture.
Umbilicus absent.

Age: Ordovician through Mississippian.
Distribution: Widespread in North America.
Comments: This genus is very common in most
limestones and shales, particularly in
the Ordovician, Silurian, and
Devonian, and occurs associated with
brachiopods and bryozoans. A genus
with a form very similar to that of
Loxonema is *Pseudozygopleura*, widespread
throughout North America from the
Mississippian through the Middle
Permian, which has no sinus, and has
transverse ribs that are limited to the
upper whorls or absent altogether.
These ribs are straight instead of
crescent-shaped.

SUPERFAMILY CERITHIACEA
(Lower Devonian to Recent)

These shells have high spires and many
whorls. In most of the more advanced
genera, the lower part of the aperture is
modified into a short, often twisted
siphonal canal.

262 Cerithium

Description: 1⅛″ (28 mm) high, ⅜″ (10 mm) wide. High, narrow spire; many flat whorls present, covered with indistinct sutures. Commonly highly ornamented in various ways, usually with revolving ribs of various sizes, which may be knobbed. Transverse ornamentation may consist of growth lines or fairly broad ribs, again with knobs or bumps. Aperture oval, with short, backward-turning channel at base; notch or short channel usually on upper side of aperture where outer lip contacts body whorl. Slightly pointed base may have revolving rows of beads. Umbilicus absent.

Age: Jurassic to Recent.

Distribution: Widespread in North America.

Comments: *Cerithium,* the cerith, is still a common marine gastropod. Many living species are found near shore in shallow water, and a few live in brackish water. *Orthonema,* widespread throughout North America in the Mississippian through the Middle Permian, is a very similar genus, though restricted to the Paleozoic. *Orthonema* has a spiral thread just below the suture, and 2 more threads or angulations below this first thread. Its aperture is rounded at the base, where *Cerithium* always has a channel. Plate 262 shows an uncharacteristically smooth species. The drawing illustrates a more typical *Cerithium.*

264 Turritella

Description: 2″ (51 mm) high, ½″ (13 mm) wide. Very high, slender spire of many whorls. Whorls convex to almost flat in profile, with upper ½ sloping in toward suture; lower ½ usually overhangs lower whorl. Ornamentation of narrow, strong revolving ridges, usually heaviest on periphery of whorls, and finer above. Ridges often beaded. Growth lines sinuous. Aperture oval, round, or square. Sutures moderately incised. Outer lip thin. Siphonal canal, notches, and umbilicus absent.

Age: Triassic to Recent.

Distribution: Widespread in North America.

Comments: *Turritella,* the turret shell, is one of the most common Mesozoic and Tertiary fossils in North America. There are innumerable species, many very valuable as guide fossils. Today, *Turritella* lives in the sediment, and filters fine particles of food from the water. *Nerinea,* found in the Jurassic and Cretaceous in the southwestern United States, has a form like that of *Turritella,* but the aperture has a short notch at its lower edge, and the columella has horizontally inclined folds, often found on the inside of the outer lip as well. *Nerinella,* from the Jurassic and Lower Cretaceous throughout North America, is like *Nerinea,* except that it has no columellar folds and does have a siphonal canal.

258 Goniobasis

Description: ½″ (13 mm) high, ¼″ (6 mm) wide. Small, spire high, base round, whorls flat to gently convex in profile. Whorls smooth or with fine revolving ribs, sometimes crossed by transverse ribs, forming bumps at intersections. Aperture oval, rounded at top. Siphonal canal and umbilicus absent.

Age: Cretaceous to Recent.

Distribution: Widespread in North America.

Comments: In Wyoming, the white shells of *Goniobasis* occur in great numbers packed within black rock. This rock is called "turritella agate" and is used to make jewelry. *Goniobasis* still lives in freshwater lakes in Florida and

Georgia. *Goniobasis* is oviparous, laying eggs that hatch outside the parent's body.

Melania, found throughout North America from the Jurassic to the Recent, is similar to *Goniobasis* and is also found in fresh water.

SUPERFAMILY STROMBACEA
(Triassic to Recent)

These are mostly big, heavy shells with a large body whorl and a narrow aperture notched at both ends.

252 Calyptraphorus

Description: 1⅜" (35 mm) high, ¾" (19 mm) wide. Spindle-shaped, with high, pointed spire and very elongate siphonal canal. In mature animals, veneer of shell deposited by mantle covers all earlier whorls from apex to base, obscuring their character. Aperture long and narrow.

Age: Tertiary to Recent.

Distribution: Atlantic and Gulf coastal plains.

Comments: The mature *Calyptraphorus* is a very distinctive shell because of its pronounced spindlelike appearance—the spire and siphonal canal are almost the same size and shape, unless broken—and the smooth covering of its shell. Before maturity, however, it looks quite different, with narrow revolving ribs and coarse transverse ribs on all whorls. These immature *Calyptraphorus* are similar to *Clathrodrillia* and some *Mangelia* species (which are found from the Eocene to the Recent on all coasts of North America), but *Calyptraphorus* is considerably larger than these, with smoother spire and a relatively larger siphonal canal.

255 Anchura

Description: 1¼″ (32 mm) high, ½″ (13 mm) wide. Spindle-shaped; spire high. Sculpture usually of transverse ribs, sometimes of fine revolving ribs. Outer lip of aperture greatly extended away from whorls, forming wing that divides at outer edge. Siphonal canal long and thin.

Age: Cretaceous.

Distribution: Widespread in North America.

Comments: The grossly extended lip of *Anchura* probably served to protect the snail's body, particularly the head, as it glided along, picking up organic matter from the sea floor. Plate 255 depicts an internal mold, or steinkern, of *Anchura* and does not show the extended outer lip. The mold formed when a dead shell was surrounded by and filled with mud. The mud hardened and the shell dissolved, leaving the steinkern showing the internal form of the shell. The drawing in the margin shows the external form.

231 Xenophora

Description: 1″ (25 mm) high, 1¼″ (32 mm) wide. Broad cone with almost straight sides and flat or depressed base. Upper surface commonly covered with foreign objects like shells, pebbles, and bits of coral; with objects removed, surface covered with hollow impressions or scars. Small aperture on base, facing down. Umbilicus covered over with shell.

Age: Cretaceous to Recent.

Distribution: Atlantic and Gulf coastal plains.

Comments: *Xenophora* is called the carrier shell today because, as it grows, it cements numerous objects onto itself. One function of this unusual decoration is doubtless camouflage: the shell is difficult to see when placed against a

background of dead and broken shells.
But the cemented decoration may serve
another purpose as well. *Xenophora*
often chooses particularly long items
with which to decorate the final whorl
of its shell, and these act as props to
hold the shell slightly off the bottom.
Xenophora feeds by picking up organic
material from the sea floor. It
apparently can feed more easily with its
aperture slightly elevated.

**SUPERFAMILY
CALYPTRAEACEA**
(Cretaceous to Recent)

Conical to limpetlike in shape, or with
a few whorls, these shells have smooth
apertures, often with a small notch on
the lower part, and a muscle flange on
the inside of the shell.

214 Crepidula

Description: 7/8″ (22 mm) high, 2⅛″ (54 mm) wide.
Boat- or slipper-shaped; coiled only
very early in life. Body whorl very
large, comprising nearly whole shell.
Ornament of growth lines; some specie
have radial ribs, which may be spiny.
Aperture very large, oval, almost ½
covered by thin, flat, shelly shelf.
Age: Upper Cretaceous to Recent.
Distribution: Coastal areas of North America.
Comments: *Crepidula,* or the slipper shell, as it is
known today, attaches its foot by
suction to a firm object, sometimes
another *Crepidula* shell, and usually
stays there throughout its life. It feeds
by filtering small bits of food out of th
water. *Crepidula* is common on all
coasts of North America, and its shells
can be found on most beaches.

SUPERFAMILY CYPRAEACEA
(Jurassic to Recent)

These are the cowries. They have egg-
shaped shells, with long, narrow
apertures and concealed spires.

---

267 Cypraea

Description: 1⅝" (41 mm) high, ⅞" (22 mm) wide.
Egg-shaped, very smooth; spire of
adult totally covered by expanded final
whorl. Side bearing aperture flattened;
aperture narrow and long, outer lip
curving inward; aperture base becomes
siphonal canal with notch. Both outer
and inner lips covered with many
small, sharp ribs.

Age: Jurassic to Recent.

Distribution: Widespread in North America.

Comments: *Cypraea,* whose common name today is
cowry, has a broad, brilliantly colored,
strikingly patterned mantle that it
carries wrapped around and completely
covering its shell. The bright colors
and patterns and the expanded mantle
may act as protective devices. When a
predator, perhaps a fish or another
gastropod, first bumps into *Cypraea,* it
senses something soft, plump, and
bright red, pink, or green. But at the
second contact, after *Cypraea* has
quickly withdrawn its mantle, the
predator touches something hard,
shiny, and brown. The confused
predator, perhaps thinking his
potential prey has escaped, goes off to
hunt for it elsewhere.

---

170 Gyrodes

Description: 1" (25 mm) high, 1⅜" (35 mm) wide.
Spire low, almost flat; final whorl large.
Whorls round and smooth except for
flattened or concave band at upper

suture on later whorls; some species
have ribs or bumps on band. Aperture
large and round. Umbilicus deep and
wide.

Age: Cretaceous.

Distribution: Widespread in North America.

Comments: *Gyrodes* is one of many fossil gastropods
having the general form called
naticoid, after the common fossil and
living genus *Natica*. Often only minor
differences in ornamentation are used to
distinguish one naticoid genus from
another.

Plate 170 shows an internal mold of
Gyrodes. The drawing shows a side view
of a complete shell.

SUPERFAMILY NATICACEA
(Triassic to Recent)

Moon snails have globular shells with
low spires, much-expanded body
whorls, and rounded apertures.

221 Polinices

Description: 1½″ (38 mm) high, 1⅜″ (35 mm)
wide. Oval to almost spherical, solid,
smooth. Spire low; final whorl very
large, round, inflated. About 4 whorls
in spire; shell suture tight. Aperture
oval; inner lip thickened by shell
deposit that also partially or completely
fills umbilicus. Outer lip smooth, not
thickened. Sinus and channel absent.

Age: Upper Cretaceous to Recent.

Distribution: Widespread in North America.

Comments: Living *Polinices* are called moon snails.
They are still major marine predators
on bivalves and other shallow-water
snails from Massachusetts and
California south. *Polinices* grasps its
prey in its enormously expanded foot
and rasps a small, beveled hole through
the shell, out of which it sucks the soft

tissues. *Polinices* lays its eggs in a distinctive circular collar composed of hardened mucus and sand grains. *Natica,* found throughout North America from the Triassic to the Recent, has a more open umbilicus than *Polinices,* and a calcareous operculum. *Lunatia,* ranging from the Cretaceous to the Recent, also throughout North America, has a thinner shell and is more high-spired than *Polinices.* A callus is sometimes present in *Lunatia.*

220 Sinum

Description: ⅝″ (16 mm) high and wide. Oval in outline, with very small spire and very broad final whorl. Spire may be almost flat, with nearly whole shell consisting of aperture. Aperture oblique, outer edge thin and sharp. Callus over inner side of aperture thin, extended to cover umbilicus completely or partly. Upper edge of aperture may form small channel where it joins body whorl.
Age: Upper Cretaceous to Recent.
Distribution: Atlantic and Pacific coastal plains.
Comments: The flat species of *Sinum* without elevated spires are commonly called baby's ears. The shell is almost totally internal in some species, is much smaller than the soft tissue, and is largely covered by an expansion of the foot.

SUPERFAMILY TONNACEA
(Cretaceous to Recent)

These shells are fairly thin, with low spires, large apertures, and a short siphonal canal at the base of the aperture.

251 Ficus

Description: 2¾" (70 mm) high, 1⅛" (28 mm) wide. Large, fig-shaped, with large, inflated body whorl and very low, almost flat spire. Sculpture of fine, close, revolving and transverse threads forms delicate cross-hatched pattern over entire shell. Aperture very elongate, reaching almost from apex to long, tapering, open siphonal canal at base. Outer lip thin, highly glazed within.

Age: Eocene to Recent.

Distribution: Atlantic and Gulf coastal plains.

Comments: *Ficus* means "fig." Other fig shells are *Priscoficus,* found on the Atlantic and Gulf coasts from the Eocene to the Recent, which has, in addition to the cross-hatching, broad revolving ribs and an outer lip with broad scallops; and *Ficopsis,* found only in the Eocene on the Atlantic, Pacific, and Gulf coasts. *Ficopsis* has a higher spire than *Ficus.*

ORDER NEOGASTROPODA
(Cretaceous to Recent)

The distinguishing feature of neogastropods is a modification of the anterior part of the mantle into a siphonal canal. The lower part of the aperture usually has a groove to hold the siphon.

SUPERFAMILY MURICACEA
(Cretaceous to Recent)

Sometimes called rock shells, these are heavy, strong shells ornamented with spines usually arranged on concentric ribs.

246 Murex

Description: 1⅜" (35 mm) high, ⅞" (22 mm) wide.
Round, inflated body whorl with
elevated spire and tapering base. Each
whorl has 3 heavy, transverse, spine-
bearing ribs marking position of former
apertures. Most species have additional
transverse ribs without spines, and
strong but narrow revolving ridges.
Aperture round, with continuous lip.
Siphonal canal long and slender,
sometimes almost completely closed
over.

Age: Tertiary to Recent (possibly also in the
Cretaceous).

Distribution: Atlantic and Gulf coastal plains.

Comments: The long siphonal canal on *Murex* and
many other neogastropods held the
siphon—a narrow, rolled extension of
the mantle. The siphon acted as a
sensor, held out in front of the head,
well protected from predators by its
shelly channel. The siphon contained
special chemical-detecting cells that
allowed the murex to sense food sources
as well as potential dangers.

244 Ecphora

Description: 2⅛" (54 mm) high, 1⅞" (48 mm)
wide. Large; squat, loosely coiled.
Umbilicus wide. Distinctive ornament
of narrow, high, coarse revolving ribs,
sometimes T-shaped in cross section;
about 4 ribs per whorl. Some species
covered with additional finer revolving
ridges. Aperture large, rounded,
extending down into short siphonal
canal. Small exhalent canal at upper,
outer edge of aperture.

Age: Cretaceous through Miocene.

Distribution: Atlantic and Gulf coastal plains.

Comments: *Ecphora* is a close, extinct relative of
Murex, and may have shared the same
feeding habits. If so, *Ecphora* would
have been a predator, boring holes in

the shells of bivalves or other snails. A
gland in the foot would have secreted a
special chemical to soften the prey's
shell. A set of tiny teeth, called the
radula, would then have rasped first the
shell and then the victim's flesh.
Cretaceous and Tertiary fossils are
commonly found with holes bored into
them, some probably by *Ecphora* and its
relatives.

248 Typhis

Description: ⅝" (16 mm) high, ⅜" (10 mm) wide.
Body whorl round, inflated; spire
elevated. Whorls have flat shoulders
and straight sides. Distinctive sculpture
of 4 heavy, sharply angled ribs on each
whorl; each rib has large, hollow spine
at shoulder. Spines look like hollow
tubes if broken. Each whorl has 4
additional spines between ribs.
Aperture oval, with continuous
elevated rim. Long, slender siphonal
canal completely roofed off from
aperture to base.

Age: Cretaceous to Recent.

Distribution: Atlantic and Gulf coastal plains.

Comments: *Typhis* is very similar in shape to
Murex, but has more rounded transverse
ribs, hollow spines only at the
shoulders, and a closed siphonal canal.

SUPERFAMILY BUCCINACEA
(Cretaceous to Recent)

Buccinaceans have sturdy shells,
usually with well-developed sculpture.
The aperture has a siphonal canal or a
notch at the bottom. The inner lip is
generally smooth.

243 Neptunea

Description: 1½" (38 mm) high, ¾" (19 mm) wide.
Large; elongate-oval, with moderately
high spire and somewhat tapered base.
Whorls have flat to sloping shoulders
and gently rounded sides. Weak to
strong, narrow revolving ribs usually
cover entire shell. Aperture oval but
drawn down to form thick, open
siphonal canal, which may be curved.
Small fold for exhalent canal sometimes
on outer lip.

Age: Cretaceous to Recent.

Distribution: Atlantic and Pacific coastal plains.

Comments: *Neptunea* has many living species,
known as neptunes, that are common
in Arctic and boreal seas. Like *Busycon*
and other whelks, *Neptunea* lays eggs.
The young spend their larval stages
within the egg case, and hatch with
complete miniature shells.

250 Busycon

Description: 5" (127 mm) high, 2¼" (57 mm) wide.
Large; heavy, pear-shaped, with low
spire. Each whorl below shoulders
completely or mostly covered by
following whorl. Sutures closed.
Sculpture various: smooth or with weak
to strong spiral lines; transverse ribs
may develop into spines at shoulders.
Aperture pear-shaped. Unthickened
outer lip, extending into long, narrow,
open canal for siphon.

Age: Upper Cretaceous to Recent.

Distribution: Widespread in North America.

Comments: The *Busycon,* or whelk, is one of the few
genera of gastropods in which the shell
may coil either to the right or the left.
The species illustrated in Plate 250
coils to the left. *Busycon* is a common
predator on all coasts today. It feeds
mainly on bivalves and gastropods,
but, unlike many gastropod predators,
does not bore holes in the shell.

Instead, it holds the prey with its foot and bangs it sharply against its own outer lip until the prey shell cracks. Then *Busycon* uses its little teeth on a long extendible proboscis to rasp out the prey tissues.

Pyropsis, found from the Upper Cretaceous through the Eocene throughout North America, has a whelklike shape, but its sculpture is predominantly spiral and sometimes beaded, and its body whorl is more inflated. Its base is contracted into a long, narrow canal, which is often broken away.

Perissolax, found in Eocene strata throughout North America, has a relatively more inflated body whorl and an aperture that is abruptly contracted and drawn down into a very narrow siphonal canal.

242 Fasciolaria

Description: ¾" (19 mm) high, ⅜" (10 mm) wide. Spindle-shaped. Whorls usually convex, sometimes shouldered. Smooth or strongly sculptured with transverse and revolving ribs. Aperture oval, expanded downward into long, twisted canal. Outer lip slightly thickened, with sinuous outline. Small, notchlike channel at junction of outer lip with body whorl.

Age: Cretaceous to Recent.

Distribution: Widespread in North America.

Comments: *Fasciolaria* are known today as tulip shells. They are common predators in shallow water from North Carolina south, and in the Gulf of Mexico, with a preference for large gastropods like *Ficus* and *Strombus.*

The marginal drawing shows a *Fasciolaria* species with a smoother shell than that of the species shown in the photograph.

SUPERFAMILY VOLUTACEA
(Cretaceous to Recent)

This group includes the olive shells. The shell shape varies and the spire is often low and largely covered by the last whorl. The aperture is long and narrow, with a notch at the base. There are well-developed spiral ridges on the columella, with the lowest one the strongest.

265 Olivella

Description: ⅝" (16 mm) high, ¼" (6 mm) wide. Olive-shaped, with short, pointed spire and long body whorl. Smooth, with slightly incised sutures. Aperture slitlike, with notch for siphon on bottom. Several short, oblique ridges on wall of inner lip.

Age: Tertiary to Recent.

Distribution: Atlantic and Gulf coastal plains.

Comments: 2 other olive-shaped genera that might be confused with *Olivella*—the olives—and that occur throughout the same areas are: *Oliva,* also called olives, and found from the Cretaceous to the Recent, which are larger, and have a thicker shell and a lower spire; and *Marginella,* the margin shells, found from the Eocene to the Recent, which are more egg-shaped, and whose spires are covered over by a secondary deposit of shell, partially obscuring the sutures.

241 Cancellaria

Description: ½" (13 mm) high, ⅜" (10 mm) wide. Oval to inflated, spindle-shaped in outline. Whorls round, body whorl inflated, spire moderately high. Equally strong revolving and transverse ribs make distinctive, cross-hatched pattern over shell. Aperture crescent-shaped, with unthickened lips. Siphonal canal short, with notch at end. Inner lip with several spiral ridges near base.

Age: Tertiary to Recent (possibly also in the Cretaceous).

Distribution: Atlantic, Pacific, and Gulf coastal plains.

Comments: Living *Cancellaria,* called nutmegs, are common on the East Coast from North Carolina south, and also on the West Coast. These animals look somewhat like nutmeg seeds because of their brown coloring, rough surface, and rounded shape.
Mataxa, found in the Upper Cretaceous in Tennessee and Texas, has a less pointed spire than *Cancellaria,* and a slightly larger siphonal canal, and lacks cross-hatched sculpture. *Aphera,* found from the Tertiary to the Recent, has a thicker, wider shelly deposit on the inner side of the aperture than that of *Cancellaria,* and a thicker outer lip.

253 Volutomorpha

Description: 3⅜″ (86 mm) high, 1¼″ (32 mm) wide. Moderately large; elongate-oval in outline, with moderately high spire of round whorls and long body whorl. Body whorl accounts for ⅘ of height. Sutures impressed, with slight shoulder just below each suture. Strong transverse ribs on spire and upper part of body whorl crossed by narrower spiral ribs, creating a beaded pattern. Aperture elongate-oval, without thickening, ending in long siphonal canal.

Age: Upper Cretaceous.

Distribution: Widespread in North America.

Comments: *Volutomorpha* is one of the most common fossils from the Cretaceous beds of the Atlantic coastal plain.

254 Volutoderma

Description: 6⅛″ (16 cm) high, 1¼″ (3 cm)
wide. Spindle-shaped. Some species
have sloping shoulders, others have
smooth spire. Spire high, with oblique
sutures; body whorl large and slender,
constituting ¾ of height. Spiral
threads or ribs usually strong; broader
transverse ribs on spire and upper part
of body whorl, often absent below.
Aperture lens-shaped, drawn out into
gracefully tapering siphonal canal.

Age: Upper Cretaceous.

Distribution: Widespread in North America.

Comments: *Volutoderma* is very similar to
Volutomorpha, but is more slender and
has oblique sutures. Plate 254 shows an
unusually long, slender species.

245 Vasum

Description: 3½″ (89 mm) high, 2½″ (64 mm)
wide. Medium to large; solid, strong,
spindle-shaped. Spire medium-high,
with 10 or more whorls. Sculpture of
spiral ribs, those on shoulder and
around middle of base bearing enlarged
spines; other ribs may have beads or
small spines. Aperture oval, extending
down into open siphonal canal.
Columella has 4 small pleats.

Age: Pliocene to Recent.

Distribution: Coastal areas of North America.

Comments: The specimen of *Vasum* that is
illustrated is from the Pliocene of
Florida, where it is fairly common.
Based on our knowledge of the living
animal, the fossil shell appears not to
have altered in the few million years
since it was alive. When alive, *Vasum*
—the vase shell—has a pure white
shell covered by a dense, black or
brown, hairy skin, and an operculum to
close off the aperture. *Vasum* is similar
to *Murex*, but has a more oval or
elongated aperture and a more open

siphonal canal, and lacks the many
small spiral threads that *Murex* has.
Perhaps most distinctively, the spines
of *Vasum* are not restricted to 3
transverse ribs as in *Murex,* but are
distributed all around the whorls.

SUPERFAMILY CONACEA
(Cretaceous to Recent)

Conaceans have a notch at the top of
the aperture where the lip connects to
the body whorl. Most conaceans have
teeth modified as harpoons, with poison
glands to immobilize prey.

249   Conus

Description: 1½" (38 mm) high, ⅝" (16 mm) wide.
Inverted-cone-shaped; very low or flat
spire forms base of cone. Whorls have
narrow, flat or slightly concave or
convex shoulders. Sides of whorls flat;
whorls overlap to form continually
sloping spire. Body whorl very large at
top, with straight sides narrowing
toward base of shell. Surface usually
smooth, but spiral grooves sometimes
present either on entire shell or on
lower part of body whorl. Aperture
long and narrow, with short channels at
top and bottom. Outer lip thin and
sharp, unthickened, straight.

Age: Cretaceous to Recent.

Distribution: Widespread in North America.

Comments: *Conus* is one of the largest of genera,
with hundreds of described species.
Today most cones, as they are
commonly called, live in the tropics.
Conus is perhaps the only gastropod
that is dangerous to humans, with a
hollow stinger in its mouth that can be
extended quickly like a spear or
harpoon. Some species in the tropical
West Pacific have a sting so venomous

that it has actually killed people. The
stinger is hurled into the prey or
intruder, and poison is pumped
through its hollow center into the
victim.

Most cone shells, like most gastropods,
have the aperture on the right. The
illustrated specimen, however, belongs
to a species almost all of whose
members are left-coiling.

263 Terebra

Description: 1″ (25 mm) high, ¼″ (6 mm) wide.
Spire solid, long, slender; whorls
numerous, evenly tapered; base round.
Whorls distinctively ornamented;
smooth or ribbed surface usually has
spiral groove or rib occurring short
distance below suture. Ornament most
pronounced in earlier whorls, becoming
weaker toward body whorl. Aperture
oval, with slit or small channel where
outer lip contacts body whorl. Aperture
ends in siphonal canal that bends back
away from aperture.

Age: Tertiary to Recent.
Distribution: Atlantic and Gulf coastal plains;
California.

Comments: *Terebra* is commonly called the auger
shell, and many subgenera and species
of it are found in southern seas. *Terebra,*
like *Conus,* has a poison gland and
stings its prey to death.

257 Clathrodrillia

Description: ¾″ (19 mm) high, ¼″ (6 mm) wide.
Slender, with spire and body whorl of
equal height. Whorls increase regularly
in size; shoulders small but
pronounced, sides convex to almost
flat. Whorls neatly sculptured, each
with 16–24 transverse ribs, narrower
than interspaces. Smaller spiral ribs

closely spaced and sometimes swollen as they pass over transverse ribs. Suture usually marked by wide spiral rib. Aperture oval and elongate, width ⅓ height. Outer lip thin, but strengthened by transverse rib near edge. Pronounced upper sinus just before outer lip contacts body whorl. Siphonal canal wide and short, with notch at base.

Age: Tertiary to Recent.

Distribution: Atlantic and Gulf coastal plains.

Comments: *Clathrodrillia* has a shell decorated with a lattice-work pattern and, as its name implies, drills into its prey. Another drill, very similar in appearance, is *Mangelia,* found from the Eocene to the Recent on the Atlantic, Pacific, and Gulf coasts. *Mangelia* can be distinguished by the absence both of the rib around the suture and the pronounced upper channel notch.

SUBCLASS OPISTHOBRANCHIA
(Mississippian to Recent; possibly also in the Devonian)

Opisthobranchs evolved from prosobranchs, and are an entirely marine group. They have a poor fossil record because the shell is commonly reduced or absent. The effects of torsion have been secondarily lost: the anus lies to the rear, as do the gills and mantle cavity, if present. Many genera have become bilaterally symmetrical. Sea slugs, sea hares, and pteropods are members of this subclass, as well as the bubble shells, a large group of shelled opisthobranchs that includes *Bulla.*

266 Bulla

Description: ½" (13 mm) high, ¼" (6 mm) wide. Small; oval to almost spherical in

outline; body whorl forms whole
exterior, completely wrapping around
all earlier whorls. Small hole at apex
with spire within. Shell smooth or
covered with spiral lines. Aperture as
long as or longer than body whorl,
narrow above, and expanded and round
below. Outer lip unthickened.
Channels and sinus absent.

Age: Jurassic to Recent.

Distribution: Atlantic and Pacific coastal plains.

Comments: *Cylichna,* found throughout North
America from the Triassic to the
Recent, and commonly called the barrel
bubble, is similar in size and shape to
Bulla, known today as the bubble shell;
but *Cylichna* has a wider opening in the
top, and also bears folds on the bottom
of the inner lip.

SUBCLASS PULMONATA
(Pennsylvanian to Recent)

The pulmonates, also descended from
prosobranchs, have no gills; they have
developed an air-breathing lung from
the mantle cavity, which is highly
vascularized. There is usually a well-
developed shell, except in the land
slugs, which have no shells.
Pulmonates inhabit both fresh water
and dry land; a few live in the sea.
Planorbis is an example of a freshwater
pulmonate.

173, 175 Planorbis

Description: ⅜″ (10 mm) high, ⅞″ (22 mm) wide
(pls. 173, 175). Disc-shaped, spire flat
or depressed. Base concave, with large,
open umbilicus. Aperture round,
oblique, with sharp, unthickened
margin. Canal and notches absent.

Age: Lower Jurassic to Recent.

Distribution: Widespread in North America.

Comments: *Planorbis* lives in fresh water. Like all pulmonates, it has lost both gills, and its mantle cavity is lined with blood vessels for gas exchange. *Planorbis* is a "form genus"—that is, it contains many similar-looking species that may not be closely related.

Plate 173 shows the top of the specimen, while Plate 175 portrays the bottom side of the shell.

CLASS ROSTROCONCHIA
(Lower Cambrian to Permian)

Until recently, rostroconchs were placed in the class Bivalvia. But a rostroconch's 2 valves grew from a single caplike larval shell, and the hingelike structure along the dorsal margin was not functional. Thus, the valves could not have opened ventrally as a bivalve's do. Rostroconchs generally have an expanded front with a large *gape* for the foot, and a short, tubular extension at the rear. Some genera are expanded at the rear. Rostroconchs may have evolved from monoplacophorans, and bivalves, in turn, probably evolved from rostroconchs.

43, 86 Hippocardia

Description: 1¼" (32 mm) long, ⅝" (16 mm) high (pl. 43); 1⅛" (28 mm) long, ¾" (19 mm) high (pl. 86). Tubelike when complete. Front elongate; with large gape; rear end extends into open tube, surrounded by hood with left and right sides. Superficially like a bivalve, but lacks true hinge. Shell layers extend across upper side. Ornamented with concentric lines and radial ribs.

Age: Middle Ordovician through Mississippian.

Distribution: Widespread in North America.
Comments: Until recently, *Hippocardia* was placed
with the bivalves on the assumption
that it had a hinge, and even teeth. It
is now clear that the "hinge" was not
functional. The gape at the front must
have been for the foot, and the one at
the rear for pumping water in and out.
Relatives of *Hippocardia* may have been
ancestral to the bivalves. If no minerals
were secreted along the dorsal margin,
the "hinge" would become flexible,
making the animal into a bivalve.
Hippocardia is now placed in a separate
class, Rostroconchia, within Mollusks.
The rostroconchs were very abundant in
the Lower Ordovician, rivalling the
bivalves, but they soon declined in
population. Plate 43 shows 2 specimens
of *Hippocardia.* The one on the left is a
rear end from which the hood, if not
broken off, would have made a flange
around the tube. This hood projects
slightly out from the center of the rear
end. The specimen on the right shows
the large gape on the anterior end from
which the foot projected. Plate 86
shows a side view of a *Hippocardia*
specimen. The marginal drawings show
a top view and a side view of another
specimen of *Hippocardia.*

CLASS BIVALVIA
(Lower Cambrian to Recent)

The most familiar bivalves are the
clams, cockles, mussels, oysters, and
scallops. Almost all bivalves have shells
composed of 2 parts, or valves. Most
are marine, but some are common in
lakes and streams.
In bivalves, the usual orientation of the
shell is with the *plane of commissure*
vertical—the position in which a
bivalve can most efficiently burrow.
The *dorsal side* of the animal is at the
hinge line, and the *ventral side* is down,

at the *lower margin*. The valves are right and left, and are generally symmetrical. Front and rear are to either side of the hinge.

The body of a bivalve is enclosed by the mantle, which has a lobe lining each valve. The mantle, which secretes new shell at the margins and also thickens the shell over the inside surface, is often attached to the shell just inside its margin, usually along a line parallel to the margin, called the *pallial line,* visible in an empty shell and often in fossils. The pallial line may have an inward bend, the *pallial sinus,* at the posterior margin. The shell is closed by 2 muscles that extend across the mantle cavity. There is usually an anterior and posterior muscle, each of which leaves a scar on the inside of the valves. In some specialized bivalves like scallops, the anterior muscle has been lost and the posterior one has become large and centrally located.

The valves are typically compressed laterally, and rounded or elongated front to rear. The shell has a hinge at the dorsal margin, bridged by a partially calcified, elastic ligament. When the closing muscles relax, the ligament acts like a spring to open the valves. In some bivalves, the ligament may be partly or entirely inside the shell; it then rests in an internal depression called a *resilifer*. Teeth are usually located inside the hinge, either on the margin or on a shelflike plate that extends from the margin. The teeth prevent rotation of the valves relative to each other. On each valve is the *umbo* (plural, *umbones*), usually a beaklike projection that is the first-formed part of the valve. In most bivalves, the umbones are inclined toward the forward end of the shell, but in some they point straight up, and in others, backward. In front of the umbones there may be a flattened, semicircular area, the *lunule*. A bivalve shell is typically thick and solid, and is composed either of aragonite or calcite, or a mixture of both. Calcite is less easily dissolved than aragonite, and most bivalves that do not burrow have an outer calcite layer for added protection.

A burrowing bivalve is usually already buried in sediment at death, and this condition decreases the chance of its being broken and dissolved.

SUPERFAMILY CTENODONTACEA

(Ordovician through Pennsylvanian)

The umbones of this group are located near the middle of the dorsal margin. The valves are nearly symmetrical along an axis from the umbones to the ventral margin. The pallial sinus and resilifer are absent.

151 Ctenodonta

Description: 1¼″ (32 mm) long, ⅝″ (16 mm) high. Large, elongate, moderately inflated. Valves equal, margins closed. Shell with some concentric sculpture; radial sculpture absent. Umbones directed slightly backward. Ligament small, immediately behind umbones, external and cylindrical. Hinge straight, with numerous short, chevron-shaped hinge teeth at approximately right angles to hinge line. Internal ligament absent. Closing-muscle scars nearly equal; pallial line simple, sinus absent. Byssal gape absent.

Age: Lower through Upper Ordovician.

Distribution: Widespread in North America.

Comments: *Ctenodonta* is similar in shape to *Nuculana*, but lacks an internal ligament and is usually larger, with the hind end less sharply drawn out.

SUPERFAMILY NUCULACEA
(Ordovician to Recent)

The valves of nuculaceans are truncate
at the rear. There are small vertical
teeth inside the *hinge margin*.

145 Nucula

Description: 1⅛" (28 mm) long, ¾" (19 mm) high.
Small, triangular to nearly circular in
outline; rear end truncated. Valves
equal, moderately inflated, internally
nacreous. Lower margins usually have
fine interlocking grooves and ridges.
Ribs usually absent; concentric
sculpture weak to strong. Umbones
narrow, turning backward. Ligament
internal. Hinge arched, with many
small teeth on both sides of umbones,
teeth transverse to hinge line. Resilifer
on hinge plate just under umbones.
Pallial sinus absent.

Age: Upper Cretaceous to Recent.

Distribution: Widespread in North America.

Comments: The illustrated specimens represent the
exterior and interior of the left valve of
Nucula. Known today as the nut clam,
Nucula is still a very common bivalve in
the marine environment, but it is easy
to overlook because of its small size.
Unlike most bivalves, it is a deposit
feeder rather than a filter feeder, with
long fleshy palps extending from its
mouth that pick up fine organic
material from the sediment. Many of
the Ordovician bivalves, including
many close relatives of *Nucula*, engage
in such deposit feeding.
This genus has recently been restudied
and separated into several genera.
Nuculoidea, formerly considered a
subgenus of *Nucula*, is found in
Devonian strata in eastern North
America. It can be distinguished from
Nucula by its triangular resilifer and
the micro-ridges on the interiors of the

lower margins. *Nuculopsis,* also formerly a subgenus, ranges from the Mississippian through the Permian throughout North America. Its valves angle farther backward than those of *Nucula,* its inner lower margins are smooth, and its resilifer is smaller.

SUPERFAMILY NUCULANACEA
(Ordovician to Recent)

The valves of the nuculanaceans are elongated at the rear, and usually show a pallial sinus inside the shell near the rear margin.

146 Palaeoneilo

Description: ½" (13 mm) long, ¼" (6 mm) high. Round to elongate. Valves equal; margins closed. Sculpture concentric, fine to coarse. Umbones point forward; shell is elongate backward. Faint to prominent rear radial groove along slope from lower back margin to umbones. Ligament external. Hinge line arched, with numerous small transverse teeth. Resilifer absent.

Age: Ordovician through Jurassic.

Distribution: Widespread in North America.

Comments: *Palaeoneilo* is in the same family as *Neilo,* a genus that first appeared in the Paleocene in coastal areas of North America, and is still living. The streamlined shape of *Palaeoneilo* probably means that the animal was a good burrower. The specimens illustrated are internal molds of the left and right valves.
The marginal drawing shows the interior of the left valve of *Palaeoneilo.*

154 **Nuculana**

Description: 1¼″ (32 mm) long, ⅝″ (16 mm) high.
Front end round; back end extended
out into sharp to blunt point. Margins
closed all around, usually smooth
internally. Concentric sculpture weak
to strong. Umbones curved slightly
backward. Hinge line arched;
numerous chevron-shaped teeth, apices
of chevrons pointing toward umbo.
Small triangular resilifer under
umbones. Pallial sinus small.

Age: Triassic to Recent.

Distribution: Widespread in North America.

Comments: On some ancient sea floors, *Nuculana*
was so abundant that its burrowing and
feeding activity made the sediment too
soft for other species to colonize. Rocks
are commonly found containing many
Nuculana shells, along with the tiny
shells of other genera that died soon
after settling, probably smothered in
the soupy sediment. *Nuculana* is still
found on all North American coasts
today, and, like *Nucula,* is commonly
known as the nut clam.
Yoldia is common throughout North
America from the Cretaceous to the
Recent, and is closely related to
Nuculana, with a similar shell. *Yoldia,*
however, has gapes both in front and
behind, and its pallial sinus is deeper
than that of *Nuculana.*

SUPERFAMILY SOLEMYACEA
(Devonian to Recent)

The 2 valves of this group are equal
in size, with the front part usually
longer than the rear. The umbones are
located near the rear, and the shell
margins gape when closed. Teeth and
lunule are absent. The ligament is
located entirely or mostly behind the
umbones. The shells are smooth or have
weak radial ornamentation.

147 Solemya

Description: ⅝" (16 mm) long, ¼" (6 mm) high. Elongate-oval to almost rectangular. Valves equal, compressed; back and front round. Ribs irregularly arranged, flat, radial. Umbones small, level with hinge margin, near back end of shell. Gape narrow front and rear. Hinge teeth absent. Ligament posterior to umbones.

Age: Devonian to Recent.

Distribution: Widespread in North America.

Comments: Living *Solemya,* commonly known as the veiled clam, is a rapid burrower. Its rounded anterior, compressed shape, and almost smooth shell aid the animal in slicing smoothly and quickly through the sediment. The specimen illustrated is a right valve.

SUPERFAMILY PRAECARDIACEA
(Middle Ordovician through Lower Mississippian; possibly also in the Upper Cambrian)

These shells are thin and oval in outline, with 2 equal muscle scars. The *hinge plate* is small or absent, and the teeth are either absent or very small and located in a row along the hinge margin. The umbones are prominent, and there is usually no pallial sinus.

---

40 Buchiola

Description: ¼" (6 mm) long and high. Small, rounded, inflated. Valves equal. Radial ribs few, flat, crossed by regularly spaced, arching growth lines. Umbones curve forward. Numerous small, vertical teeth under umbones.

Age: Silurian through Upper Devonian.

Distribution: Widespread in North America.

Comments: *Buchiola* is often found preserved with

both valves still attached but spread out, their convex sides down. The *Buchiola* specimen illustrated is an internal mold of the right valve. 2 similar genera in the same family as *Buchiola* are: *Praecardium*, from the Devonian of New York, Michigan, and Ontario, which has fewer radial ribs and less prominent concentric growth lines; and *Cardiola*, from the Upper Silurian of Indiana and Illinois (and perhaps other states as well), which is larger than *Buchiola*, and has very coarse, nodelike ribs and a flat, triangular area under the umbones.

SUPERFAMILY ARCACEA
(Lower Ordovician to Recent)

Arks usually have large, heavy, trapezoidal shells that are equal in size, highly inflated, and covered with radial ornamentation. A flat area that may be grooved rises above the hinge line.

46 Barbatia

Description: 1⅜" (35 mm) long, 1" (25 mm) high. Elongate. Valves equal, slightly inflated, ends rounded or nearly angular. Small radial ribs crossed by concentric growth lines give finely beaded appearance. Umbones point forward. Many small teeth on curved hinge; largest teeth farthest from umbones, angled obliquely. Large area under umbones, showing former position of ligament, crossed by growth lines parallel to hinge line. Gape at lower margin for byssus.

Age: Jurassic to Recent (possibly also in the Triassic).

Distribution: New Jersey, Maryland, Georgia, Florida, Alabama, Mississippi, Arkansas, Louisiana, Texas, Wyoming,

New Mexico, Utah, Washington, Oregon, and California.

Comments: Living *Barbatia* is a byssate epifaunal nestler—that is, it attaches with byssal threads to a firm substrate with some protection from currents and waves, in a depression or a crevice between coral heads, or between coral branches. It may also attach to the roots of marine plants. Its elongated shape and long, strong hinge are well adapted to withstand strong currents and predators in this environment.

47 Anadara

Description: 2½" (64 mm) long, 1½" (38 mm) high. Round to almost trapezoidal. Valves not quite equal. Umbones large, pointing forward; moderately to strongly inflated. Surface covered by strong radial ribs that interlock at lower margin. Hinge slightly arched, bearing numerous small teeth, generally transverse to hinge, becoming larger and more oblique away from umbones. Area between umbones and hinge narrow, elongate, with grooves. Byssal gape absent.

Age: Upper Cretaceous to Recent.

Distribution: Atlantic, Pacific, and Gulf coastal plains.

Comments: Living *Anadara*, commonly known as arks, are either free-living (usually burrowing shallowly) or attached by a thin byssus. The fossil forms probably had similar life habits.

149 Parallelodon

Description: ¾" (19 mm) long, ⅜" (10 mm) high. 2 or more times as long as high; some species have winglike back end. Valves equal, inflated. Ornamented with concentric growth lines that may become sheetlike, and usually with radial ribs. Umbones very near front end, pointing forward. Hinge line long, straight, with several small teeth diverging from beneath umbones, and 2–4 elongate lateral teeth. Byssal gape on lower margin with lateral sinus in shell margin at gape. Margins smooth internally.

Age: Devonian through Upper Jurassic.

Distribution: Widespread in North America.

Comments: The name *Parallelodon*, meaning "parallel teeth," refers to the long side teeth at the hind end of the hinge. *Parallelodon* was a filter feeder that attached its byssus to a firm object such as a rock, a coral, or another shell. The *Parallelodon* specimen illustrated is the internal mold of a right valve. *Nemodon*, found in the upper Cretaceous throughout North America, is closely related and similar in appearance to *Parallelodon*, but has anterior teeth almost parallel to the hinge line.

42 Cucullaea

Description: 3⅜″ (86 mm) long, 3⅛″ (79 mm) high. Valves rounded, inflated, heavy, with long, straight hinge line. Margin closed. Surface ornamented with strong radial ribs. Umbones near center, bent over hinge line. Area between umbones and hinge large, with chevron-shaped growth lines. Hinge straight, tooth row slightly arched. Middle teeth small, vertical; side teeth large, angled obliquely. Byssal gape and pallial sinus absent.

Age: Lower Jurassic through Paleocene.

Distribution: Atlantic and Gulf coastal plains, British Columbia, Oregon, and California.

Comments: *Cucullaea* probably lived partially buried in sediment, with only its posterior margin above the surface. The heavy shell, inflated shape, and heavy ribs would have helped to stabilize it. Living in this position would have protected it from being washed away, knocked over, eaten, or baked and dried at low tides if it lived above the low-tide line.

The *Cucullaea* specimen in Plate 42 still has some of the shell preserved, but *Cucullaea* fossils are often only internal molds. The marginal drawings show the interior and exterior of a left valve. *Idonearca*, widespread throughout North America in the Jurassic through the Cretaceous, is so similar to *Cucullaea* that it is often included as a subgenus. It has much finer radial ribs than *Cucullaea*.

128 Glycymeris

Description: 1⅝″ (41 mm) long, 1½″ (38 mm) high. Almost circular. Valves equal, heavy, moderately inflated. Margins ribbed. Surface smooth or radially ribbed. Umbones central, straight to pointing slightly backward. Hinge heavy, with arched row of numerous teeth, smaller toward center and sides, vertical or chevron-shaped with angles of chevrons pointing toward middle of hinge. Area between umbones and hinge broad and triangular, with diverging growth lines.

Age: Lower Cretaceous to Recent.

Distribution: Widespread in North America.

Comments: *Glycymeris* reached a maximum abundance and diversity in the Cretaceous, but is still common today from North Carolina south, where it is known as the bittersweet clam. It is free-living on current-swept bottoms with coarse sediment. Some living species burrow shallowly, with their commissures vertical; other species lie on the surface horizontally and are moved around by the currents.

SUPERFAMILY MYTILACEA
(Mississippian to Recent)

The valves of mussels are equal in size, with the umbones near the front end of the shell and pointing forward. The hinge margins may be smooth or have a few small teeth. The forward muscle scar is small or absent, and the pallial line may have a small pallial sinus. The ligament is supported by a calcified ridge.

Mytilus

Description: 1⅝″ (41 mm) long, 3½″ (89 mm) high. Almost teardrop-shaped; rear side has slight keel. Valves equal, thin, moderately inflated; smooth or, more rarely, with radial ribs. Growth lines may be prominent. Umbones pointed, terminal, inclined forward. Teeth few, small, weak, close to umbones.

Age: Oligocene to Recent.

Distribution: Atlantic and Pacific coastal plains.

Comments: Living *Mytilus,* the mussel, is a very successful genus today, with large populations on all North American coasts. It attaches to firm objects by a short, strong byssus, and individuals tend to cluster together, with shells attaching to each other. Living *Mytilus* is particularly common on rocks, but is also found on marsh grass and pilings. *Brachydontes* is a similar genus, also common, ranging from the Jurassic to the Recent throughout North America. It has external forking ribs and internal ribs on its margins.

155 Promytilus

Description: ½″ (13 mm) long, 1⅛″ (28 mm) high. Shell thin, elongate, smooth. Valves inflated, particularly at terminal umbones. Small front lobe set off by broad groove extending from umbones to ventral margin. Hinge line smooth; lateral byssal sinus present; teeth absent.

Age: Mississippian through Permian.

Distribution: Widespread in North America.

Comments: *Promytilus* is closely related to the living mussel, *Modiolus,* that is common throughout North America in Mesozoic and Cenozoic rocks. *Modiolus,* however, is more rounded in front than *Promytilus,* with the umbones not quite terminal. *Promytilus* probably lived similarly to *Modiolus.* Illustrated is the exterior of a left valve of *Promytilus.*

SUPERFAMILY PINNACEA
(Mississippian to Recent)

Pinnaceans are called pen shells; they have triangular or fan-shaped shells, with the valves generally equal in size and shape. The shell's umbones are at its apex. The lower margin has a large gape.

159 Pinna

Description: 2″ (51 mm) long, 5⅞″ (149 mm) high. Large; wedge- to ham-shaped, thin-shelled, gaping at back end. Usually has radial ridges, and sometimes scales. Umbones terminal. Hinge long; teeth absent. Interior has 2 lobes of pearly shell material extending from umbones to mid-shell.

Age: Jurassic to Recent.

Distribution: Widespread in North America.

Comments: *Pinna* is a member of the family of pen shells. A genus similar to *Pinna* and also common throughout North America is *Atrina,* another pen shell that ranges from the Middle Jurassic to the Recent. *Atrina* has a more rounded

back end than *Pinna*, and a single
pearly lobe extending ²⁄₃–³⁄₄ of the
total length.

Both *Atrina* and *Pinna* are common
today in tropical and subtropical
marine grass beds. They burrow with
the back end up, exposed above the
substrate. Illustrated are 2 external
molds of *Pinna* specimens, with parts of
the internal mold clinging to one of
them.

SUPERFAMILY AMBONYCHIACEA

(Middle Ordovician through Lower
Jurassic; possibly also in the Lower
Ordovician and the Upper Jurassic)

Ambonychiaceans have triangular or
4-sided shells, with umbones near the
forward end of the hinge. The pallial
line is a series of pits, and there is no
sinus. The forward muscle scar is
reduced or absent.

39 Ambonychia

Description: ⁷⁄₈″ (22 mm) long, 1¼″ (32 mm) high.
Mussel-shaped; umbones skewed
toward front. Valves equal, irregular in
outline, thin, inflated, covered with
coarse radiating ribs. Umbones
terminal. Hinge has 2 or 3 small teeth
below umbones, and side teeth at rear
hinge margin. Gape for byssus just
below umbones; lateral byssal sinus
present; pallial line pitted and
complete.

Age: Middle and Upper Ordovician.

Distribution: Common in eastern North America;
also found in Wyoming.

Comments: *Ambonychia* belongs to a highly
successful lower Paleozoic family, and
includes at least 40 named species. In
life, *Ambonychia* probably lay on its

front commissure with its byssus attached to hard surfaces. Many *Ambonychia* shells are preserved with edrioasteroids attached. When the growth of *Ambonychia* shells was slowed or stopped by storms, cold weather, injury, or other forms of disturbance, the slowing of the process resulted in a shallow groove concentric with the margin, such as the one evident in the left valve exterior illustrated.

165 Myalina

Description:	2¾″ (70 mm) long, 3⅛″ (79 mm) high. Obliquely oval, but variable in outline. Rear margin sometimes drawn out into "wing." Right valve slightly smaller and less convex than left. Commonly ornamented only with growth lines. Umbones pointed and terminal. Teeth absent. Pallial line complete and pitted.
Age:	Lower Mississippian through Upper Permian.
Distribution:	Widespread in North America.
Comments:	Based on its similarity to living bivalves, such as certain mussels, *Myalina* is thought to have buried partially in soft substrates, with the byssus attached to shell debris and the "wing" exposed to the current. *Myalina* may also have lived completely exposed on firm substrates.

SUPERFAMILY PTERIACEA
(Middle Ordovician to Recent)

Pteriaceans have valves of unequal size, with the right usually less convex than the left. The umbones point forward and are located near the front end of the shell, with the ligament behind them. The right valve has a byssal notch below the hinge.

156 Leptodesma

Description: 1⅜" (35 mm) long, ⅞" (22 mm) high. Obliquely oval to almost triangular, with large rear "wing" and smaller front auricle. Valves not equal, with left umbo more prominent than right, left valve more convex. Radial ornamentation absent. Umbones point toward front; hinge has few large teeth.

Age: Middle Silurian through Upper Permian.

Distribution: Widespread in North America.

Comments: Like *Myalina, Leptodesma* may have burrowed with the back "wing" exposed, or may have been completely exposed and attached by the byssus to a firm substrate. *Leptodesma* now includes the genus *Leiopteria* as a subgenus. *Ptychopteria,* a genus similar to *Leptodesma,* is widespread throughout North America from the Silurian through the Lower Permian. It differs from *Leptodesma* in having radial ornamentation on the left valve, and sometimes on the right valve.

41 Oxytoma

Description: ¼" (6 mm) long and high. Obliquely oval; left valve varies in convexity and has protruding umbo; right valve flat or slightly convex. Rear "wing" elongate. Front auricles small; deep notch just below right front auricle. Comblike row of small teeth on lower side of notch. Umbones point forward. Smooth or with strong radial ribs; ornament deeper on right valve than on left. Hinge line long; hinge teeth absent. Pallial line discontinuous, broken up into series of pits.

Age: Upper Triassic through Upper Cretaceous.

Distribution: Widespread in North America.

Comments: *Oxytoma* may have been adapted to live attached above the sea floor where currents are strong. It can swing freely on the byssus; the long rear "wing" orients the shell so that the front faces into the current. Water enters the shell at the front and lower margins, passes over the gills, and flows out at the rear margin. Plate 41 shows many *Oxytoma* shells probably swept together by currents. Some of the shell material is broken away to show internal molds. The marginal drawing shows the exterior of a left valve.

158, 162 Inoceramus

Description: 3⅜" (9 cm) long, 6¼" (16 cm) high (pl. 158); 2⅜" (60 mm) long, 3" (76 mm) high (pl. 162). Round to oval to trapezoidal in outline; feebly to strongly inflated in profile. Left valve slightly to greatly more convex than right valve. Ornamentation of faint to strong concentric ridges; some species have faint to strong radial ribs. Umbones are prominent and point forward. Hinge line long and straight; teeth absent, but numerous small, vertical ligament pits present.

Age: Lower Jurassic through Upper Cretaceous.

Distribution: Widespread in North America.

Comments: *Inoceramus* is characteristic of the Jurassic and Cretaceous, and some species grew to enormous size: many in the western United States reached a length of 4' (1.2 m). The very large specimens must have lain on the sediment unattached, their broad, flat, thin shells preventing them from sinking into the mud. The smaller species may have been free-swinging from a byssus, or attached in clusters to shell debris and to each other, like living *Mytilus*. The illustrations show molds of the exterior and interior (pl. 158) and the right valve exterior (pl. 162).

SUPERFAMILY PECTINACEA
(Silurian to Recent)

Most pectinaceans, or scallops, are circular in outline. The valves are unequal in size, with one more convex than the other. Small extensions of the *hinge line* to the front and rear of the umbones form auricles. There is only 1 muscle scar. Most genera have a notch for the byssus below the right forward auricle. The sculpture is usually radial.

121 Pterinopecten

Description: 1″ (25 mm) long, ¾″ (19 mm) high. Scallop-shaped; valves oblique, with lower rear margin somewhat elongate. Valves only slightly convex. Rear auricle poorly differentiated but longer than front auricle. Slitlike notch for byssus under front auricle on right valve. Fairly fine radial ribs, with new ribs arising in grooves between older ribs. Resilifer on hinge margin absent; series of chevron-shaped grooves below umbones.

Age: Upper Silurian through Upper Devonian.

Distribution: Widespread in North America.

Comments: *Pterinopecten,* a fossil scallop, is commonly found in Devonian beds with strophomenid brachiopods, and may have had a life habit very similar to theirs. *Pterinopecten* probably attached by its byssus to shell debris, with the hinge line resting on the surface and the lower margin elevated off the substrate. The strophomenids were very similar to *Pterinopecten* in shell shape and also lay on the bottom with the hinge down and the anterior margin elevated. Plate 121 shows internal molds of a left and right valve of *Pterinopecten.*

Dunbarella, found throughout North America in the Mississippian through the Pennsylvanian, has a shape similar to that of *Pterinopecten,* and was formerly included in the same genus. *Dunbarella* can be distinguished by the forking ribs on its right valve. *Lyriopecten,* from the Middle and Upper Devonian, is another common early pectinid found in New York, Pennsylvania, Maryland, and West Virginia. *Lyriopecten* has a smaller, more rounded front auricle than that of *Pterinopecten.*

---

34 Aviculopecten

Description: 1″ (25 mm) long and high. Scallop-shaped. Right valve smaller than left, and less convex. Auricles large, rear auricle as long as or longer than more rounded front auricle. Front auricle on right valve has byssal notch below. Ornamentation of radial ribs that increase in number by forking on right valve, and arising in grooves on left valve. Umbones upright or skewed slightly forward. Triangular, oblique resilifer on inner hinge margin below umbones. Teeth absent. Single closing-muscle scar present.

Age: Lower Mississippian through Upper Permian.

Distribution: Widespread in North America.

Comments: *Aviculopecten,* like *Pterinopecten* a fossil scallop, probably preferred to live off the bottom in areas with a moderate current. On a reef there would have been many objects to attach to, such as corals, crinoids, and upright bryozoans. *Annuliconcha,* found in the midwestern and southwestern United States from the Lower Mississippian through the Upper Permian, resembles *Aviculopecten* in shape, but has strong concentric growth ridges, and its radial ribs are weak and interrupted.
Fasciculiconcha, from the Middle and Upper Pennsylvanian of the Midwest

and Texas, has fine lines or grooves on the strong radial ribs of the left valve, but not on the right valve. *Euchondria,* ranging throughout North America from the Lower Mississippian through the Upper Permian, has a more oval resilifer, and numerous small pits along the inside of the hinge area.

37   Pseudomonotis

Description: 1⅛" (28 mm) long, 1" (25 mm) high. Round to almost triangular in outline. Valves very different in shape: right valve flat or concave, left valve moderately inflated; lower rear margin extended into more or less pronounced "wing." Ornamentation on left valve of irregular radial ribs, commonly with roughened surfaces. New ribs arise in grooves. Hinge margin short; small, rounded auricle front and rear; byssal notch in right valve in young, closed in adults.

Age: Mississippian through Upper Permian.

Distribution: Widespread in North America.

Comments: *Pseudomonotis* attached itself by a byssus when young, but then cemented its right valve to some firm surface and lost the byssus. Cementation usually occurred in areas near shore with strong waves and currents. *Pseudomonotis* may have been the ancestor of the most important group of living cemented bivalves—the oysters. In Plate 37 *Pseudomonotis* is shown cemented to a brachiopod shell. Its ventral margin seems to be situated to take maximum advantage of feeding currents created by the brachiopod. Since both animals probably ate the same food—tiny plants and animals filtered from sea water—the brachiopod would probably have fared better without the bivalve cemented to its shell.

35 Chlamys

Description: 2½" (64 mm) long, 2⅞" (73 mm) high. Scallop-shaped. Left valve usually more convex than right. Round or higher than long, usually somewhat oblique. Inner lower margins scalloped. Auricles well developed, usually large; front auricle sometimes longer than rear one. Front auricle on right valve with byssal notch below. Radial ribs fine to coarse, often crossed by concentric lines, with spines or beads at junctures. Umbones central and straight. Hinge line straight, with triangular resilifer in middle. Numerous small teeth present in young, absent in adult. Variable number of grooves and ridges radiate from resilifer and function as teeth. Single closing-muscle scar large, round, dorsal.

Age: Triassic to Recent.

Distribution: Widespread in North America.

Comments: The specimen of *Chlamys* illustrated in Plate 35 belongs to the subgenus *Argopecten*. Members of this subgenus have nearly equally inflated valves, unusually large auricles, and fine concentric lines that loop across the ribs. *Chlamys* was formerly considered a subgenus of *Pecten,* but now has 34 subgenera of its own. Like *Pecten* it is a member of the scallop family. Some species are free-living, lying on the sea floor and swimming by clapping the valves together, usually in response to a threat from a potential predator, like a starfish. Others nestle in crevices and fissures, attached by the byssus. The ribs strengthen the shell, and also make it more difficult for currents or predators to dislodge it.

36 Chesapecten

Description: 5½" (140 mm) long, 5⅛" (130 mm) high. Shells very large, thick;

scalloped-shaped, slightly longer than high in adults. Moderately inflated to fairly flat; left valve more convex than right. Auricles well developed, about equal in length, with notch under front auricle deep to shallow. Sculpture of strong, wide ribs on both valves; width of ribs about equal to width of interspaces; usually 10–16 ribs but 4–23 may be present; ribs crossed by other spiny or scaly concentric ribs. Inner margins scalloped. Closing-muscle scar single, large, rounded.

Age: Miocene through Pliocene.

Distribution: Atlantic coastal plain.

Comments: *Chesapecten* is the very first North American fossil to be described and illustrated. In 1687, Martin Lister published "Historiae Conchyliorum, Liber III," which contained a drawing and description of a *Chesapecten* shell, but Lister did not give it a name. *Chesapecten* is named for Chesapeake Bay because outcrops along the shores of the bay contain the best collecting sites for the genus.

Chesapecten lived with its flatter right valve resting on the sea floor in depths from a few feet to 131′ (40 meters). It could undoubtedly swim like *Chlamys* and *Pecten*. At various times, *Chesapecten* has been placed in the genera *Chlamys*, *Pecten*, and *Lyropecten*, but is now considered distinctive enough to warrant its own genus. Plate 36 shows the exterior of a right valve. *Chesapecten* differs from *Chlamys* in having fewer and heavier unbranching ribs. It is also different from *Pecten*, another scallop, which is found as a fossil in deposits of Upper Eocene to Recent age on all North American coasts. The right valve of *Pecten* is more convex, the ribs are not scaly or spiny, and the ribs on the right valve are wider than the interspaces.

SUPERFAMILY ANOMIACEA
(Cretaceous to Recent; possibly also in the Permian)

Anomiaceans are mostly sessile, each attaching for life by a *byssus,* which may be calcified in adults and extended through a hole in the lower valve. The shells are irregular in outline and the valves are not equal in size. There are no true teeth on the hinge.

124 Anomia

Description: 1¼″ (32 mm) long, ⅞″ (22 mm) high. Irregularly rounded, thin-shelled; valves unequal. Right valve usually flatter, with hole near umbo. Left valve larger, convex. Fine radial lines, sometimes crossed by concentric lines, may be present. Resilifer present; teeth absent. Right valve has central closing-muscle scar; left valve has 4 scars in central area—3 for byssus and 1 for closing-muscle.

Age: Cretaceous to Recent.

Distribution: Widespread in North America.

Comments: Young *Anomia* attach the right valve with a byssus to some firm surface. As the shell grows, the margins grow around the byssus, finally forming a hole in the right valve; at the same time, the byssus becomes calcified and pluglike. *Anomia* are often found on beaches, and are called jingle shells. Illustrated are the exterior and interior of the left valve of *Anomia. Paranomia,* found in the Upper Cretaceous of eastern North America, has a similar shape and is closely related. It differs from *Anomia* in that the hole in the right valve is closed in adults, and the left valve has only 1 byssal retractor scar. The surface is sculpted with small, spiny ribs.

SUPERFAMILY LIMACEA
(Mississippian to Recent)

The valves of limaceans are usually the
same size; oval, rounded, or almost
triangular in outline; and usually
higher than long. The shells are often
extended obliquely in the lower front.
There may be 2 auricles or the front
auricle may be reduced or absent. The
umbones are well separated and there is
a triangular *ligament pit* in both valves.
If teeth are present, they are small,
weak, and vertical. The interior of the
valves has only 1 muscle scar. The shell
is smooth or covered with radial
ornamentation.

38 Lima

Description: 1⅛″ (28 mm) long, ¾″ (19 mm) high.
Obliquely oval, with lower rear margin
extended. Valves equal, inflated, with
slight gape in front and sometimes rear
margins. Ornamented with radial ribs
that may be scaly. Prominent umbones
point forward. Hinge line straight;
auricles well defined but small; front
auricle slightly smaller. Resilifer
central; teeth absent. Interior closing-
muscle scar single, large.

Age: Jurassic to Recent.

Distribution: Atlantic and Gulf coastal plains, the
Midwest and midcontinent.

Comments: *Lima,* the file shell, was particularly
abundant in the Jurassic and
Cretaceous, but is still common in
warm seas. Most *Lima* are free-living,
although some normally attach with a
byssus in small fissures. *Lima* has long
sensory tentacles on the outer edges of
the mantle which cannot be fully
withdrawn into the shell. Some species
build nests of byssal thread around
themselves, probably to help protect
the tentacles. Most living species can
"swim" to some extent by clapping

their valves together, but this ability is not as developed in *Lima* as it is in *Pecten* and *Chlamys*.

SUPERFAMILY OSTREACEA
(Upper Triassic to Recent)

Ostreaceans live on the sea floor and, except for a few species, cement the left valve to a firm surface. The valves are unequal in size and composed of thin, overlapping sheets of shell. There are no teeth on the hinge. The ligament is divided into 3 areas: a wide central resilifer in the middle of the hinge area, with a smaller area to each side. Inside the valves there is no mantle line, except in 1 genus, where the line is broken.

138 Pycnodonte

Description: 4¾" (121 mm) long, 4⅜" (111 mm) high. Irregular in outline, but tending to circular. Valves unequal; right valve smaller, flat to concave, with flange forming a shelf around valve; left valve highly convex, with umbo high and curved over hinge. Surface uneven, sometimes with weak to well-defined radial ribs. Right valve commonly has sharp radial gashes. Small to large attachment area on left valve umbo. Ridges and grooves at hinge line function as teeth.

Age: Cretaceous through Miocene.

Distribution: Atlantic and Gulf coastal plains.

Comments: Plate 138 shows one specimen of *Pycnodonte* with the valves hinged; the right valve is forward. The other specimen is an isolated right valve. *Pycnodonte* was formerly considered part of the genus *Gryphaea*, which is now known to be restricted to the Triassic and Jurassic. *Gryphaea* is rare in North

America, but very common in Europe. Another genus that was included in *Gryphaea* until recently is *Texigryphaea*, restricted to the Cretaceous and found in Kansas, Texas, and the southwestern United States. *Texigryphaea* is more elongated vertically than *Pycnodonte*, and has a rear flange set off from the rest of the shell by a deep groove. In Texas, in particular, shell banks can be found containing countless individuals. These banks probably formed when storms washed *Texigryphaea* ashore, depositing the shells either on beaches, where they were covered over by sand, or in underwater depressions that were later filled in with sand. Such activity occurs today along the east coast of North America, with genera like *Spisula*, the surf clam, when winter storms wash thousands of individuals onto the shore. These accumulate in depressions behind the beach. If the sea level continues to rise, the accumulations will be covered by sediment, and consequently will have a good chance of becoming fossil shell banks in the future.

211 Exogyra

Description: 3½″ (89 mm) long, 4″ (102 mm) high
Massive, heavy. Left valve deep; right valve flat, lidlike. Umbo of left valve curved backward into a coil; right umbo flat, but also coiled. Large, flat or concave area on umbo of left valve where shell was cemented. Concentric growth sheets on both valves; radial ribs may be present. Ribs may have bumps on crests.

Age: Cretaceous.
Distribution: Atlantic coast; Gulf coastal plain to Utah.
Comments: *Exogyra* is one of the most distinctive and common of Cretaceous bivalves. It probably cemented to a firm surface

when young and then, as it grew larger, broke free and lay on the sediment. The large, heavy shells were not easily destroyed after death, so many have survived as fossils. There are many species, and the ornament on them is highly variable. However, *Exogyra* is an easily recognized genus because all species have the distinctive backward-coiling umbo on the left valve. It is the left valve of *Exogyra* that is illustrated in Plate 211.

212, 213 Ilymatogyra

Description: ⅝″ (16 mm) long, 1¼″ (32 mm) high (pl. 212); ⅜″ (10 mm) long, ⅝″ (16 mm) high (pl. 213). Small. Left valve begins as tight, corkscrewlike spiral with narrow radial ribs, then opens into deep cup; right valve small, flat, caplike, countersunk. Concentric growth sheets on both valves. Attachment area absent.

Age: Upper Cretaceous.

Distribution: Oklahoma, Texas.

Comments: *Ilymatogyra* is 1 of only 2 genera of oysters that never attached in any stage of life. It was formerly included in the genus *Exogyra*. The strange, snail-like coiling habit of *Ilymatogyra* was probably an adaptation for living on a soft, muddy bottom without the means to burrow. The ancestors of *Ilymatogyra* were probably oysterlike, and lacked the ability to burrow. In order to keep its valve margins above the sediment surface, *Ilymatogyra* grew upward as its umbones became more deeply buried. The 2 illustrations of *Ilymatogyra* show a single left valve (pl. 212) and a bedding plan with many specimens of *Ilymatogyra* (pl. 213).

157 Crassostrea

Description: 1½" (38 mm) long, 2¼" (57 mm) high. Variable in outline, but usually high, slender, spatula-shaped, with parallel front and rear margins. Surface rough with concentric growth sheets, sometimes frilled on free ends. Rounded radial ribs may be present, usually on left valve. Left valve usually deeper and larger than right, with deep cavity under umbo. Ligament pit high, triangular, with many cross-grooves and ridges outside ligament margins.

Age: Lower Cretaceous to Recent.

Distribution: Widespread in North America.

Comments: The deep umbonal cavity beneath the hinge plate of the left valve of *Crassostrea* identifies it as a genus that does not incubate its young, and distinguishes it from those oysters, including *Ostrea,* that keep eggs inside the mantle cavity for fertilization and development of the larvae. *Crassostrea virginica* is the common edible oyster of the Atlantic coast. Even before the evolution of human beings, *Crassostrea* was a choice food item for many members of the marine world, including fishes, starfishes, and many gastropods. Living *Crassostrea* is better able to defend itself than its ancestors were because it cements its shell to a substrate, unlike its ancestors, which attached by a byssus. Without a byssus, *Crassostrea* can close its valves more tightly. In fact, the closure is especially tight because a thin, flexible fringe around the upper valve molds itself onto the margin of the lower valve. This tight seal also allows oyster to withstand prolonged periods of exposure to air. *Crassostrea* is probably the fastest-growing and largest of all living oysters.

160 Ostrea

Description: 4¾″ (121 mm) long, 5⅛″ (130 mm) high. Variable in shape, but usually round to teardrop-shaped with narrow hinge. Left valve slightly convex, never deeply cupped. Right valve usually smaller, flatter. Left valve has wide, irregular ribs; both valves have sheetlike concentric bands. Sheets frilled, delicate. Ligament area triangular, usually longer than high. Commissure usually flat, but may be zigzagged.

Age: Cretaceous to Recent.

Distribution: Widespread in North America.

Comments: The absence of a deep cavity under the left umbo in *Ostrea* and related oyster genera means that the eggs were fertilized and brooded inside the shell. Another oyster, similar to *Ostrea*, that incubated its young is *Lopha*.

164 Agerostrea

Description: ⅜″ (10 mm) long, ⅞″ (22 mm) high. Outline sickle-shaped. Both valves flat, lacking ribs or folds in area inside margins. Margins with as many as 20 narrow folds, up to 1⅝″ (41 mm) high, ending in sharp or rounded points. Small auricles to sides of hinge sometimes present. Closing-muscle scar comma-shaped, situated close to hinge.

Age: Upper Cretaceous.

Distribution: Widespread in North America.

Comments: *Agerostrea* has previously been identified as *Ostrea* larva.
Rastellum is a similar genus widespread in the Middle Jurassic through the Upper Cretaceous. *Rastellum* may be shaped like *Agerostrea*, or may be more triangular. It has strong to weak radial ribs, which end in as many as 100 or more interlocking zigzags on the commissure.

163 Lopha

Description: 3″ (76 mm) long, 3½″ (89 mm) high.
Round to oval or paw-shaped; usually
higher than long; valves convex and
almost equal in size; some species
slightly crescent-shaped in outline,
with slight posterior auricle. Surface
covered with 6–50 or more angular to
rounded ribs that produce zigzag
margin. In mature specimens, margin
thickens as valves stop increasing in
height. Attachment area present. Some
species have shelly, spinelike claspers
that grow out at intervals from ribs to
lock around stems or branches. Tiny
bumps present inside valves, especially
near margins.

Age: Triassic to Recent.

Distribution: Widespread in North America.

Comments: Living *Lopha* are mostly tropical, and
occasionally subtropical. Some
members of the genus are specialized to
attach to the stems of sea fans. The
claspers grow out at the margins from
the crests of ribs and around the stems,
locking *Lopha* to its host. The zigzag
margin of *Lopha* is very similar to the
margins of many brachiopods, like
Enteletes or *Rhynchotrema*, and must have
served the same function: to increase
the area of the opening between the
valves without increasing the distance
between them. This zigzag margin
reduces the size of the particles that can
pass into the mantle cavity. Bits of food
and animals that are too large to eat
can more easily be prevented from
entering.
The extreme thickening of the shell at
the margins, which has occurred on the
specimen illustrated, must function as
defense against predators that either
bore into the shell, like the moon snail
and the oyster drills, or try to break the
shell at the margins, like the whelks.
Lopha was formerly called *Alectryonia*.

SUPERFAMILY MODIOMORPHACEA
(Lower Ordovician through Lower Permian; possibly also in the Upper Permian)

Modiomorphacean shells tend to be elongated front to rear, with the umbones located near the front end but not terminal. There may be radially diverging teeth under the umbones, with some side teeth to the rear, beginning under the umbones. The ligament is not supported by a calcified ridge.

150 Modiolopsis

Description: 1¼" (32 mm) long, ⅝" (16 mm) wide. Transversely elongated, with small umbones close to front end; highest at posterior. Valves equal, thin, ornamented with concentric lines. Margins closed. Teeth absent. 2 closing-muscle scars: front scar deep, back scar large, faint.

Age: Middle and Upper Ordovician.

Distribution: Widespread in North America.

Comments: *Modiolopsis* includes at least 163 named species. The specimen illustrated is an internal mold of the right valve. *Modiomorpha* is in the same family as, and very close in appearance to, *Modiolopsis*. *Modiomorpha* lived during the Middle Silurian through the Lower Permian in eastern and midwestern North America. It is difficult to distinguish the two genera externally, but *Modiomorpha* has a single large, wedge-shaped tooth on the hinge of the left valve, and a corresponding socket on the right valve.

SUPERFAMILY TRIGONIACEA
(Devonian to Recent; possibly as early as the Lower Ordovician)

The shells of trigoniaceans may be triangular, oval, or 4-sided. The rear of the shell is usually truncated, with a ridge running from the umbo to the lower rear margin. The surface of the shells may be smooth, but is more commonly ornamented with radial or concentric ribs. The left valve typically has a strong middle tooth under the umbo and a weaker one on each side. The right valve usually has 2 strong teeth, and sometimes a weaker third tooth. The small muscle scars are located near the teeth. There is no pallial sinus.

---

140 Schizodus

Description:	1½″ (38 mm) long, 1¼″ (32 mm) high. Almost triangular to square in outline, longest along front-rear axis. Profile widest at umbones, tapering to sharp wedge at rear. Valves equal, smooth or with concentric lines. Right valve has 1 tooth below umbo; left valve has 2 or 3. Umbones high, pointing forward. 2 closing-muscle scars; sinus absent on pallial line.
Age:	Mississippian through Permian.
Distribution:	Widespread in North America.
Comments:	The shape of *Schizodus* appears to be adapted for burrowing. Plate 140 shows a right valve of *Schizodus*.

---

87 Scabrotrigonia

Description:	1⅛″ (28 mm) long, ¾″ (19 mm) high. Semicircular to crescent-shaped in outline; inflated; shells thick. Front rounded, with extended, angular lower rear margin. Umbones nearly terminal

pointing backward. Ribs narrow, high, beaded, radiating from large escutcheon, also crossed by ribs. Escutcheon set off by grooves; ribs cross grooves, changing angle and producing chevron pattern. Right valve has 2 large, grooved teeth under umbo, left valve has 3. Pallial line simple.

Age: Middle Jurassic through Upper Cretaceous.

Distribution: Widespread in North America.

Comments: *Scabrotrigonia* may have burrowed shallowly, with the escutcheon almost horizontal at the surface. The ribbing on the flanks probably helped to stabilize the shell in the sand or mud and that on the escutcheon may have inhibited the scouring action of currents. *Scabrotrigonia* is a member of the distinctive family Trigoniidae. Trigoniids are much more common in Europe than in North America, but their distribution indicates that they evolved in the Pacific during the Triassic, spread to western South America, then to North America, and finally to Europe. The family Trigoniidae used to consist of the single genus *Trigonia,* which has now been divided into more than 40 genera and numerous subgenera. Almost all are distinguished by umbones similar to those of *Scabrotrigonia,* strong ornamentation on the flanks—usually either radial or oblique—and a different pattern of ornamentation on the dorsal surface. Left and right valve exteriors and a left valve interior of *Scabrotrigonia* are illustrated in Plate 87. The drawing depicts a dorsal view.

SUPERFAMILY LUCINACEA
(Silurian to Recent)

Lucinaceans have equal-size valves, usually rounded in outline but sometimes oval or almost triangular. The surface is often smooth, but may have concentric sculpture. Many genera have a fold on the shell from the umbones to the lower margin, both in front and in the rear. The umbones are usually small and point forward or straight up. There are 1 or 2 teeth under the umbo in the right valve and 2 teeth in the left valve. There is no pallial sinus.

125 Codakia

Description: ⅝" (16 mm) long and high. Round; valves equal, profile moderately compressed on both sides. Valves almost symmetrical along line from umbo to lower margin. Ornamentation of concentric lines or latticelike sculpture. Umbones almost central, pointing slightly forward. Hinge under umbones has 2 teeth in each valve. Ridgelike side teeth well developed in front of umbo, small or absent behind. Lunule deep and narrow. Front closing muscle scar elongated, mostly within pallial line. Pallial sinus absent. Inner margin smooth.

Age: Upper Jurassic to Recent.

Distribution: Widespread in North America.

Comments: The species of *Codakia* illustrated is a member of the subgenus *Claibornites*. The illustration shows the exterior of right valve and the interior of a left valve.
Diplodonta, a genus similar to *Codakia,* is found in the Paleocene to the Recent on all North American coasts, and is known today as the diplodon. It differs by being smaller and having, under the umbones, a groove in the rear tooth of the right valve and in the forward tooth of the left valve. In *Diplodonta,* the outer edges of the closing-muscle scar are continuous with the pallial line. *Lucina,* found throughout North

America in the Upper Cretaceous to the Recent, is a very similar bivalve in the same family as *Codakia*. *Lucina* is less well rounded than *Codakia*, with a shallow groove marking off a posterior dorsal area. The inner margin is finely toothed. *Lucina* has only 1 siphon that takes water and wastes from the mantle cavity. The long foot forms a mucus-lined tube to the surface, and this is used to bring water containing food into the shell. *Lucina* still lives on the east coast of North America south of Cape Cod; the many lucine species living today are all fairly shallow burrowers in marine sediments. *Codakia*, also called the lucine, still lives on the East Coast south of North Carolina, in the Gulf of Mexico, and off California. It burrows in sand and is a common filter feeder near shore.

127 Paracyclas

Description: ¾" (19 mm) long, ⅝" (16 mm) high. Round, thin-shelled, sometimes with oblique, very shallow furrow from umbones to lower rear margin. Valves moderately inflated, covered with concentric lines. Umbones small, low, pointing forward. Ligament in deep groove; hinge line short. 2 small teeth on right valve just below umbo. Lunule absent; pallial line lacks sinus.

Age: Devonian.

Distribution: Widespread in North America.

Comments: *Paracyclas* is abundant in Devonian rock. The specimen illustrated is a left valve.

SUPERFAMILY CARDITACEA
(Devonian to Recent; possibly also in the Ordovician)

Carditaceans may be triangular, heart-shaped, trapezoidal, or mussel-shaped. Radial sculpture usually dominates, and the interior shell layer has straight radial ribs. There is a strong rib or a change of angle on the posterior slope. The umbones point forward. Each valve has 2 teeth under the umbones; side teeth are separated from umbonal teeth by a gape.

44 Venericardia

Description: 3⅞" (98 mm) long, 4⅛" (105 mm) high. Large; thick, heavy, triangular to round, with inflated, heart-shaped profile. Inside margin with widely spaced ridges and grooves. Radial ribs evenly spaced in young, becoming flattened and wider in adults. Umbones high, pointing forward. 2 large, elongated teeth under umbo, the posterior tooth much elongated; hinge plate deep. Pallial line simple.

Age: Paleocene through Eocene (possibly also in the Upper Cretaceous).

Distribution: Widespread in North America.

Comments: *Venericardia* is a member of the Carditacea, or cockle, superfamily. Most cockles burrow shallowly near shore, where they are easily dislodged by surf and currents. However, they have a strong foot and can quickly reburrow. Illustrated in Plate 44 are the exterior of a right valve and the interior of a left valve of *Venericardia*.

SUPERFAMILY CRASSATELLACEA
(Lower Ordovician to Recent)

Crassatellaceans are either triangular, trapezoidal, or rounded. The shell is smooth or covered with concentric sculpture, and the umbones are pointed and inclined forward. 2 or 3 teeth are present on the hinge plate under the umbones in each valve, and lateral teeth are also present. The pallial sinus, if present, is very faint.

---

137 Lirodiscus

Description: 1" (25 mm) long, ¾" (19 mm) high. Small, thick, oval, compressed, elongate front to rear. Some species have grooves from umbo to rear ventral margin, forming rear fold. Margins closed, small teeth on insides. Concentric sculpture present. Umbones small, forward-pointing, flattened. 2 teeth under umbones in left valve and 1 in right valve. Strong, nearly equal closing-muscle scars. Pallial line simple, sinus absent.

Age: Eocene.

Distribution: Atlantic, Pacific, and Gulf coastal plains.

Comments: *Lirodiscus* may have lacked siphons, and if so, could not burrow deeply below the surface. The shells are thick and small and not easily broken or dissolved, a characteristic that increased their chances of being preserved as fossils. The specimens illustrated in Plate 137 are the exterior and interior of the right valve of *Lirodiscus*. *Astarte* is a similar genus in the same family. *Astarte* ranges from the Jurassic to the Recent, and is widespread in North America. The shape of its shell is more triangular than that of *Lirodiscus*, with prominent umbones and more inflated valves.

143 Bathytormus

Description: 1½″ (38 mm) long, ⅞″ (22 mm) high. Almost triangular in outline, but with lower rear margin extended, in some species narrowed and slightly truncated. Abrupt change in angle forms ridge from umbo to rear margin. Moderately inflated, covered with concentric ribs. Margin interiors may have small nodes. Umbones small, close together. Hinge plate has 2 oblique teeth under umbo in each valve, with very large pit behind teeth extending to lower margin of hinge plate and pushing teeth forward; elongate, ridgelike teeth in front and rear also present. Closing-muscle scars deep; sinus absent on pallial line.

Age: Upper Cretaceous to Recent.

Distribution: Widespread in North America.

Comments: Plate 143 shows the interior of the left valve and the exterior of the right valve (top), and the exterior of the left valve (bottom). The outer shell layer has worn off on parts of the valve exteriors, revealing the radial ribs on the inner layer.

A genus similar to *Bathytormus* is *Crassatella,* which ranges from the Middle Cretaceous through the Miocene, and is also found throughout North America. The valves of *Crassatella* are less transversely elongated than those of *Bathytormus,* and the animal has a heavier hinge plate, with a smaller pit behind the teeth that does not extend to the margin of the hinge plate.

SUPERFAMILY CARDIACEA
(Upper Triassic to Recent)

The shells of cardiaceans are usually rounded or oval, but may be trapezoidal. The sculpture is typically radial, with a change of pattern on the rear slope. There are 2 conical teeth under the umbones in both valves; the teeth in the left valve are unequal in size, and the teeth in the right valve may be somewhat fused. The side teeth are distant from the umbonal teeth, and the anterior side teeth may be absent.

---

45 "Cerastoderma"

Description: 3½" (89 mm) long and high. Large, round or obliquely oval; heart-shaped in profile. Ribs numerous, elevated, either smooth or with fine concentric lines that appear scalloped as they cross ribs. Umbones curve forward. 2 teeth under umbones; anterior tooth in left valve and posterior tooth in right valve large and elevated. Hinge line relatively long and arched. Inside margin coarsely ribbed; pallial sinus absent.

Age: Upper Oligocene to Recent.

Distribution: Atlantic coastal plain.

Comments: *"Cerastoderma"* is placed in quotation marks because this group of bivalves is in the process of being reclassified, and this genus will probably soon receive a new name. Plate 45 shows the interior of a left valve and the exterior of a right valve of *"Cerastoderma."* The ribs have been broken or worn off over much of the right valve—probably before the animal died—giving it a ringed appearance. *"Cerastoderma,"* is in the same family—the Cardiidae—as *Cardium,* the cockle, and was formally placed as a subgenus in *Cardium.* Now the genus *Cardium* has been redefined, and is restricted to Europe and Africa. Other similar members of the Cardiidae family are *Granocardium* and *Laevicardium. Granocardium* is found in the Cretaceous throughout North America. It may have spines on the ribs, and the spaces between the ribs

are filled with pits, rows of small
spines, or 1–3 smaller ribs.
Laevicardium, another cockle found
along the coasts of North America in
Eocene to Recent deposits, has less well
developed ribs that are noticeably
weaker on the rear slope.

100 Protocardia

Description: 2¾″ (70 mm) long, 2½″ (64 mm)
high. Rounded, inflated; umbones
nearly central. Posterior slope covered
with radial ribs, rest of shell with
concentric ribs. Hinge long, slightly
arched; 2 teeth under umbones,
anterior ones larger and curved upward.
Pallial line nearly complete, sometimes
with small sinus near posterior closing-
muscle scar. Inside margin toothed.

Age: Upper Triassic through Upper
Cretaceous.

Distribution: Widespread in North America.

Comments: Plate 100 shows a specimen of
Protocardia with its valves still
articulated and closed. The animal
probably died while still buried or the
valves would have opened when the
closing muscles rotted. The left valve is
at the top, and the radial ribbing,
restricted to the posterior slope, is
clearly visible. This radial ribbing may
mark the area commonly exposed when
Protocardia burrowed, and may have
reduced the scouring action of water on
the shell more than the concentric
ribbing on the rest of the shell could
have done.

SUPERFAMILY MACTRACEA
(Upper Cretaceous to Recent)

Mactraceans have thin, triangular to
oval shells. The umbones are centrally
located and point forward. Each hinge

plate has a socketlike pit to hold an internal ligament. In the right valve there are 2 teeth under the umbo; in the left valve there is a single inverted-V-shaped tooth in front of the ligament pit. There is almost always a large pallial sinus. Shell surfaces are smooth or have concentric sculpture.

132 Rangia

Description: 2⅛" (54 mm) long, 2⅜" (60 mm) high. Thick-shelled; almost triangular in outline. Valves equal; inner margins smooth. Umbones inflated, tilted forward. Shells smooth or with growth lines present. 1 tooth under umbo in left valve, 2 in right valve; strong, curved, ridgelike side teeth to front and rear of umbones. Resilifer under umbones broad, deep. Pallial sinus small, deep.

Age: Miocene to Recent (possibly also in the Paleocene).

Distribution: Atlantic, Pacific, and Gulf coastal plains.

Comments: Living *Rangia* is an estuarine genus found in areas where salinity is low and fluctuating. It occurs from Chesapeake Bay to Florida and in the Gulf of Mexico. Shown in Plate 132 are the interior of a right valve and the exterior of a left valve of *Rangia*.
A close relative of *Rangia* is *Spisula*, common in the fossil record from the Paleocene to the Recent in eastern and western North America. A living species, the Atlantic Surf Clam *(Spisula solidissima)*, is the most important commercial clam on the East Coast, and occurs from Nova Scotia to North Carolina. *Spisula* can be distinguished from *Rangia* by its larger, broader resilifer, curved lateral teeth, and greater length relative to height. *Cymbophora*, found in the Upper Cretaceous of the east and west coasts of

North America, is another close
relative. *Cymbophora* has a slight gape
between the valves in the rear,
umbones that point almost straight up,
and a much smaller resilifer than that
of *Rangia*.

SUPERFAMILY TELLINACEA
(Upper Triassic to Recent)

Tellinaceans are rounded or, more
commonly, triangular or elongated
front to rear. The shells are smooth or
have concentric or, rarely, radial
sculpture. On the hinge under each
umbo are 2 teeth, often cleft. The side
teeth are commonly well developed.
The pallial line connects the muscle
scars and has a deep sinus.

144 Tellina

Description: 2¼" (57 mm) long, 1⅛" (28 mm)
high. Elongate-oval; valves compressed
and slightly unequal, left valve more
convex than right. Front round, rear
angular and gaping. Umbones placed
slightly to rear of middle. Oblique fold
from umbo to lower rear margin.
Smooth, or with weak concentric
sculpture present. Each valve has 2
small teeth under umbo; right rear and
left front teeth have deep groove in
center. Right valve has elongate side
teeth to front and rear of umbones;
elongate teeth indistinct on left valve.
Interior margins smooth. Pallial sinus
wide and deep; closing-muscle scars
connected by pallial line.

Age: Tertiary to Recent (possibly also in the
Cretaceous).

Distribution: Widespread in North America.

Comments: Living tellins have 2 very long, thin
siphons that are not connected. The
animals burrow well below the surface

with the left valve up and the commissure horizontal. In most tellins the rear end is bent slightly to the left, where the shell gapes. The siphons are extended through this gape. 1 siphon reaches up to the surface, where it picks up organic material and sucks it down to the shell. The other siphon discharges water and wastes below the surface. Shown in Plate 144 are exteriors of the right and left valve, and an interior of the right valve.

SUPERFAMILY CORBICULACEA
(Middle Jurassic to Recent; possibly also in the Lower Jurassic)

Corbiculacean shells vary in shape from triangular to oval. Their ornamentation is usually restricted to growth lines. There are 3 or fewer teeth on each hinge plate under the umbo, and a large central tooth in the right valve. There may be a small pallial sinus.

129 Corbicula

Description: ¾″ (19 mm) long, ⅝″ (16 mm) high. Outline triangular, but with round lower margin; or more oval, extended front to rear. Valves equal, inflated, heart-shaped in profile; inner margins smooth. Faint to pronounced concentric sculpture present. Umbones slope forward. 3 cleft teeth under umbo in each valve; to front and rear of cleft teeth are long, ridgelike teeth with grooves. Pallial line may have small sinus. Closing-muscle scars nearly equal.

Age: Upper Cretaceous to Recent.

Distribution: Central and western North America.

Comments: Living *Corbicula,* like the Asiatic Clam (*Corbicula fluminea*), are found in fresh to partly-salt water. Fossils are also found in sediments that must have been marine. The Asiatic Clam was introduced into the United States from Japan, and now sometimes appears in such numbers that it clogs drainage systems. Introduced species such as this one sometimes become overabundant and can turn into economic liabilities.

SUPERFAMILY VENERACEA
(Lower Cretaceous to Recent)

Veneracean shells tend to be thick and rounded, triangular, or oval. Ornamentation, if present, is almost always concentric. The umbones point forward and are located toward the front; there are usually 3 teeth under the umbo in each valve. The middle umbonal tooth in both valves is commonly thicker than the front umbonal tooth. The rear side teeth are weak or absent, and the front side teeth may also be absent.

131 Pitar

Description: 1½″ (38 mm) long, 1⅛″ (28 mm) high. Oval or rounded-triangular in outline, moderately inflated, with umbones pointing forward. Smooth or with fine concentric lines or ribs. Each valve has 3 teeth under umbones; right valve has strong tooth to rear, 2 teeth forward, with the front tooth very thin; left valve has large tooth to rear and thinner teeth forward. Ridgelike teeth to front of umbones well developed; side teeth absent to rear of umbones. Inner margins smooth; pallial sinus pointed, reaching to center of valves, or sinus short and rounded.

Age: Eocene to Recent.

Distribution: Atlantic, Pacific, and Gulf coastal plains.

Comments: Living *Pitar* are common along all North American coasts. The exterior of a right valve is shown in Plate 131. *Aeora*, found throughout North America in the Upper Cretaceous, looks very similar to *Pitar*. The main difference is in the teeth: *Aeora* has clefts in the outer 2 teeth of the left valve and a cleft in the back tooth of the right valve. The pallial sinus of *Aeora* is deep, rounded, and ascending. *Aphrodina*, widespread throughout North America from the Lower Cretaceous through the Upper Eocene, is closely related to *Pitar* and has the same general shape. Its teeth are similar to those of *Aeora*, but it has a deep, descending pallial sinus.

126 Dosinia

Description: 2⅝″ (67 mm) long and high. Almost round, with umbones pointing forward. Moderately inflated and lens-shaped in cross section. Inner margins smooth. Surface covered with incised concentric lines, usually evenly spaced. Teeth located on anterior of broad, platformlike hinge plate. 3 teeth under each umbo; some cleft, the others very thin and separated by deep sockets. Ridgelike teeth in front of umbones roughened. Lunule small and deeply incised. Large pallial sinus makes sharp angle that points to front margin just below umbones.

Age: Lower Eocene to Recent.

Distribution: Atlantic and Pacific coastal plains.

Comments: *Dosinia* is common today, living as a filter feeder on the Florida coast and in the West Indies, where it is known as the disc shell. The fossil specimens shown in Plate 126 are an exterior and an interior of a left valve. *Dosiniopsis*, found in eastern and western North America in the Paleocene through the Eocene, is very similar, but has a

broad, deep channel above the rear
tooth of the right valve. Its ridgelike
front teeth are roughened.

Cyprimeria

Description: 1⅛" (28 mm) long, ⅞" (22 mm) high.
Almost circular in outline; very
compressed in profile. Ornamentation
of weak concentric lines and very faint
radial traces. Umbones point forward.
3 teeth under umbo of each valve.
First front tooth on left valve and first 2
teeth on right valve are entire; other teeth
are grooved or bisected. Ridgelike
teeth present to rear of umbones,
absent in front of umbones. Lunule
absent; escutcheon deep. Inner margins
smooth. Pallial sinus very shallow.

Age: Cretaceous.

Distribution: Atlantic and Gulf coastal plains; New
Mexico.

Comments: The marginal drawing shows the
exterior of a left valve of *Cyprimeria*.
Cyprimeria is in the same family as the
fossil and living genus *Dosinia*, and is
very similar. Like *Dosinia*, *Cyprimeria*
was probably a filter feeder burrowing
just below the surface in shallow water,
with the rear margin up.
Clementia, found in the Eocene to the
Recent on the east and west coasts of
North America, is in the same family
as *Cyprimeria* and looks very similar.
Clementia is more elongated front to
rear and has a thinner shell than
Cyprimeria, its 2 posterior left teeth are
bisected, there are no ridgelike teeth,
and the pallial sinus is angular and
much longer than that of *Cyprimeria*.

Chione

Description: ¾" (19 mm) long, ⅝" (16 mm) high.
Outline triangular, but with rounded

lower margin and umbones pointing forward. Strong concentric sculpture commonly has erect frills and, in some species, radial ribs. Hinge plate large; 2 or 3 teeth under umbo of each valve: clefts in central tooth of left valve and 2 rear teeth of right valve. Side teeth absent. Lunule and escutcheon well defined. Inner margins ribbed; pallial sinus short and angular.

Age: Oligocene to Recent.

Distribution: Widespread in North America except in the Midwest.

Comments: *Chione* is still a very common clam genus today. East Coast species like the Cross-barred Venus *(Chione cancellata)* and the Gray Pygmy Venus *(Chione grus)* are found from North Carolina south; West Coast species such as the Common Californian Venus *(Chione californiensis)* are found from California south. They live in shallow water and burrow just below the surface. The concentric sculpture or frills help to hold their shells in the sediment. The genus *Chione* has been divided into several subgenera, one of which bears the same name as the genus itself. The marginal drawing shows the subgenus *Chione,* which is a very common fossil in Miocene and Pliocene sediments along North American coasts. This *Chione* probably lived very close to shore in shallow water, where even mild storms could throw the shells onto the beach or into lagoons where they would be buried.

136 Lirophora

Description: ¾″ (19 mm) long, ⅝″ (16 mm) high. Almost triangular; thick; moderately inflated. Umbones bluntly pointed, tilted forward. Ornamented with large, heavy concentric folds that may be flat or rounded. Hinge plate heavy. 2 large teeth under each umbo; rear tooth grooved in left valve; both teeth grooved in right valve; side teeth absent. Interior has pallial sinus and fine ribs on margin.

Age: Oligocene to Recent.

Distribution: Atlantic, Pacific, and Gulf coastal plains.

Comments: Plate 136 shows the exterior of a right valve of *Lirophora,* and the interior of a left valve whose lower margin is broken. *Lirophora* is very similar to *Chione,* and until recently was considered a subgenus of *Chione.* The strong concentric sculpture on *Lirophora* is distinctive, however, and these groups have now been separated. *Lirophora* still lives on North American coasts, south from North Carolina and southern California. It is usually found in shallow water, and burrows just below the surface.

135 Mercenaria

Description: 3¾" (95 mm) long, 3⅛" (79 mm) high. Large, heavy, triangular to oval in outline. Umbones relatively small, pointing forward. Shell smooth or covered with strong concentric growth lines, sometimes expanded as frills; faint ribs may be present. Lunule small. Hinge plate large, heavy, roughened to rear of umbones. 3 teeth in each valve under umbones; middle tooth in left valve and posterior 2 teeth in right valve are cleft. Inner margin finely ribbed; pallial sinus short, angular.

Age: Oligocene to Recent.

Distribution: Atlantic and Gulf coastal plains.

Comments: The specimen illustrated in Plate 135 is the exterior of a right valve of *Mercenaria tridacnoides.* This is a fairly young fossil—only a few million years old—and the shell material has not been altered.

Living *Mercenaria* is an abundant clam,

harvested commercially along the eastern coast of North America. It has many common names: Hard Shell Clam, Little Neck, Cherry Stone, Chowder Clam and Quahog. *Mercenaria* is a common fossil in Miocene, Pliocene, and Pleistocene deposits, and apparently has been abundant in the same area since the late Tertiary. There are 2 living North American species of *Mercenaria*, *M. mercenaria*, the northern species, and *M. campechiensis*, the southern species. *Mercenaria* lives in shallow water, often even above low-tide line, buried just below the surface. Older individuals may live with their posterior margins protruding above the sand or mud.

SUPERFAMILY MYACEA
(Upper Jurassic to Recent)

The valves of myaceans are unequal in size and usually elongated front to rear. There is usually a wide gape between the valves at the rear end. The surface of the shells is smooth or has concentric ornamentation. The hinge on the left valve generally has a socket for the internal ligament, and hinge teeth are usually absent. The interior margins on the valves are smooth, and there is almost always a well-developed sinus.

130, 141 Mya

Description: 2⅞" (73 mm) long, 1⅞" (48 mm) high (pl. 130); 1⅝" (41 mm) long, 1⅜" (35 mm) high (pl. 141). Outline oval, elongate front to rear, with wide rear gape, sometimes with truncated rear. Valves thin, chalky. Surface has concentric growth lines. Hinge teeth absent, but left valve has projecting semicircular and concave plate to hold

internal ligament. Inner margins
smooth. Pallial sinus well developed,
covering almost ½ of shell. Lunule and
escutcheon absent.

Age: Oligocene to Recent.

Distribution: Atlantic coastal plain.

Comments: *Mya* is the common soft-shell clam of
the East Coast. It burrows deeply near
shore, and has long siphons to reach the
surface. These siphons are so large that
they cannot be withdrawn into the
shell. The animal is protected not by
its shell, but by its habitat well below
the surface. Even though the valves are
thin and fragile, *Mya* is often preserved
as a fossil because of its deep burrowing
behavior. When it dies, it stays buried
instead of being washed ashore, where
destruction is more likely.
Plate 130 shows the interior and
exterior of a left valve of *Mya*. Plate
141 shows the interiors of left and right
valves and the exterior of a left valve.
This species has a truncated rear.
Hiatella, found widespread in North
America in the Upper Jurassic to the
Recent, is a nestling bivalve which
nevertheless has a shape similar to that
of *Mya. Hiatella* has a rather irregular
outline, truncated posteriorly, but with
the posterior end wider than the
anterior. *Hiatella* lacks the internal
ligament plate but does have a few very
weak teeth. *Hiatella* is also known as
Saxicava.

139 Corbula

Description: 1" (25 mm) long, ⅞" (22 mm) high.
Small; thin; triangular in outline;
profile inflated. Right valve larger,
more convex; left valve fits inside
margins of right valve. Front end
rounded; back end usually drawn out
into beak, with ridge from umbo to
lower rear margin. Smooth or
concentrically ribbed. Right valve with

1 strong tooth in front of resilifer and 1 behind. Margins smooth internally.

Age: Cretaceous to Recent.

Distribution: Widespread in North America.

Comments: *Corbula* inhabits marine environments. It attaches by a byssus to objects on the surface when young, but loses the byssus with age, and burrows into the sediment. *Corbula* is very sluggish, taking days to burrow just below the surface. Not surprisingly, it prefers to live in mud, where currents are less strong and the bottom more stable compared to areas where the sediment is coarser. Shown in Plate 139 are the right and left exteriors and the right interior of *Corbula*.

The genus *Ursirivus*, found in the Upper Cretaceous throughout North America, used to be included in *Corbula* but is now considered a separate genus. *Ursirivus* lacks the ridge from the umbo to the rear margin, and the 2 valves are almost equal in size. A common species is *U. pyriformis*, which is found in Wyoming, Utah, and Idaho, and lived in fresh and brackish water.

The specimen illustrated belongs to the subgenus *Bicorbula*. This subgenus is characterized by fairly weak sculpture. Most other subgenera of *Corbula* have much stronger concentric ribs.

SUPERFAMILY HIATELLACEA
(Permian to Recent)

The shells of hiatellaceans are elongated front to rear, and either oval or 4-sided. There is a narrow to wide rear gape. The hinge has 1 or 2 weak teeth, and the pallial sinus is large.

142 Panopea

Description: 6¼" (16 cm) long, 3⅜" (9 cm) high.
Thick-shelled, valves equal, outline
oblong to almost rectangular. Valves
gape slightly in front, widely in rear.
Margins smooth inside. Surface covered
with concentric growth lines. Umbones
very low, positioned near front; hinge
with 1 tooth in each valve. Pallial sinus
very deep.

Age: Lower Cretaceous to Recent.

Distribution: Atlantic and Gulf coastal plains,
Washington, Oregon, and California.

Comments: Very large living *Panopea* are found on
the West Coast, where they are called
geoducks (pronounced gooey-ducks),
collected for food, and considered a
delicacy because they are hard to catch.
Panopea burrows very deep, up to 3'
(1 m) below the surface. Its shell does
not cover the soft parts, but because it
lives so deep, it is well protected from
predation by most animals except very
determined humans. *Panopea* is called
Panope in some books, and is illustrated
in Plate 142 by the interior of the left
valve and the exterior of the right
valve.

SUPERFAMILY EDMONDIACEA
(Upper Devonian to Upper Permian)

Edmondiacean shells have valves that
are equal in size, inflated, and egg-
shaped when closed. The hinge plate
thin and lacks teeth.

123 Edmondia

Description: ⅞" (22 mm) long, ¾" (19 mm) high.
Valves equal, elongate-oval in outline
evenly inflated. Margins closed all
around. Umbones small, pointing
forward, situated ¼ to ⅓ distance fr

front margin. Irregular concentric ridges or growth lines present. Ligament was external and supported by small, elongate shelves along each side of ligament pit. Hinge plate has internal parallel ridge. Teeth absent. 2 equal closing-muscle scars; shallow pallial sinus.

Age: Upper Devonian through Upper Permian.

Distribution: Widespread in North America.

Comments: *Edmondia* burrowed and fed by filtering organic material from sea water. It is illustrated in Plate 123 by an internal mold of the right valve.

Cardiomorpha, found in the Mississippian throughout North America, is similar to *Edmondia*, but has more prominent umbones set closer to the front, and a smoother surface. The internal characteristics are not known for this genus, but the supposition is that it is closely related to *Edmondia*.

Grammysioidea, found in the Silurian through the Upper Devonian in eastern North America and Nevada, has the same shape as *Edmondia*, although it is more closely related to *Grammysia*. The umbones are larger than those of *Edmondia*, and jut farther forward, and a single radial groove from the umbones to the rear margin may be present.

SUPERFAMILY PHOLADOMYACEA
(Middle Ordovician to Recent)

Pholadomyacean valves are equal in size and egg-shaped when joined; or more elongated, especially to the rear. Many shells have a large rear gape.

148 Orthonota

Description: 1⅜" (35 mm) long, ⅜" (10 mm) high. Extremely elongate. Umbones small, flat, very near front end. Upper and lower margins straight and parallel. Rear gape slight. At least 1 fold runs obliquely from umbones to lower rear margin. Low concentric ridges narrow in front of folds, wider in back. Teeth absent.

Age: Middle Ordovician through Middle Devonian.

Distribution: Widespread in North America.

Comments: The shape of *Orthonota* is very similar to that of the living razor clam *Ensis*. Judging from *Ensis, Orthonota* may also have been a very fast, deep burrower, escaping from its Paleozoic predators the way *Ensis* often escapes today. The specimen illustrated is an external mold of both valves of *Orthonota. Cymatonota,* found in the Upper Ordovician in eastern North America, has a shape almost identical to that of *Orthonota,* but lacks the radial folds and concentric ribs. The more elongate species of *Modiolopsis* are also similar to *Orthonota* but again lack the distinctive ribbing and folds. They are also relatively higher at the rear margin.

122 Grammysia

Description: 2" (51 mm) long, 1⅛" (28 mm) high. Oval in outline, very inflated, with 1 or 2 shallow to deep grooves running from umbones to middle of lower margin. Concentric ornamentation of lines or undulations. Umbones large, curving in and forward, located toward front in middle of upper margin. Teeth on hinge absent. Lunule and escutcheon well defined. Inner margin with fine radial lines. Pallial sinus absent.

Age: Upper Silurian through Upper Devonian.

Distribution: Widespread in North America.

Comments: A famous discovery concerning *Grammysia* was made early in this century when J. M. Clarke exposed a bedding surface of Mt. Marion sandstone in New York State and found about 400 starfishes *(Devonaster)* and numerous specimens of *Grammysia.* By tapping on a *Grammysia* with a hammer, Clarke removed bits of rock and then found a starfish just below, with its mouth up, perfectly positioned, as though killed in the act of devouring the *Grammysia.* Thus, probably even in the Devonian, starfishes were feeding on bivalves. An internal mold of both valves of *Grammysia* is shown in Plate 122, the left valve at the top.

153 Pholadomya

Description: 2½" (64 mm) long, 1" (25 mm) high. Oval, thin, but with high, broad, round umbones near front, pointing forward. Valves equal, inflated, especially in front. Ornamentation of radial ribs, faint in upper rear, and crossed by concentric lines or ribs. Surface covered with tiny, delicate bumps. Teeth absent. Deep sinus on pallial line.

Age: Upper Triassic to Recent.

Distribution: Widespread in North America.

Comments: Living *Pholadomya* are found in the deep sea, but fossil *Pholadomya* inhabited shallow waters. Since *Pholadomya* preferred to burrow in mud, fossils of it found in sandy sediments were most likely washed there after death.

This is a fairly common pattern when fossil organisms have close living relatives. The living species may be found only in the deep sea, but the fossils obviously lived near shore in shallow water, judging from the kind of rock in which they are found and the other fossils in the assemblage.

The organisms apparently became extinct near shore for reasons that did not affect the deeper-water forms. A general lowering of sea level could have caused this phenomenon.

CLASS SCAPHOPODA
(Ordovician to Recent)

Scaphopods are also called tusk shells. The shell is shaped like a small tusk, open on both ends, and slightly curved. The surface of the shell is generally covered with fine growth lines and longitudinal ribs. The class is probably now at its maximum diversity, with more species alive today than there were fossil species for all of the Tertiary.

340 Dentalium

Description: 1⅞" (48 mm) long, ¼" (6 mm) wide. Long, narrow, slightly curved tubes, open on both ends; cross section rounded or polygonal, or some species square at apex, changing to rounded or polygonal at aperture. Usually has 4–20 narrow longitudinal ribs, but some species smooth or with close, fine, deeply engraved longitudinal lines near apex. Opening at apex may have small slit or notch.

Age: Middle Triassic to Recent.

Distribution: Widespread in North America.

Comments: Scaphopods are entirely marine, still present worldwide, and commonly called tusk shells. In life, the shell is oriented with the broad aperture down in the sediment and the apex above the surface. The animal pumps water in and out of the apex for respiration, and feeds by picking up small bits of food from the sediment with delicate tentacles. The Paleozoic scaphopods were formerly included in the genus *Dentalium,* but have now been reclassified. One such scaphopod is *Plagioglypta,* found from the Upper Devonian through the Upper Cambrian throughout North America. *Plagioglypta* has a fairly long and broad slit in the apex opening. It is covered with fine, oblique, encircling wrinkles, in some species over the entire surface, and in some only on the area near the larger aperture. *Prodentalium* is another Paleozoic genus that ranges from the Devonian through the Pennsylvanian throughout North America. *Prodentalium* is covered by oblique growth lines and fine longitudinal ribs, generally with a slightly zigzag alignment.

CLASS CEPHALOPODA
(Upper Cambrian to Recent)

Living cephalopods include the squids, cuttlefish, octopods, and *Nautilus,* the only living nautiloid. *Nautilus* alone has a fully developed shell, and this living form is most similar to the majority of fossil cephalopods, almost all of which had external shells. *Nautilus* has a bilaterally symmetrical shell that coils over its head in a single plane. The shell is divided into *chambers* by partitions, called *septa,* at right angles to the long axis of the shell. The area in which a septum fuses to the inside of the shell is called the *suture.* The last chamber—the *body chamber*— is larger than the rest and holds the soft parts of the animal. A tubular extension of the mantle passes back through all the septa. Where the tube passes through the septa, there are backward-projecting extensions of the septa called *septal necks.* The tube is enclosed by rings, called *connecting rings,* between the septal necks. The mantle tube, the septal necks, and the connecting rings together constitute the *siphuncle.*

The first fossil cephalopods had straight shells, but in most other respects were probably very similar to *Nautilus.* The main difference is that most of the straight shells contain calcareous deposits—in the chambers, the siphuncle, or both—that made the shells heavier. The nature of these deposits is important in classification. Fossil cephalopods may be ornamented with concentric ribs and longitudinal lines, ribs, or bumps. In some cephalopod shells concentric and longitudinal ornamentation are both present, with nodes in the form of bumps where 2 ridges of ornamentation intersect.

SUBCLASS ENDOCERATOIDEA
(Lower through Upper Ordovician; possibly also in the Middle Silurian)

The shells of this early group of cephalopods are mostly long and straight; but some are short, and a few others, slightly curved. The siphuncle is relatively large, and probably held a substantial portion of the internal organs. There are never calcareous deposits inside the chambers. The distinctive feature of this group is the presence of *endocones,* siphuncular deposits in the form of calcareous cones secreted in the rear parts of the siphuncle. A few endoceratoids have siphuncular deposits that are sheetlike, radially arranged in the siphuncle, and extended longitudinally.

ORDER ENDOCERIDA
(Lower through Upper Ordovician;
possibly also in the Middle Silurian)

This order has the characteristics of the
subclass; the siphuncular deposits are
always of the endocone variety.

341 Endoceras

Description: 11¾" (30 cm) long, 3¼" (8 cm)
wide. Large, heavy, straight, conical;
nearly cylindrical. Sutures simple and
straight. Concentric ribs variable from
low and closely placed to well elevated
and distinctly separate. Siphuncle
ventral, large, with long septal necks
that reach across segment to previous
septum. Thick, strongly calcified
connecting rings line inside of
siphuncle, forming double wall.
Calcareous deposit inside siphuncle of
hollow cones that open toward front
and are stacked inside siphuncle like
paper cups in a dispenser.

Age: Middle and Upper Ordovician.
Distribution: Widespread in North America.
Comments: *Endoceras* is known only from
fragmentary remains. The siphuncle is
the strongest portion of the shell, and
often the only part preserved. The
family Endoceratidae, named for the
genus *Endoceras,* includes the largest
known Paleozoic invertebrate fossils.
The shells of some endoceratids are 30'
(9 m) long.

The specimen illustrated is an internal
mold, with the apex broken off. The
fossil has been cut open on the right
side to show the siphuncle. The
marginal drawing shows a longitudinal
section of the shell of *Endoceras.*
Endoceras was among the first shelled
predators, and occupied a new niche in
the food chain. Food must have been
abundant for *Endoceras,* which had no
enemies. *Cameroceras,* a Middle and

Upper Ordovician cephalopod found i
Quebec, Vermont, New York,
Ontario, Ohio, Indiana, Illinois,
Manitoba, Minnesota, and Missouri,
closely resembles *Endoceras,* but lacks
concentric ribs.

Vaginoceras, found in the Middle
Ordovician of eastern North America,
also looks like *Endoceras,* but, when
seen in cross section, has 2 concentric
rings within the calcareous deposits
lining the siphuncle.

SUBCLASS ACTINOCERATOIDE.
(Middle Ordovician through
Pennsylvanian)

The shells of actinoceratoids are
generally medium to large, straight
cones. A few genera have curved shells
The siphuncle is typically very large,
and is expanded where passing throug.
a chamber and contracted where
passing through a septum. The septal
necks are flared out and bent toward
the outer walls of the cone. In early-
formed parts of the shell, the siphuncl
may occupy almost the entire chamber
in later chambers, the siphuncle may
be ½ the diameter of the shell. Almos
all actinoceratoids have calcareous
deposits in their siphuncles. These
deposits are arranged so that the space
between deposits constitute a canal
system radiating from a small central
tube. Many actinoceratoids also have
calcareous deposits in the chambers.
The only order in this subclass is the
order Actinocerida.

342 Actinoceras

Description: 2¾" (70 mm) long, ¾" (19 mm) wide
Large, straight, conical, slightly
spindle-shaped; decreasing in diamete

from just behind body chamber to aperture. Nearly circular to circular in cross section. Sutures straight, transverse. Siphuncle unusually large, strongly expanded between septa, located in ventral part of cone. Septal necks long, with ends curved out toward walls of cone. Connecting rings strongly calcified, gently inflated. Deposits in segments near apex fill siphuncle except for irregular, narrow central canal and narrow radiating canals. Calcareous deposits also commonly present in chambers on both sides of septa.

Age: Middle Ordovician through Lower Silurian.

Distribution: Widespread in North America.

Comments: The heavy calcareous deposits inside the rear chambers of the shell of *Actinoceras* acted to balance the body of the animal in the forward body chamber. Without this ballast, the rear chambers, which were filled with gas, would have been too light, and the end of the shell would have pointed obliquely upward. Almost all of the Paleozoic cephalopods with straight, conical shells balanced their shells with such deposits to increase efficiency when they swam or crawled. *Armenoceras*, a genus similar to *Actinoceras*, is found in the Middle Ordovician through the Upper Silurian of Arctic America, Anticosti Island (Quebec), New York, Ontario, Michigan, Indiana, Tennessee, Iowa, and Manitoba. The septal necks of *Armenoceras* are bent sharply back against the septum, and are inconspicuous.

In the specimen of *Actinoceras* illustrated, the septa and calcareous deposits in the siphuncle are clearly visible. The marginal drawing shows an internal mold. The upper part of the fossil is the filling of the body chamber.

299 Gonioceras

Description: 2¾″ (70 mm) long, 2½″ (64 mm) wide. Large, straight, with strongly flattened ventral surface; dorsal surface moderately convex, with sharply angular flanks. Aperture contracted. Sutures sinuous, with broad dorsal and ventral lobes, narrower rounded saddles, and pointed lateral lobes. Siphuncle large, situated near center of cone; segments short, inflated; rear segments filled with calcareous deposits, leaving only narrow canal.

Age: Middle Ordovician.

Distribution: Widespread in North America.

Comments: *Gonioceras* means "angle-horn," and refers to the sharp angles on the sides the cone. The wide, low, flattened she is similar in shape to a flounder, and some paleontologists have suggested that *Gonioceras,* like the flounder, might have spent most of its time lyir quietly in wait on the sea floor, and reaching out quickly for passing food. However, the fact that *Gonioceras* established a wide geographic range in a short period of time and has a shell weighted with calcareous deposits suggests that it was a very mobile animal. The photograph of *Gonioceras* shows a natural longitudinal section in which the septa and inflated siphuncle with calcareous deposits are visible. The marginal drawing illustrates a whole cone, and depicts the sutures that would show only after the outer shell was removed.

SUBCLASS NAUTILOIDEA
(Upper Cambrian to Recent)

The form of the shell ranges from straight and long to tightly coiled. Th siphuncle is commonly slender and straight, and may be moderately large In some nautiloids it has slightly bowed connecting rings. The septal necks are mostly straight, but may be recurved. The straight or slightly curved cones commonly have calcareous deposits in the chambers; there may be deposits in the siphuncle as well, but these are different from the endocone deposits in the Endoceratoidea and the deposits forming the elaborate canal system in the Actinoceratoidea. The siphuncle in most nautiloids is near the center of the cone, but it may be more dorsal in some genera. The sutures are most commonly straight or slightly wavy, but in a few of the more advanced nautiloids the sutures are strongly bent.

ORDER ORTHOCERIDA
(Lower Ordovician through Upper Triassic)

The orthocerids are the most common of the straight or slightly curved Paleozoic cephalopods. The siphuncle is generally straight, and may be empty or filled with calcareous deposits. The septal necks are commonly straight and the connecting rings are thin. The surface of the orthocerid shell may be smooth or it may be ornamented with transverse and, in some cases, longitudinal ribs.

320 Michelinoceras

Description: 2″ (51 mm) long, 1⅛″ (28 mm) wide. Long, slender, straight cone, circular in cross section, with smooth surface. Body chamber very long. Sutures straight and transverse. Siphuncle relatively small, central or near center of cone, lacking calcareous deposits. Septal necks straight; connecting rings cylindrical, delicate. Calcareous deposits completely fill early chambers and cover septa of later ones.

Age: Lower Ordovician through Upper Triassic.

Distribution: Widespread in North America.

Comments: All long, straight cephalopods with simple siphuncles used to be called *Orthoceras*. However, a closer look convinced paleontologists that no North American species fit the type, and now *Orthoceras* is an appropriate name mostly for European species. Many of the North American cones formerly called *Orthoceras* are now called *Michelinoceras*. Others are placed in the genera *Pleurorthoceras*, found in the Upper Ordovician of Ohio and Manitoba, and *Kionoceras*, from the Middle Ordovician through the Lower Permian throughout North America. *Pleurorthoceras* has connecting rings that are slightly bowed, making the siphuncle segments inflated, and has calcareous deposits only on the outside walls of the chambers. The outside of the shell of *Kionoceras* is longitudinally fluted, and is sometimes covered with additional less conspicuous, long, transverse lines or threads. The siphuncle segments may be straight, as in *Michelinoceras*, or slightly inflated, as in *Pleurorthoceras*. The specimen of *Michelinoceras* illustrated is an internal mold. The large section on the right is a mold of 2 chambers and part of the body chamber. The narrow section on the left is the internal mold of the siphuncle. The marginal drawing is a reconstruction of *Michelinoceras* showing the shell and an internal mold, with calcareous deposits in the chambers and siphuncle.

344 Dawsonoceras

Description: 6¼″ (16 cm) long, 1¼″ (3 cm) wide. Straight to slightly curved, long, narrow cone, circular or almost circular in cross section. Conspicuous ribs mark positions of chambers. Outer surface of shell shows straight or wrinkled growth lines and, in some species, irregular longitudinal ribs. Siphuncle small, almost central, with short septal necks constricting siphuncle at septal openings; siphuncle straight between septa.

Age: Silurian (possibly also in the Middle Devonian in Ontario).

Distribution: Widespread in North America.

Comments: The photograph and the marginal drawing show typical *Dawsonoceras* fossils: internal molds formed after death when the cephalopod's empty shell filled with and was surrounded by mud, which eventually hardened into rock. In time, the shell dissolved away. When the rock enclosing the fossil broke, splitting along the now empty areas once occupied by the cephalopod's shell, the shell filling came free. The surface of the shell filling, or mold, shows an impression of the inside surface of the shell. In some cases the rock around an internal mold will show an impression of the outside of the shell that it once enclosed. This is called an external mold. Molds are the most common means of preservation for Paleozoic cephalopods since, in the hundreds of millions of years since the original animals' deaths, most shell material has dissolved.

ORDER ONCOCERIDA
(Middle Ordovician through Mississippian)

Most oncocerids have shells that are short, blunt, curved cones, but some are long and straight and a few are coiled. The siphuncle is generally slender, but may be large and inflated. In primitive genera the septal necks are straight, but in more advanced forms they flare out and bend toward the outer shell walls in the anterior chambers. The siphuncles of many oncocerids have calcareous deposits in the form of radially arranged longitudinal sheets. In many oncocerid the body chamber is expanded and the aperture constricted.

208 Oncoceras

Description: 3⅛" (79 mm) long, ¾" (19 mm) wide. Short, blunt cone expanding to greatest diameter near base of body chamber, contracting to aperture. Shell compressed laterally, slightly curved, with ventral side convex, dorsal side concave, except at base of body chamber, where diameter greatest. Aperture oval, with sinus in ventral edge. Sutures straight, transverse. Siphuncle small, near ventral wall, empty; segments narrowly spindle-shaped, generally longer than wide; connecting rings thin.

Age: Middle and Upper Ordovician.

Distribution: Arctic America, Anticosti Island (Quebec), New York, Ontario, Wisconsin, Illinois, Manitoba, Minnesota, Wyoming.

Comments: 2 fairly common genera are similar to *Oncoceras. Beloitoceras,* found in the Middle and Upper Ordovician of Arctic America, Ontario, Wisconsin, Illinois, Minnesota, Iowa, and Wyoming, is more inflated than *Oncoceras* in back of the body chamber, and the dorsal side of its cone is straighter over the body chamber. *Augustoceras,* found in the Middle and Upper Ordovician of Kentucky and Ohio, is almost triangular in cross section, and has oblique sutures, faint, slightly wrinkled growth lines, and a siphuncle with slightly inflated segments and calcareous deposits.

The photograph of *Oncoceras* shows the rear part of a shell with the body chamber broken away.

ORDER DISCOSORIDA
(Middle Ordovician through Middle Devonian; possibly also in the Upper Devonian)

Most discosorids have short, conical shells that are straight or slightly curved. The siphuncle is greatly expanded between the septa. The septal necks are strongly recurved and the ends of the necks almost touch the rear sides of the septa. The connecting rings may be very thick, or the siphuncle lined with calcareous deposits, or both. Calcareous deposits may also be present in the chambers, but these are thin in most genera.

207 Phragmoceras

Description: 1¾" (44 mm) long, ⅞" (22 mm) wide. Rapidly expanding, laterally compressed, open coil. Body chamber also compressed laterally. Aperture strongly contracted, opening ventrally into narrow, oval to almost triangular hole, with long slit above connecting to larger, circular opening at dorsal margin. Small siphuncle at ventral edge; segments inflated with thick connecting rings. Calcareous deposits in siphuncle resemble lining of small beads in cross section.

Age: Middle Silurian.

Distribution: Anticosti Island (Quebec), New York, Ontario, Ohio, Indiana, Wisconsin, Illinois, Manitoba.

Comments: The contracted and peculiarly shaped aperture of *Phragmoceras* seems to serve as protection for its soft tissues. The head and arms probably extended out of the large dorsal part of the aperture, and the funnel out of the smaller ventral part. The funnel could then face to the rear and blow out jets of water to drive the animal forward. The chambers did not contain calcareous deposits, as they usually do not in coiled shells, since balance is not a problem for them.

Most *Phragmoceras* fossils are internal molds of the body chamber. The aperture shows as a rough, rounded area where the circular opening was, and a lower, oval to triangular area, also roughened. The areas are rough because the fossil has been broken in these places. The photograph of *Phragmoceras* shows an internal mold, with a few chambers to the right and the body chamber to the left. The marginal drawing shows the shell from the front.

ORDER TARPHYCERIDA
(Lower Ordovician through Upper Silurian)

These earliest and most primitive of the coiled cephalopods are most abundant in the Lower Ordovician. The shells of tarphycerids are loosely coiled, with all the *whorls* in 1 plane. When the animal was mature, the body chamber diverged away from the coil. In some forms, only the first part of the shell is coiled, and most of the chambers form a long, straight section. The sutures are simple. The siphuncle varies in position within the chambers. Septal necks are straight and connecting rings are layered.

204 Bickmorites

Description: 8″ (20 cm) long, 1¾″ (4 cm) wide. Long, narrow, coiled cone with whorls not touching; circular or slightly compressed laterally in cross section. Body chamber straight and tubular, or slightly curved. Sutures transverse, almost straight, with slight lateral lobes. Siphuncle slightly ventral to center, tubular, straight, with straight septal necks and thin-walled connecting rings. Calcareous deposits in siphuncle and chambers absent. Ornamentation of prominent concentric ribs inclined toward rear at ventral surface, outlining ventral sinus in aperture. Mature specimens have additional fine concentric and longitudinal lines.

Age: Upper Ordovician through Middle Silurian.

Distribution: Widespread in North America.

Comments: *Barrandeoceras,* found in the Middle Ordovician of Quebec, New York, and Ontario, is very similar to *Bickmorites,* but its whorls, although not impressed, are in contact.

The illustrated internal mold of *Bickmorites* shows sculpture because the inside of the shell was ribbed, as well as the outside. The sutures show very faintly.

ORDER NAUTILIDA
(Lower Devonian to Recent)

Most nautiloids from the mid-Paleozoic to the Recent belong to this order. The shell is commonly coiled in a single plane, but in some genera it is simply curved.

247 Cooperoceras

Description: 5¼″ (133 mm) long, 2⅝″ (67 mm) wide. Loosely coiled cone, with whorls touching but not impressed. Whorls flattened ventrally and laterally.

Umbilicus wide, with hole through middle. Sutures with rounded ventral, lateral, and dorsal lobes. Septa simple, concave toward body chamber. Siphuncle is simple, straight tube; deposits absent. Ornamentation of narrow, shallow groove on middle of ventral surface; last whorl has sinuous ribs on sides and long, slender, hollow spines to each side of ventral surface. Spines project to sides and curve toward earlier parts of whorl.

Age: Lower Permian.

Distribution: Widespread in western North America.

Comments: *Cooperoceras* is especially common in Texas. Its spines are very distinctive, but since no spiny cephalopods are living today, we can only guess at their function. Since they project laterally, they could have been used to stabilize *Cooperoceras* when it lay on the sea floor. They may also have served for defense, making *Cooperoceras* a larger and less palatable mouthful for a predator. They clearly had a disadvantage, however, since they reduced the streamlining of the shell.

200 Temnocheilus

Description: 1¼" (32 mm) long, ⅜" (10 mm) wide. Disc-shaped coil, with whorls touching but only slightly impressed. Umbilicus wide, with hole through center. Ventral margin broad, strongly flattened, with sides converging toward umbilicus. Aperture has deep sinus on ventral edge. Suture has shallow ventral, lateral, and dorsal lobes. Siphuncle small, almost central, straight, lacking deposits. Ventrolateral shoulders ornamented with single row of large, longitudinally elongated bumps.

Age: Mississippian through Permian.

Distribution: Kentucky, Illinois, Missouri, Kansas, Texas, Colorado.

Comments: *Metacoceras* is similar to *Temnocheilus*, but its shell is not as flattened ventrally, and its bumps are less pronounced. *Metacoceras* is found from the Pennsylvanian through the Permian in Ohio, Illinois, Missouri, Kansas, Oklahoma, and Texas.

195 Centroceras

Description: 6¼" (16 cm) long, 1½" (4 cm) wide. Coiled cone with few whorls that touch but barely overlap, and rapidly expanding diameter; umbilicus wide, with hole through center. Body chamber ½ length of whorl. Whorl section 4-sided, compressed laterally, with sharp shoulders to sides of ventral surface and at umbilicus. Sides oblique, convergent toward narrow, slightly convex ventral surface. Suture has shallow ventral lobe; broad, shallow lobe near umbilicus in mature part of shell. Siphuncle tubular, empty, located near ventral surface. Ornamentation of fine concentric ribs and grooves; earlier whorls have small bumps at shoulders.

Age: Middle Devonian.

Distribution: Canada, New York, Ohio.

Comments: *Centroceras* was formerly considered to range in age from the Devonian through the Pennsylvanian, but the genus has recently been restricted in definition, and consequently also in age.

The specimen in the photograph is an internal mold, showing the long body chamber and the sutures. The marginal drawing shows the shell of a *Centroceras* specimen.

189 *Eutrephoceras*

Description: 2⅛" (54 mm) long, 1¾" (44 mm) wide. Almost spherical, tightly coiled, with outer whorl completely covering inner whorls. Whorls broadly rounded ventrally and laterally. Umbilicus very small or absent. Aperture has broad, shallow, rounded sinus on ventral margin. Sutures almost straight. Siphuncle small, variable in position. Surface smooth except for fine growth lines.

Age: Upper Jurassic through Miocene.

Distribution: Widespread in coastal areas and western interior of North America.

Comments: *Eutrephoceras* is in the same family as the only living nautiloid cephalopod genus, *Nautilus,* called the Chambered, or Pearly, Nautilus. *Nautilus* lives in the southwestern Pacific Ocean around coral reefs, and ranges down to 1800' (550 m). It can swim backward or forward by expelling jets of water from its funnel. *Nautilus* has been studied extensively, but there are still many unanswered questions about its biology and life habits.

The specimen illustrated has been filled with mud, but some of the original shell material is preserved, particularly on the septa.

SUBCLASS AMMONOIDEA
(Devonian through Cretaceous)

Ammonoids are a large group particularly abundant in the late Paleozoic and Mesozoic. Most have tightly coiled shells, and many are ornamented with ribs, bumps, and spines.

Ammonoids differ from nautiloids most conspicuously in the nature of the sutures, which in ammonoids are commonly folded into complex patterns. The *suture patterns* are very important in classification. In this book, a diagram of the suture pattern is given with the description of each ammonoid genus. The suture pattern diagrams usually extend to where the whorl is covered by a later whorl, but in some books patterns may include the inner side of the whorl; the diagrams here show only the exposed part of the suture. To orient the viewer, an arrow is included that points toward the aperture and passes through the suture where the middle of the *venter* would be. The venter is the outermost part of a whorl. The part of the pattern that is convex in the direction of the aperture on the venter is called the *ventral saddle,* and the small concave area in the middle is called the *ventral lobe.* To the sides of the ventral saddle are *lateral saddles* and *lobes,* which may also be highly folded.

Except for some early, primitive forms, ammonoids also differ from nautiloids in having siphuncles that are near the outer margin of the whorls, and septal necks directed forward. Some ammonoids have a ridge along the venter, called a *keel.*

ORDER ANARCESTIDA
(Middle and Upper Devonian)

The anarcestids are the ancestral stock from which all other ammonoids evolved. Like those of the nautiloids, the septal necks of anarcestids are directed backward, and the *umbilicus* may be perforated. The sutures commonly have 3–4 lobes, but there may be fewer or more. The lobes and saddles are not complexly folded as in more advanced groups.

337 Bactrites

Description: ½" (13 mm) long, ⅛" (3 mm) wide.
Long, straight, very slender cone, with
angle at apex of cone less than 10°.
Circular or slightly oval in cross
section. Aperture has ventral sinus.
Sutures almost straight, with only
small ventral lobe. Septa simple,
transverse. Narrow siphuncle in contact
with ventral wall.

Age: Devonian through Permian (possibly
also in the Silurian).

Distribution: New York, Pennsylvania, Maryland,
Virginia, West Virginia, Ontario,
Ohio, Michigan, Arkansas, Oklahoma,
Texas, Mackenzie Basin (Northwest
Territories), Alberta.

Comments: Although not closely related, *Bactrites*
is very similar to *Michelinoceras*.
However, the siphuncle of *Bactrites* is at
the ventral margin, not near the center
of the cone as in *Michelinoceras*, and
there are never any deposits in its
chambers. *Bactrites* and its close
relatives have been proposed as
ancestors of all other ammonoids and
belemnoids because all have the
siphuncle located ventrally.
3 fragments of *Bactrites* are shown in
Plate 337. The marginal drawing is a
reconstruction of a complete shell.

Suture:

Manticoceras

Description: 3⅜" (86 mm) long, 1⅛" (28 mm)
wide. Disc-shaped, somewhat
compressed laterally. Surface smooth
except for fine, sinuous growth lines.
Umbilicus moderately large. Aperture
triangular or oval. Sutures simple, with
divided ventral saddle and large first

lateral saddle. Siphuncle small, ventral, marginal.

Age: Upper Devonian.

Distribution: Widespread in North America.

Comments: The marginal drawing shows a front view of a complete internal mold. Early ammonoids like *Manticoceras* tend to be smooth-surfaced, with very little ornamentation. This shell form undoubtedly increased their efficiency in swimming. *Agoniatites* is an even earlier ammonoid than *Manticoceras*, found in the Middle and possibly the Lower Devonian, and widespread in North America. The shape of *Agoniatites* is almost identical with that of *Manticoceras*, but its sutures show only 3 lobes, 1 lobe in the middle of the venter and 1 broad, low lobe on each side of the whorls.

Acanthoclymenia, like *Manticoceras*, is another Upper Devonian ammonoid found throughout North America. It is smaller than *Manticoceras* and has a wider umbilicus, and a single deep, rounded lobe on the venter.

Suture:

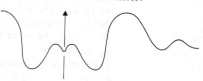

ORDER GONIATITIDA
(Middle Devonian through Upper Permian)

Almost all goniatitids have septal necks directed forward. The sutures are generally sharply angular, without intricate folding. Usually there are 8 lobes around the whorls, but in primitive forms there are fewer lobes, and in advanced forms there are more. A few genera have complexly folded lobes.

192 Tornoceras

Description: 1⅜″ (35 mm) long, ⅝″ (16 mm) wide.
Disc-shaped, with whorls flattened
laterally and strongly impressed
dorsally. Venter may be rounded,
flattened, or angular. Umbilicus
completely closed, or small, open, and
funnel-shaped. Sutures simple, with
small lobe dividing fairly high ventral
saddle, and much higher, rounded
saddle just before umbilicus. Siphuncle
small, ventral, and marginal, with
septal necks pointing forward.
Ornamentation absent except for fine,
sinuous growth lines, forming shallow
sinuses on venter and sides of shell.
Shape of growth lines reflects shape of
aperture.

Age: Middle and Upper Devonian.

Distribution: New York, Pennsylvania, Maryland,
Virginia, West Virginia, Ontario,
Ohio, Michigan, and Iowa.

Comments: The nature of the outer shell surface is
generally important in identification,
but not as important as the sutures,
which are hidden by the shell. It is
therefore sometimes necessary to peel
off some of the outer shell to expose the
suture pattern.
The specimen illustrated is preserved as
an internal mold, showing the suture
pattern clearly.

Suture:

191 Imitoceras

Description: 3⅜″ (86 mm) long, 1⅞″ (48 mm)
wide. Globular to disc-shaped; surface
smooth. Umbilicus closed. Suture
pattern simple, with 8 undivided
saddles around outside of whorls; only
4 of these show, 2 on each side.

Age: Upper Devonian through Middle Permian.

Distribution: Widespread in North America.

Comments: The *Imitoceras* fossil illustrated is an internal mold. None of the original shell material has been preserved. This most common state of preservation for cephalopods fortunately gives adequate information for identification. When the shell filled with mud or sand after the death of the animal, all of the chambers and the siphuncle, as well as the body chamber, were usually filled. The filling hardened to rock and the shell dissolved away, leaving spaces around the outside of the fossil as well as between each chamber where the septa used to be. When the rock that enclosed the fossil is broken open, the entire mold often lifts neatly out. The suture pattern, the single most important characteristic for identifying ammonoids, is usually clearly seen because it is outlined on the surface of the mold by the spaces that formerly held the septa.

Agathiceras is an ammonoid very similar to *Imitoceras,* both in form and suture pattern. *Agathiceras* occurs in British Columbia, Oklahoma, and Texas in the Lower Pennsylvanian through the Middle Permian. Its shell has revolving ribs, and the venter has a saddle (instead of a lobe) that is indented. 2 short prongs extend from the edges of the indentation.

Suture:

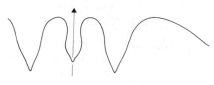

---

183 Shumardites

Description: 3⅜" (86 mm) long, 2" (51 mm) wide. Globular, smooth, with medium-sized umbilicus. Concentric ribs on inner side of shell at every ¼ turn of whorl form grooved impressions across internal mold of whorl. Sutures have ventral lobe with 2 prongs and 5 lateral saddles that are simple, knoblike, and separated by deep lobes that may be divided.

Age: Upper Pennsylvanian.

Distribution: Texas, New Mexico.

Comments: The grooves on the internal mold under the outer shell of *Shumardites* are called varices, and represent the thickening of the lip of earlier apertures. The shell's growth must have been discontinuous, with long pauses during which this thickening of the lip occurred, followed by rapid growth forward to a new aperture position. *Shumardites* was named for Benjamin F. Shumard (1820–1869), an American paleontologist who studied the geology of Wisconsin, Minnesota, Iowa, Missouri, Texas, and Oregon. The photograph shows a specimen in which half of the outer whorl is broken away. The marginal drawing shows a complete specimen of *Shumardites*.

Suture:

181 Waagenoceras

Description: 1¼″ (32 mm) long, 1⅛″ (28 mm) wide. Almost globular, coiled, with smooth surface. Umbilicus wide. Suture line, with 4–8 pairs of lobes, arcs out toward aperture; lobes and saddles strongly divided.

Age: Middle Permian.

Distribution: Texas.

Comments: Many of the Paleozoic ammonoids, like *Waagenoceras,* have no ornamentation and are so similar in shape to other genera that the only reliable way to tell them apart is through the study of their sutures. Ammonoid sutures tend to become increasingly complex through the Paleozoic and Mesozoic. *Waagenoceras* has sutures of medium complexity compared to those of other ammonoids.

The illustration shows 2 *Waagenoceras* fossils, both broken.

Properrinites, a cephalopod of the Lower Permian in the southwestern United States, has a suture pattern very similar to that of *Waagenoceras,* but its whorls are more compressed. *Perrinites,* from the Middle Permian of the western United States, also has more compressed whorls than *Waagenoceras,* but in its suture pattern the saddles are more divided.

Suture:

184, 190 Muensteroceras

Description: 2⅛" (54 mm) long, ⅜" (10 mm) wide. Disclike, but variable from compressed to globular. Umbilicus moderately wide, with whorls overlapping only slightly. Sutures simple. Lobe on venter narrow, with parallel sides. Surface smooth.

Age: Mississippian.

Distribution: Michigan, Indiana, Missouri.

Comments: *Goniatites* is a very similar genus found in the Upper Mississippian throughout North America. The only reliable way to distinguish *Muensteroceras* from *Goniatites* is by comparing the sutures on the venter. *Goniatites* has a high saddle with a tiny lobe in the middle; *Muensteroceras* has a small, divided

saddle sunken into a deep lobe.
2 sides of the same specimen of
Muensteroceras are illustrated: Plate 190
clearly shows the suture pattern, and
Plate 184, the well-preserved shell.

Suture:

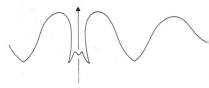

187, 188   Gastrioceras

Description: ⅝″ (16 mm) long, ⅜″ (10 mm) wide
(pl. 187); 1¼″ (32 mm) long, ¾″ (19
mm) wide (pl. 188). Disc-shaped to
almost globular, with whorls
overlapping only slightly. Umbilicus
wide, showing all whorls. Umbilical
side of whorls has pronounced ribs that
fade out toward venter. Sutures simple;
tiny lobe on venter has sharp prongs
pointing toward aperture.

Age: Pennsylvanian.

Distribution: Widespread in North America.

Comments: Both photographs of *Gastrioceras* show
crushed internal molds, with the
original shell material showing white at
the sutures. *Pseudoparalegoceras,* found
in the Pennsylvanian of the
southwestern United States, resembles
Gastrioceras externally, but its sutures
have an additional sharp lobe just
before the umbilicus.

Suture:

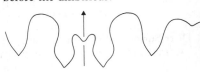

ORDER CERATITIDA
(Permian through Triassic)

Almost all Triassic ammonoids are ceratitids. They commonly have well-developed ornamentation, and the sutures tend to be more elaborate than in the goniatitids. In a few genera the sutures have simple lobes and saddles, but in most genera, sutures have complexly folded lobes and smooth saddles. Sutures of other genera have both lobes and saddles that are complexly folded. The shells are regularly coiled in 1 plane.

185 Meekoceras

Description: 3⅛″ (79 mm) long, ¾″ (19 mm) wide. Disc-shaped, compressed, with flattened venter. Umbilicus narrow or moderately wide. Body chamber short. Sutures have many small folds in bases of lobes; saddles rounded and simple. Surface smooth or with weak, broad ribs on sides.

Age: Lower Triassic.

Distribution: Idaho, Utah, Nevada, California.

Comments: Wherever *Meekoceras* is found, paleontologists are able to identify the stratum as Lower Triassic. This makes *Meekoceras*, like most ammonoids with their short age ranges, a very good guide fossil.
In the specimen illustrated, some of the original shell material is preserved, but it is broken away over the body chamber.

Suture:

ORDER LYTOCERATIDA
(Jurassic through Cretaceous)

The shells of the lytoceratids are loosely coiled and commonly have rounded whorl sections. In many genera the whorls do not touch throughout the shell; in others the shell is coiled into a spire, like that of a gastropod. The sutures tend to have only a few lobes and saddles, but these are very intricately folded.

209 Oxybeloceras

Description: 1⅜" (35 mm) long, ¼" (6 mm) wide. U-shaped, with the 2 straight, parallel shafts pressed closely together. Shell begins as tiny, tight coil, forms 1 open coil about ½" (13 mm) wide, then forms long, straight shaft that doubles back on itself. Ornamentation of strong ribs that may be oblique; 2 rows of small spines on ribs on outer side of shafts.

Age: Upper Cretaceous.

Distribution: Widespread in North America.

Comments: It is difficult to imagine how the peculiar shape of *Oxybeloceras* was adaptive, and what mode of life the animal observed. Some paleontologists think that it drifted in the ocean. Since the empty chambers were filled with gas, the head and arms would have hung down in the early stages of the animal's life; but after the final bend in the shell was secreted, the aperture, head, and arms would have faced up. Other paleontologists think *Oxybeloceras* crawled on the ocean floor.
This genus is usually found either as an isolated shaft or as 2 shafts connected by a 180° curve. Both shafts of the specimen illustrated are broken. *Oxybeloceras* has been called both *Ptychoceras* and *Solenoceras,* and is very similar to both these genera. *Ptychoceras*

found in the Lower Cretaceous of British Columbia and California, lacks the initial coil and begins as a straight shaft, with weaker ribbing that is restricted to the outer surface. *Solenoceras* is found in the Upper Cretaceous throughout North America. It lacks the final open coil in the young shell, and has a constricted aperture with a collar around it. *Hamites* is another genus with a similar form, found throughout North America in the Lower Cretaceous. It usually has 3 parallel shafts that are well separated; some species form an open coil. It has ribs, typically strong, fine, and dense, but never has spines on the ribs.

Suture:

---

338 Baculites

Description: 3¾" (95 mm) long, 1¼" (32 mm) wide. Long and straight, except for a few tight initial whorls. Sutures very complex, with saddles and lobes folded, and resulting lobes folded. Surface smooth or with sinuous growth lines following shape of aperture, which is extended as umbo on dorsal side.

Age: Upper Cretaceous.

Distribution: Widespread in North America.

Comments: *Baculites* fossils from the Pierre shale of Wyoming and South Dakota are often very beautiful: the shell is filled with dark, shiny rock, and the elaborate suture design stands out in pearly white with an iridescent sheen.

Suture:

210 Turrilites

Description: 1¼" (32 mm) long, ⅝" (16 mm) wide. Shell coiled into tight, high spire, with all whorls in contact (in old age, final whorl may be out of contact). Sutures very complexly folded. Ribs present but weak, with 4 bumps on each. Ribs may disappear in final whorl, with only 3 bumps remaining.

Age: Middle Cretaceous.

Distribution: Western United States.

Comments: *Turrilites* looks like a large gastropod, but its cephalopod identity is clear from the internal chambers and suture pattern. Similarity of shell form, however, probably means similarity of life habit, and *Turrilites* might have lived like a gastropod, crawling on the sea floor and picking up bits of food from the sand or mud.

Helicoceras, found in the Cretaceous of the western United States, has the same spiral shape as *Turrilites,* but can be distinguished because its whorls are separate at all stages.

Suture:

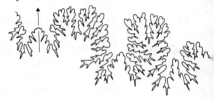

ORDER AMMONITIDA
(Lower Jurassic through Upper Cretaceous)

Most ammonitids have thick shells and strong ornamentation. The sutures of most ammonitids are very complexly folded on both the lobes and saddles. All ammonitids are tightly coiled.

194, 199 Scaphites

Description: ⅜″ (10 mm) long, ⅛″ (3 mm) wide
(pl. 194); ¾″ (19 mm) long, ¼″ (6
mm) wide (pl. 199). Hooklike,
compressed, or very inflated. Shell
grows first in tight coil, then straight,
then back upon initial coil, with
aperture almost touching initial coil.
Umbilicus tiny, with whorls strongly
overlapping. Aperture commonly has
thickened rim. Sutures complexly
folded and refolded. Concentric ribs
extend to edge of ventral surface,
become bumpy and thick, and split
into weaker secondary ribs. Ribs on
side and venter equal in mature shell.
Body chamber has bumps on side.

Age: Middle Cretaceous.

Distribution: Widespread in North America.

Comments: *Hoploscaphites* is closely related to
Scaphites and very similar, but it has a
shorter shaft: its shell does not grow
out straight as far as that of *Scaphites*
before turning back to make the hook.
Hoploscaphites may also differ by having
strong bumps on the sides near the
periphery of the shaft and hook. Plate
194 shows juvenile specimens that have
been crushed on a bedding plane. The
marginal drawing shows a
reconstruction of one of these crushed
specimens.

Suture:

196 Hoploscaphites

Description: 4⅝″ (117 mm) long, 1⅝″ (41 mm)
wide. Disc-shaped; compressed with
flat sides to inflated with convex sides.
Venter flat or rounded; whorls tightly

coiled up to body chamber, which grows away from coil; shaft grows out away from coil, then hooks in so aperture inclines toward coil. Venter has elongated bumps; sides may also have bumps.

Age: Upper Cretaceous.

Distribution: Widespread in North America.

Comments: The specimen of *Hoploscaphites* illustrated is preserved with the original shell material. This kind of preservation results in a beautiful fossil, but may actually make identification more difficult than the preservation of the more common internal mold fossil. The outer shell preserved in this specimen covers the suture pattern, the most important feature for identification. Because ammonoids may be very similar in general form and ornamentation, often only the suture pattern can distinguish one species, or even one genus, from another. Sometimes the outer shell must be chipped away over at least 1 septum before identification can be confirmed.

Suture:

182 Desmoceras

Description: 10″ (25 cm) long, 2⅛″ (5.4 cm) wide. Disc-shaped, inflated; cross section of whorl rounded to almost square. Sutures complex. Outside of shell covered with strong, wavy ribs, widely spaced, with dense lines or weak, rounded ribs between them.

Age: Cretaceous.

Distribution: Texas, California.

Comments: The specimen illustrated is an internal mold of *Desmoceras* from which the shell material has completely dissolved away. This kind of preservation usually shows the suture pattern clearly. The complete pattern seen in the mature parts of the shell and exposed on the outside of the whorls is shown in the suture diagram. As ammonites grew, the sutures changed, starting simply and becoming more complex with age. Therefore it is important to compare such a drawing only with the pattern of a mature shell.

Suture:

193, 197 Placenticeras

Description: 9½″ (24 cm) long, 2⅜″ (6 cm) wide (pls. 193, 197). Disc-shaped, compressed laterally, with strongly overlapping whorls. Umbilicus small, about ⅐ shell diameter. Venter flattened, very narrow, smooth or with faint ribs, and with center groove. Sides slightly convex. Bumps along umbilicus in early whorls move up the middle of the sides on later whorls. Suture complex, with 3 prominent lateral lobes and 6–7 smaller lobes. Weak sculpture of faint ribs present in young stages, absent or obscure in later stages.

Age: Upper Cretaceous.

Distribution: Widespread in North America.

Comments: *Placenticeras* has one of the most complex suture patterns of all ammonoids. No cephalopods alive today have folded sutures, so paleontologists have puzzled over the function of such sutures for years, and there are various interpretations. The sutures are the traces of the junction of the septa and the outer shell—folded sutures mean fluted septa, and these

undoubtedly strengthened the shell.
The specimen illustrated in Plate 196
has been cut through the whorls,
exposing the septa and chambers.
Following the death of the animal,
chert crystals grew around the septa,
making them appear layered.

Suture:

203   Goodhallites

Description: 3″ (76 mm) long, ⅞″ (22 mm) wide.
Disc-shaped, compressed, narrower at
venter, with high ventral keel; coils
loose; umbilicus wide. Ornamentation
of weak, fine, close ribs that branch on
inner whorls; ribs strong on outer
whorl. Bumps around umbilicus and on
sides of venter become stronger with
age. Suture pattern complex; saddles
generally squarish, symmetrical, deeply
and sharply indented.

Age: Lower Cretaceous.

Distribution: Widespread in North America.

Comments: *Mortoniceras,* which is the same age, has
the same distribution, and is in the
same family as *Goodhallites,* is very
similar. *Mortoniceras* has stronger ribs
and lacks a keel in some species. The
saddles on the sutures of *Mortoniceras*
are more rounded in outline, and less
sharply incised.
Goodhallites is sometimes placed as a
subgenus in the genus *Prohysteroceras.*

Suture:

201 Metoicoceras

Description: 2″ (51 mm) long, ⅝″ (16 mm) wide.
Disc-shaped, compressed, whorls
strongly overlapping. Umbilicus small,
sometimes bordered by bumps. Venter
usually flattened or grooved, bordered
by distinct bumps where ribs cross.
Sutures relatively simple. Ribs faint if
umbilical bumps lacking, strong if
bumps present.

Age: Middle Cretaceous.

Distribution: Western North America.

Comments: Most Mesozoic ammonoids are
ornamented, commonly with ribs and
bumps like *Metoicoceras,* but sometimes
also with elaborate spines and frills.
These ornaments would clearly have
reduced streamlining, and might have
limited the animals' swimming
capacity. Some experts think their
highly ornamented shells mean that the
animals spent most of their lives
crawling on the sea floor.
The specimen illustrated is an internal
mold that shows the sutures clearly.
The ribs are also visible, since they
were present on the inside of the shell.

Suture:

186 Prionocyclus

Description: 2¾″ (70 mm) long, 1″ (25 mm) wide.
Disc-shaped, slightly compressed, with
flattened sides. Whorls only slightly
overlapping; umbilicus large. Venter
broad, with continuous narrow keel in
center that is not bumpy. Sutures fairly
simple for an ammonoid. Ribs well
developed, irregular in size and length
on outer whorl, bending toward
aperture as they approach keel. Bumps
present at ventrolateral margin and

fainter over sides.

Age: Middle Cretaceous.

Distribution: Western interior of the United States.

Comments: Although the age of *Prionocyclus* is given as Middle Cretaceous, the animal is actually found only in a single stage of that period—the Turonian—that lasted a few million years. Thus, when *Prionocyclus* is found in a rock, the age of that rock is pinpointed with remarkable precision.

Suture:

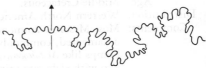

202   Texanites

Description: 3⅞″ (98 mm) long, 1¼″ (32 mm) wide. Disc-shaped, compressed, and high-whorled. Sides flat but converging to narrow, flat venter with small, continuous keel. Sutures fairly simple, with nonfolded ventral saddle and 2 additional saddles. Ribs strong, straight, and simple, with 5 bumps per rib on each side.

Age: Upper Cretaceous.

Distribution: Texas.

Comments: In North America, *Texanites* is found only in Texas. Worldwide, however, it occurs in Europe, Africa, Madagascar, southern India, and Japan. Many ammonoids show a similarly wide distribution. The *Texanites* specimen illustrated in Plate 202 is an internal mold.

Suture:

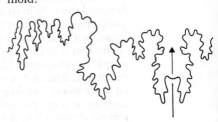

198 Sphenodiscus

Description: 3⅜" (86 mm) long, ¾" (19 mm) wide.
Disc-shaped, compressed, with strongly
overlapping whorls. Umbilicus small.
Venter sharply angled. Sutures have
many small, branching lobes and
saddles. Surface commonly smooth, but
covered with bumps in a few species.

Age: Upper Cretaceous.

Distribution: Widespread in North America.

Comments: *Sphenodiscus* is streamlined, and must
have been a good swimmer.
The specimen illustrated still has most
of the original shell material preserved,
but in one area the shell has been
chipped away to show the sutures.

Suture:

SUBCLASS COLEOIDEA
(Mississippian to Recent)

With the exception of the single
nautiloid genus *Nautilus,* all living
cephalopods belong to the subclass
Coleoidea. Included here are the
squids, cuttlefish, and octopods, as well
as the extinct belemnoids. Only the
belemnoids have a good fossil record.
The shells of coleoids are internal and
much reduced in size.

ORDER BELEMNOIDEA
(Mississippian through Eocene)

Fossil belemnoids are known
principally from the calcareous internal
shell, which is divided into 2 parts: a
solid, heavy *rostrum,* or guard, that
served as ballast, and a chambered
section with a tonguelike projection.

339 Belemnitella

Description: 6¾" (17 cm) long, ⅝" (2 cm) wide. Long, narrow cone, or guard, with blunt apex; solid except for inverse conical cavity at broad end extending into cone ⅓ distance to apex. Short, deep furrow, beginning near margin of cavity, extends down outside of cone toward apex; cone commonly splits longitudinally along this furrow. Growth layers like tree rings visible inside cavity and when cone is sectioned. Cone usually light brown with glassy appearance.

Age: Cretaceous.

Distribution: Widespread in North America.

Comments: The 2 specimens shown illustrate the state of preservation in which *Belemnitella* is usually found: with the hollow parts of the cones broken off. *Belemnitella* was a squidlike animal, probably closely related to the ancestors of modern squids and cuttlefish. The shell was internal, but with 2 parts. The part usually found as a fossil is the guard, described above. Another part of the shell, the phragmocone, is the equivalent of the external shells of cephalopods like *Bactrites*. The phragmocone fit into the cavity in the guard, and had septa and a ventral siphuncle inside. It projected forward from the guard as a slender, beaklike blade. Squid and cuttlefish have an internal shell like the phragmocone, but no guard. Although most *Belemnitella* fossils are only guards, in some belemnoid fossils from Germany 10 tentacles and the body form have been clearly preserved as thin carbon films, and confirm that belemnoids like *Belemnitella* were similar to squids.

Phylum Annelida
(Precambrian to Recent)

Annelids are worms with distinct heads, segmented bodies, and unsegmented tails. They are divided into 3 classes: the Polychaeta; the Oligochaeta, which include the familiar earthworms; and the Hirudinia, or leeches. Since oligochaetes and leeches have no hard parts and do not secrete calcified tubes, their fossil records are very poor; almost all annelid fossils are polychaetes.

CLASS POLYCHAETA
(Cambrian to Recent)

In marine sediments today, polychaetes are usually the most abundant and diverse animals visible to the naked eye. They build temporary or permanent tubes in the sediment, burrow through it, or live free on the surface.

Many polychaetes secrete calcified tubes, or have chitinous teeth that are easily fossilized. The fossil teeth are called *scolecodonts* and are always very tiny. Polychaete tubes are made from many different materials; those that fossilize best are composed of calcium carbonate, but others are made from hardened mucus—sometimes called parchment—or from sand or other small grains stuck together with a chitinlike cement. Most polychaete fossils are traces of burrowing activity in soft sediment, although sometimes burrows in hard materials are preserved. In a few cases it is possible to identify the genus that made the burrows in hard material by examining the shape of the burrows; but it is almost never possible to specify the genus of an animal that burrowed in soft sediment, because many other wormlike animals also burrow in sediment. It is not possible to identify most trace fossils even to phylum.

469 Serpula

Description: 1¼" (32 mm) long, ⅟₃₂" (1 mm) wide. Long, slender, coiled or contorted, irregularly tapering tube; usually attached to shell, at least at lower end. Surface ornamented with small concentric ridges.

Age: Silurian to Recent.

Distribution: Widespread in North America.

Comments: *Serpula* is an uncommon fossil before

the Cretaceous, but quite common afterward, and still common worldwide. The worm that lived in the tube was a "feather-duster," with a crown of feathery tentacles that it used to strain food out of the water. Many shells, living and dead, along the North American coast are encrusted with *Serpula*. Sometimes *Serpula* completely overgrows its host with a large mass of tubes.

206 Hamulus

Description: ¾" (19 mm) long, ¼" (6 mm) wide. Conical, elongated, gently curved tube, with 3–7 longitudinal ribs. Inner surface smooth. Small, immature tubes commonly attached; larger, mature tubes have usually broken away.

Age: Cretaceous.

Distribution: Widespread and common in the Atlantic and Gulf coastal regions.

Comments: Since *Hamulus* is extinct and the soft parts have never been found, we can only speculate about the kind of animal that lived in the tube. The construction of the tube, however, is basically like that of *Serpula* and *Spirorbis*, so *Hamulus* was most likely a similar worm, with a feather-duster crown of tentacles and a filter-feeding habit.

Spirorbis

Description: ¹⁄₃₂" (1 mm) long, ¹⁄₅₀" (0.5 mm) wide. Tiny, snail-like tube, coiled in flat spiral, cemented to a surface. Tube may coil to right or left. Ornamented with concentric lines or ridges; some species have bumps and spines.

Age: Ordovician to Recent.

Distribution: Widespread in North America.

Comments: Living *Spirorbis*, called the spiral tube worm, is a tiny, segmented worm with a set of feather-dusterlike tentacles that it uses to strain particles of food from the water. The animal builds its tube on almost any flat surface. *Spirorbis* is particularly common on eelgrass or turtlegrass in shallow water. The tubes look like tiny white spots unless seen under a microscope. As fossils, they are often found still cemented to mollusks and brachiopods. The marginal drawings show the exterior of 2 *Spirorbis* shells.

205 Rotularia

Description: ⅜" (10 mm) long, ⅛" (3 mm) wide. Small, coiled, snail-like tube, usually cemented to a surface. Tube approximately same diameter throughout its length, but ending in restricted aperture that extends out from coil. Outer surface of tube smooth or with concentric wrinkles. Some species have 1 or 2 keels on outer edge of tube.

Age: Upper Cretaceous through Eocene.

Distribution: Widespread in North America.

Comments: In the specimen illustrated, the end of the tube, which would have extended out from the coil, has broken off, so the restricted aperture is not visible. The keel can be seen on the outer coil. The tube is very similar to that of *Serpula*, except that it is coiled more tightly. It also resembles that of *Spirorbis*, except that *Spirorbis* is smaller and lacks the extended terminal part of the tube and the restricted aperture. All 3 of these tubes are placed in the family Serpulidae. If this classification is correct, then the occupant of the *Rotularia* tube was a small worm with a feathery crown of tentacles around its head, 1 of which was modified as a plug to seal off the tube.

Small Tubes of Uncertain Affinities
(Cambrian to Recent)

Many small fossil tubes cannot be classified because nothing is known about the animals that lived in them; the tubes themselves have no characteristics that definitely relate them to any known group.

ORDER TENTACULITIDA
(Lower Ordovician through Upper Devonian)

Tentaculitids are small, narrow, straight, conical tubes that reached their maximum diversity and abundance in the Middle Devonian. The tubes are composed of calcium carbonate, and vary in size from less than $\frac{1}{32}''$ to $3\frac{1}{8}''$ (1 to 79 mm) long. The apex ends either in a point or a small bulb. In most genera, the part of the tube near the apex is divided into chambers by *septa*. The tube wall is layered, not cellular like that of the cornulitids. All tentaculitids are found in marine rocks, and are most common in rocks that probably formed in shallow lagoons.

346 Tentaculites

Description: $2\frac{3}{8}''$ (60 mm) long, $1\frac{3}{4}''$ (44 mm) wide. Small, narrow, straight cone, with coarse rings more uniform toward apex. Area between rings usually has concentric ringlets or lines. Inner wall of cone also has rings. Walls thick, multi-layered, pierced by tiny radiating canals. Interior toward apex has slightly concave septa sealing off sections of tube. Apex closed and bluntly pointed.

Age: Lower Silurian through Upper Devonian (possibly also in the Lower Ordovician).

Distribution: Widespread in North America.

Comments: *Tentaculites* continues to elude classification. At various times, it has been considered a hydroid, a mollusk, tube-dwelling annelid, the spine of a brachiopod or echinoid, and the arm of a crinoid. Valid arguments can be made against all of these relationships. The shell is radially symmetrical, and thus was probably not dragged on the sea floor, for which mode of locomotion a flattened side would have been more efficient. While it is unlikely that the shell was carried flat on the bottom, it may have been carried upright, that is, with the apex pointing up and the head of the animal down. The animal may have fed on bottom detritus or small animals in the sediment.

Tentaculites almost certainly did not burrow, since the shell is never found oblique to bedding planes. *Tentaculites* may have been free-swimming, or have floated in the plankton like the pteropods, a group of tiny gastropods that float near the surface of the ocean. *Tentaculites* is quite often found clustered together in enormous numbers on bedding planes, the shells oriented by the currents that existed when the animals were alive. *Tentaculites* is most abundant in Silurian and Devonian limestones and shales.

ORDER HYOLITHIDA
(Lower Cambrian through Middle Permian)

Hyolithids are small, bilaterally symmetrical tubes. In cross section, most hyolithids are flattened on 1 side, with the other side arched. A few genera have tubes that are round or square in cross section, with ornamentation that varies from simple growth lines to longitudinal or concentric ribs. Some genera have small, closely spaced septa across the tube near the apex.

347 Hyolithes

Description: 1⅜" (35 mm) long, ½" (13 mm) wide. Tube oval, conical, tapering, triangular in cross section; having concentric growth lines, but lacking exterior ribs.

1 side of tube rounded, the other flattened, broad. Margin of flattened side projects somewhat beyond opposite margin. Growth lines on dorsal side bend upward in center, following shape of aperture. Interior has platforms across lower section. Operculum may be preserved as small plate with concentric growth lines that fits over opening at top of tube. Exceptionally well-preserved specimens show 2 slender, armlike structures curving out and down from sides of aperture.

Age: Lower Cambrian through Middle Permian.

Distribution: Widespread in North America.

Comments: Most paleontologists now put *Hyolithes* with the mollusks, as a separate class. The soft parts of the animal have never been found, but may have been similar to those of a snail or a cephalopod. *Hyolithes* may have crawled along the sea bottom, pulling its shell behind it. However, its mode of life has also been interpreted as sedentary (with the apex of the tube thrust into the sediment) or pelagic (with the animal swimming freely). Perhaps different species experienced different modes of life. The role of the "arms" is unclear, and many suggestions have been made about them: that they helped to prop the shell up off the surface; that they held the operculum open; or that they were supports for winglike flaps that helped to move the animal along the bottom in much the same way that skates and rays move along, flapping the extended sides of the body.

Hyolithes is abundant in the Cambrian, but fairly rare afterward. A common mode of preservation is the one illustrated: many shells, mostly broken at the apex and aperture, sorted according to size and clustered together by water currents. The opercula and "arms" are missing.

FAMILY CORNULITIDAE
(Middle Ordovician through Mississippian)

There are 4 genera of cornulitids. All are small to medium-size tubes that may be bent in various ways, are ringed, and taper to a point at 1 end. The walls of the tubes are thick, with a cellular structure, and composed of calcium carbonate. The length of the tube varies from ¼″ to 3⅛″ (6 to 79 mm); the width of the opening, from ⅛″ to ¾″ (3 to 19 mm). Cornulitids are almost always found in calcareous rock, especially limestone. They are usually attached to calcium carbonate shells or skeletons, and may be found isolated or in clusters.

345 Cornulites

Description: ¾″ (19 mm) long, ⅛″ (3 mm) wide. Small, trumpet-shaped tube, circular in cross section; gently tapering, small end usually bent; closed at lower end and either wholly or partly adhering to some foreign object such as a shell. Walls thick; prominent rings present, larger tubes often have longitudinal lines as well.

Age: Middle Ordovician through Middle Devonian.

Distribution: Common and widespread in North America.

Comments: The young *Cornulites* almost certainly travelled as free-swimming larvae. When they settled, they soon attached to some firm object, usually a coral, gastropod, bryozoan, or brachiopod. They then formed a tube that grew upward. The tube was smooth on the outside at first, but soon developed prominent rings and, in old age, longitudinal lines; just what kind of animal lived in it is not clear. The shape of the tube suggests a polychaete annelid, since many living polychaetes occupy very similar-looking tubes. The structure of the walls in *Cornulites* is different, however. The walls contain large, rounded or oval cellular cavities with thin walls. This structure most resembles the skeletons of stromatoporoids and those of some calcareous hydroids or the large, extinct group of foraminiferans called fusulines (single-celled animals related to amoebas). For these reasons, most paleontologists prefer not to assign *Cornulites* to a phylum.

There are 2 other North American genera of small tubes similar to *Cornulites*. *Conchicolites*, found throughout North America from the Upper Ordovician through the Lower Devonian, usually has tubes attached in clustered masses to the host. The walls of the tubes are thinner and the rings more rounded than in *Cornulites*. *Cornulitella*, found from the Middle Ordovician through the Mississippian throughout North America, is a solitary tube like *Cornulites*. It is attached along all of 1 side, and is somewhat flattened on that side. It never develops longitudinal lines, and its rings are usually more rounded than those of *Cornulites*.

Cornulites fossils are most often found in limestone and calcareous shale.

Phylum Arthropoda
(Cambrian to Recent)

The arthropods are the most diverse and abundant of all animal phyla. Included are the terrestrial scorpions, spiders, millipedes, centipedes, and insects, and the aquatic horseshoe crabs, shrimps, lobsters, and crabs. The extinct trilobites and eurypterids also belong here.

The distinguishing features of arthropods are an external skeleton, or *exoskeleton*, of chitin—calcified in some groups—a segmented body, and jointed appendages, 1 pair for each segment in primitive genera. Many of the more advanced genera have lost appendages on some segments. In the head region several segments are fused together, and appendages are usually modified as sense organs or for eating. The exoskeleton of an arthropod functions as a support for the soft tissues and a place for the muscles to attach much the same as the internal skeleton of a vertebrate does.

The fossil record of arthropods does not do justice to their diversity and abundance through time because most arthropods have exoskeletons that are not calcified. Those whose exoskeletons are calcified, the trilobites and some crustaceans, have very good records; but the insects, which are both terrestrial and uncalcified, are relatively rare as fossils.

There is a growing opinion among arthropod experts that the process of developing an exoskeleton and jointed appendages arose 3 times in the history of life, and that arthropods should therefore be divided into 3 phyla on this basis. Many biologists and paleontologists have already accepted this conclusion. The 3 proposed phyla of arthropods are the Uniramia, the Crustacea, and the Chelicerata. The Uniramia are named for the single branch (*ramus*) of the leg. This group includes the millipedes, centipedes, insects, and onychophorans. All Uniramia have jaws formed from the tip of a whole appendage. Of this group the insects alone have a significant fossil record, and will be the only class considered in this guide. Because they are rare as fossils, insects are treated in a general essay that follows the accounts of fossil genera. The Crustacea have legs with 2 branches, and jaws formed from inner segments of legs. The Chelicerata— scorpions, spiders, and horseshoe crabs —have a pair of *pincers* in front of the mouth, and the jaws move transversely in primitive genera. In more advanced genera the jaws move longitudinally. The position of the trilobites is presently unclear, and this group may eventually be considered a fourth phylum.

In this guide these 4 major groups of arthropods are treated as superclasses, consistent with the *Treatise on Invertebrate Paleontology*.

SUPERCLASS
TRILOBITOMORPHA
(Cambrian through Permian)

The trilobitomorphs include 2 classes, the trilobitoids and the trilobites. The trilobitoids are primitive, trilobitelike arthropods with thin, organic shells. They are known almost entirely from the Middle Cambrian Burgess Shale—found in a single locality in British Columbia—and are not treated here.

CLASS TRILOBITA
(Lower Cambrian through Permian)

Trilobites are the first unequivocal arthropods to appear in the fossil record. Their shells at the base of the Cambrian mark the beginning of the Paleozoic Era; they became extinct at the end of the Paleozoic. During these 345 million years, more than 1500 genera, with about 4000 species, evolved. For a group this large and long-lasting, trilobites show unusual consistency of form, especially notable if one compares them to crustaceans, which have evolved a tremendous variety of forms. The contrasting conservative nature of trilobite evolution may be explained as the failure of trilobites to develop specialization in the form and function of their legs. Thus trilobites could never exploit more than a limited number of ecological niches.

Trilobites are named for the 3 longitudinal lobes of their bodies. A raised middle lobe begins at the head and runs down into the tail, and a flatter lobe is present at each side of the middle one. The trilobite body is also divided into 3 transverse sections: the *head,* the *thorax,* and the *tail.* The dorsal, or upper, part of the skeleton was calcified, but most of the lower, or

ventral, part, including the legs, was
not, so most trilobite fossils are dorsal
only.

The head shield is flat or domed and
generally semicircular. In the center is
a raised lobe called the *glabella*. The
glabella usually shows signs of
segmentation, with transverse grooves.
To each side of the glabella are the
eyes, usually crescent-shaped and
convex toward the sides of the head.
Many trilobites have a border around
the head shield, set off by a shallow
groove. The lateral-rear corners of the
head shield commonly terminate in
backward-directed spines. Like all
arthropods, trilobites had to molt in
order to grow, so many have *facial
sutures* across the head shield where the
skeleton would split to allow the
animal to crawl out of its old covering.
The position of the facial sutures is
important in classification.

To the rear of the head is the thorax,
composed of 2–40 or more jointed
segments. Each segment has a raised
portion in the middle (the *middle lobe*)
and flattened and grooved portions,
called *pleura,* to the sides. The pleura
may terminate in points or spines, or
be rounded. Some segments may have
unusually long spines.

The tail lies to the rear of the thorax,
and is composed of a variable number
of segments that have been fused into
plate. The central lobe is almost alway
raised, and the pleura are defined by
grooves. There may be a border aroun
the tail similar to the head border. Th
tail may have 1 or more marginal
spines projecting backward or to the
sides.

The underside of the trilobite skeleton
was almost never preserved, since it w
not calcified, but a few rare
preservations indicate that there was a
pair of antennae on the head, and a
series of legs, a pair on each segment.
The legs had 2 parts—a long,

segmented part for walking, and another branch bearing a fine fringe of bladelike filaments, or *gills*, for respiration. All the legs appear to have been of the same structure, differing only in size. The head bore 3 or 4 pairs, and there was a pair on all body segments except the last, which sometimes bore a set of antennalike extensions.

Most Postcambrian trilobites could roll themselves up into a ball by bringing the ventral side of the tail into contact with the ventral side of the head. In the Cambrian, only the agnostids could do this. The others were prevented by the pleura from doing more than curling slightly. Some phacopids had "tooth and socket" structures for locking the tail shield under the head shield.

Most trilobite fossils are shed skeletons. In those trilobites where facial sutures crossed the head shield, the molts lack part of the head shield where it split at the suture line. Often the fossils are further disjointed, with the head shield, thorax, and tail shield all separated. Many trilobite fossils are *internal molds* of the dorsal skeleton. Since the skeleton was very thin, it is sometimes difficult to tell an internal mold from a skeleton.

ORDER AGNOSTIDA
(Lower Cambrian through Upper Ordovician)

The agnostids are a small but important order, quite different from other trilobites. All agnostids are very small, usually less than ⅜" (10 mm) long. The head and tail are connected by only 2 or 3 thoracic segments. Most agnostids were blind and lacked facial sutures.

268 Ptychagnostus

Description: ¼" (6 mm) long, ⅛" (3 mm) wide. Tiny, oval, with symmetrical, semicircular head and tail shields, both with prominent borders. Glabella small, oval, but slightly narrower at front; divided transversely into 2 lobes with front lobe much smaller than rear lobe. Eyes and facial sutures absent. Thorax has 2 segments. Tail shield has long, triangular, unsegmented middle lobe extending to border.

Age: Middle Cambrian.

Distribution: Newfoundland, New Brunswick, Montana, and western North America

Comments: *Ptychagnostus* is representative of an important order of trilobites called the agnostids. The agnostids are minute— all smaller than ½" (13 mm) in size— with head and tail shields connected by only 2–3 thoracic segments. The head and tail are often so similar that it is difficult to tell front from rear. Almost all agnostids were sightless and had no facial sutures.
The marginal drawing shows a head and tail of *Ptychagnostus*.

ORDER REDLICHIIDA
(Lower and Middle Cambrian)

This order may contain the most primitive of all trilobites. The head shield is relatively large and semicircular, usually with spines at the lateral-rear corners. In most genera, the glabella is divided into lobes by transverse furrows, and the eyes are elongated crescents. There are many thoracic segments and a tiny tail shield. If the top of the head has facial sutures, they cross the front margin to either side of the glabella, follow the eye ridges, and cross the lower margin to the inside of the corner spines.

292 Olenellus

Description: 3" (76 mm) long, 2¼" (57 mm) wide.
Large, oval, nearly flat, with large,
semicircular head shield and tiny tail
shield. Glabella has parallel sides, is
rounded in front, and reaches to narrow
border; 3 pairs of lateral furrows that
bend backward divide glabella into
lobes. Lateral-rear corners of head
shield extended into short or long,
backward-trending spines. Eyes large,
crescent-shaped, set close to glabella.
Facial suture marginal. Thorax has 18–
44 or more segments; first 14 wide,
remaining ones narrow. Wide segments
marked by furrows on pleura, which
end in backward-directed spines. 3rd
segment larger than others, with longer
spines. 15th segment has prominent
center spine pointing to rear, overlying
and partly covering narrow segments
that begin abruptly at 16th segment.
Tail shield is small, flattened plate
with rounded outline; length
approximately equal to preceding 3–4
segments combined.

Age: Lower Cambrian.

Distribution: Widespread in North America.

Comments: *Olenellus* appears to be a primitive
trilobite, mainly because of its
wormlike lower thorax and rudimentary
tail. In many fossils of *Olenellus* the
wormlike segments on the thorax are
lost, and for years paleontologists
thought the 15th segment with the
long spine was really the tail shield.
Olenellus has facial sutures that follow
the outer margins of the head shield
and thus are not evident on most
fossils. This kind of facial suture is
called protoparian. Some classifications
of trilobites have been based mainly on
the position of the facial sutures, but
most experts now consider this method
of classification inadequate, since the
same kind of facial suture has evolved
more than once.
Peachella, found in the Lower Cambrian

in the southwestern United States, is like *Olenellus* except for its more slender glabella and shorter, more bluntly rounded spines at the lateral-rear corners of the head shield. *Wanneria*, found in Lower Cambrian strata throughout North America, looks very much like an *Olenellus* specimen that has lost its lower thorax segments. However, *Wanneria* has a more expanded front lobe on the glabella than *Olenellus*, tiny bumps in the middle of the first 15 thoracic segments, and a small tail shield with a double lobe. *Wanneria* also lacks the enlarged 3rd thoracic segment.

293 Paedeumias

Description: 1½″ (38 mm) long, ⅞″ (22 mm) wide. Almost triangular, with large, broad, semicircular head shield and narrow, rapidly tapering thorax. Glabella narrow, not reaching to border of head shield, with rounded front lobe and 3 more lobes behind, set off by transverse furrows. Small ridge extends vertically from front lobe to border. Eyes large and crescent-shaped, set close to glabella. Facial sutures entirely marginal. Thorax has fairly wide, prominent middle lobe. Pleura relatively short, with prominent middle furrow, ending in spines that bend abruptly backward; 3rd segment has larger pleura and very long spines. 15th segment is very long, straight spine, partly covering numerous small segments. Tail inconspicuous.

Age: Lower Cambrian.

Distribution: Vermont, Pennsylvania, Nevada, British Columbia.

Comments: Occasionally trilobite fossils show evidence of damage and repair. A notable specimen of *Paedeumias* from British Columbia looks as though a predator took a bite out of the right side of the animal's thorax, nipping the 4th–6th segments, which then healed but remained shorter than the others. *Paedeumias* has many of the same characteristics as *Olenellus,* but can be distinguished by the median ridge that runs from the glabella to the anterior margin. *Paedeumias* also has a narrower, more tapering thorax than *Olenellus.* The specimen illustrated in Plate 293 is an external mold of *Paedeumias.* After death, the animal must have been covered with mud, which later hardened into rock. The shell disintegrated, but left the impression of its surface in the rock. The front part of the head is not well preserved, and the ridge connecting the glabella to the border is obscured. The long spine of the 15th segment and the lower part of the tail are missing. The marginal drawing shows a complete *Paedeumias* specimen.

---

288 Holmia

Description: 6¼" (16 cm) long, 4⅜" (11 cm) wide. Almost oval; semicircular head shield somewhat wider than thorax and much larger than tail shield. Glabella broadly rounded in front and expanded laterally; transverse furrows divide glabella into 4 lobes; 5th lobe has small spine in middle. Head shield surrounded by fairly wide border. Lateral-rear corners have large spines; smaller spines located on rear margin, midway between outer corners and center of glabellar base. Eyes large, crescent-shaped, located opposite glabellar middle lobe. Facial sutures at margins of head shield. Thorax has 16 segments; small, backward-directed spines in center of each segment, becoming longer near tail. Each pleuron ends in fairly short, backward-directed spine. Tail shield consists of 2

segments and rounded, terminal
portion. Outer surface covered with fine
crosshatch pattern.

Age: Lower Cambrian.

Distribution: Widespread in North America.

Comments: *Callavia*, a trilobite restricted to the
Lower Cambrian of eastern North
America, has many of the same
characteristics as *Holmia*. However, the
thorax of *Callavia* is wider relative to
the head shield than that of *Holmia*, the
front lobe on the glabella is more
pointed, and the spine on the last
glabellar lobe is very long, extending
halfway down the thorax.

When *Holmia* enrolled, curling its tail
under its body, its small spines would
have projected in a threatening
manner, protecting the animal against
danger.
The marginal drawing shows a
reconstructed specimen of *Holmia*.

294 Paradoxides

Description: 10½″ (27 cm) long, 6¼″ (16 cm)
wide. Large. Head shield broad,
semicircular; thorax long, narrow; tail
tiny. Glabella prominent, expanding
forward, reaching front margin. Border
well developed, particularly wide at
sides of shield. Lateral-rear corners
extend down in long, heavy spines that
reach to mid-thorax or farther. Eyes
moderately large, crescent-shaped.
Facial sutures extend from anterior
margin at sides of glabella, along
ridges over eyes, and down to rear
margin midway between corner spines
and base of glabella. Thorax long,
narrow, evenly tapered; pleura deeply
furrowed, ending in backward-curving
spines. Middle lobe raised, well
defined, extending down into very
small tail shield. Tail shield has flat,
unfurrowed, spatula-shaped area around
middle lobe.

Age: Middle Cambrian.

Distribution: Widespread in eastern North America.

Comments: Some species of *Paradoxides* were giants among Cambrian trilobites, reaching lengths of 18″ (45 cm).

Plate 294 shows a specimen that has been partially reconstructed. The "fossil" above the fracture that runs across the head shield is a plaster reconstruction of the missing part. The plaster glabella has not been made to reach the border furrow, as it should.

296 Bathynotus

Description: 2″ (51 mm) long, 1″ (25 mm) wide. Rectangular, elongate, with extremely long spines protruding from corners of semicircular head shield. Glabella large, broad, strongly convex, tapering toward front, and nearly straight across front end. Glabella reaches edge of front border and is divided into 4 lobes by transverse furrows; last lobe has bump or spine in center. Eyes very large, ¾ length of glabella. Facial sutures intersect front margin above glabella but follow inside curve of eyes to sides of midline, and intersect rear margin just inside bases of corner spines. Thorax has 13 segments with small bump in center of each; middle lobe very wide in anterior segments, tapering toward tail. As middle lobe decreases in width, length of pleural spines increases regularly from 4th to 11th segment; spines on 11th segment much longer than others. Tail shield broad, semicircular, smaller than head shield; 1 or 2 segments in middle lobe, which ends before margin. Pleural furrows absent. Skeleton surface has grainy texture.

Age: Lower and Middle Cambrian.

Distribution: Widespread in eastern North America.

Comments: Although the specimen illustrated in Plate 296 has lost its head, the thorax and tail are well preserved. The spines on the 11th thoracic segment are extremely long, and may mark the position of the animal's internal sex organs. No one knows how trilobites reproduced, but the process may have been similar to that of the horseshoe crab living today. If so, the female trilobite would have deposited eggs, perhaps stored in the enlarged thoracic segment, in a small hole in mud or sand. The male trilobite would have followed closely to fertilize the eggs with sperm. Horseshoe crabs cover over their nests, but trilobites may have left the nests to be covered by drifting sand.

The marginal drawing shows a complete specimen of *Bathynotus*.

ORDER CORYNEXOCHIDA
(Lower through Upper Cambrian)

Corynexochids are elongated and oval in outline. The head is semicircular, with well-developed lateral-rear spines. The glabella has straight, parallel sides, but may expand at the anterior end. The facial sutures cross the front margin to the sides of the glabella, follow the eye ridges, and cross the rear margin of the head shield inside the base of the lateral-rear spines. The thorax has 5−11 segments and the pleura end in spines. The tail, which is about the same size as the head, or smaller, may have a smooth margin or marginal spines.

274, 279 **Olenoides**

Description: 1⅝″ (41 mm) long, 1⅜″ (35 mm) wide (pl. 274); 1½″ (38 mm) long; ⅞″ (22 mm) wide (pl. 279). Oval; head and tail shields semicircular, almost equal in size and shape. Glabella has straight, almost parallel sides, but expands slightly toward front and reaches border. Eyes small, elongate, narrow. Lateral-rear corners of head shield extend down into spines. Facial sutures cross rear margin just inside of bases of corner spines. Thorax consists of 7–8 equal-size segments; middle of each segment may have bumps or backward-curving spines. Pleura have well-marked furrows. Tail shield has 5–11 middle lobe segments; pleura have distinct grooves within and between segments; margin has 4–8 pairs of fairly short spines, usually of equal length. Surface of skeleton has fine, bumpy texture.

Age: Middle Cambrian (possibly also in the Upper Cambrian).

Distribution: Widespread in North America.

Comments: Well-preserved specimens of *Olenoides* from the Burgess shale of British Columbia usually show the ventral surface with all the legs preserved, as well as 2 antennae in front and 2 antennalike appendages at the tail. Plate 274 shows only the head and part of a few thoracic segments of *Olenoides*. The left facial suture is clearly visible, but that part of the head shield outside the suture has broken away on the right side. The impression of the left lateral-rear spine is visible. Plate 279 shows an internal mold of a molted skeleton, with the outer edges of the head shield absent.

Kootenia, in the same family as *Olenoides* and very similar to it, is found throughout North America in the Cambrian. *Kootenia* can be identified by the absence of grooves within the pleura on the tail shield, and also by the corner spines of the head shield set closer to the thorax. *Marjumia,* found in the Middle Cambrian of the Rocky Mountains, is not closely related to *Olenoides,* but does look very similar to it. *Marjumia,* however, has a glabella

that is narrower at the front and ends
well before the border, 14 segments in
the thorax, and only 1–4 pairs of
spines on the tail shield.

284 Ogygopsis

Description: 3¼" (83 mm) long, 1¾" (44 mm)
wide. Oval; head and tail shields both
semicircular, but head shield slightly
larger. Glabella prominent, smooth,
with straight sides reaching to anterior
border. Moderately large spines extend
back from lateral-rear corners of head
shield. Ridges extend from front edges
of fairly small eyes to glabella. Facial
sutures cross front border well to sides
of glabella, follow curve of eyes, and
intersect rear margin just inside
bases of corner spines. Thorax has 8
segments, prominent but narrow
middle lobe, and pleura with grooves
parallel to transverse edges. Ends of
pleura bluntly pointed but not
extended into spines. Tail shield large,
with middle lobe clearly segmented
and extending almost to rear border;
spines absent. Tail pleura have grooves
within and between segments.

Age: Middle Cambrian.

Distribution: Rocky Mountains.

Comments: Plate 284 shows a specimen of *Ogygopsis*
preserved with the outer edges of the
head shield beyond the suture lines
missing. *Ogygopsis,* like all trilobites,
had to shed its outer skeleton in order
to grow. When this molting occurred,
the head shield split along the facial
sutures, and the animal crawled out of
its old skeleton. The part of the head
shield inside the sutures remained with
the thorax and tail, and the outer parts
of the head shield were scattered. Most
Ogygopsis fossils are molted skeletons
with incomplete head shields. Since a
trilobite molted 20 or more times
during its lifetime, the molted skeleton

is the most common fossil of this class. The marginal drawing shows a complete specimen of *Ogygopsis*.

282 Bathyuriscus

Description: ½″ (13 mm) long, ¼″ (6 mm) wide. Almost oval, tapering from wide, semicircular head shield to smaller tail shield. Front lobe of glabella bulges out slightly, touching border; 3–4 pairs of transverse furrows reach partway across glabella. Short, backward-directed spines at lateral-rear corners of head shield. Facial sutures cross front margin to sides of glabella, curve along eyes, and cross rear margin inside bases of corner spines. 7–9 segments in thorax, which is slightly narrower than head shield, and tapers only slightly to tail. Middle lobe well defined; pleura end in points or very short spines. Semicircular tail shield has 7 middle lobe segments that reach almost to prominent border, and pronounced pleural grooves within and between segments.

Age: Middle Cambrian.

Distribution: Widespread in North America.

Comments: *Bathyuriscus* lacks distinctive, specialized features, and so may be confused with other genera. The specimen illustrated has a poorly preserved head, with the corner spines broken away. The marginal drawing shows a complete specimen of *Bathyuriscus*.

Anoria is the same age and has the same distribution and basic shape as *Bathyuriscus,* but the facial sutures curve in front of the glabella before intersecting the front margin, and the thorax has a small bump in the middle of each segment and especially long, backward-directed spines on the 5th segment. *Glossopleura,* also the same age as *Bathyuriscus,* and with the same distribution, has a smoother glabella and a relatively larger, smoother tail shield with a particularly wide border.

303 Orria

Description: 2¾″ (70 mm) long, 1¼″ (32 mm) wide. Oval; head shield semicircular, slightly wider and shorter than tail shield. Glabella narrow, long, sides parallel, reaching to narrow anterior border; 4 pairs of indistinctly defined transverse furrows. Eyes small, set close to rear of glabella. Lateral-rear corners of head shield extended as short, backward-directed spines. Facial sutures intersect front margin close to sides of glabella, extend down along ridges over eyes, and intersect rear margin just inside of bases of corner spines. Area of head shield outside of sutures covered with irregular network of raised lines. Thorax has 9 segments, straight sides, and narrow middle lobe set off by broad furrows to each side; tapers evenly to tail. Tail shield long, containing 8–9 middle lobe segments, with strong grooves within and between pleura. Middle lobe ends well before border.

Age: Middle Cambrian.

Distribution: Western North America.

Comments: Plate 303 shows an internal mold of an incomplete and broken specimen of *Orria* that is probably a molted skeleton. The marginal drawing shows a complete specimen of *Orria*. The shape of *Orria* is very similar to that of *Ogygopsis*, but *Ogygopsis* has facial sutures that intersect the front margin well to the sides of the glabella, and a middle lobe that extends almost to the end of the tail shield.

285 Zacanthoides

Description: 2¼" (57 mm) long, 1⅜" (35 mm)
wide. Outline almost triangular; head
shield very wide; body tapering to
small, narrow tail. Glabella nearly
rectangular, ending before front border;
partially crossed by 4 pairs of furrows.
Spines on lateral-rear corners of head
shield; small spines on rear margin.
Eyes large, extending nearly to rear
margin. Facial sutures intersect front
margin opposite sides of glabella, curve
outward across border, then down to
beginning of lobe over eyes and along
lobe to end, when sutures make sharp
bend to intersect rear margin below
bases of spines. Thorax has 9 segments;
pleura have oblique furrows and sharp
points or spines. Some species have
long center spine on middle lobe of 1st
and 8th segments; others have spines
on all middle lobe segments. Tail
shield small, containing middle lobe
with 4–8 segments that extends nearly
or completely to rear margin. Pleura
directed backward and extended into
several pairs of marginal spines.

Age: Middle Cambrian.

Distribution: Widespread in North America.

Comments: *Holmia* looks like *Zacanthoides*, but has
entirely marginal facial sutures, shorter
eyes, and a more bulbous glabella than
Zacanthoides.

The specimen of *Zacanthoides* in Plate
285 appears to be the underside of a
molted skeleton with part of the tail
missing. The trilobite crawled out of
its old skeleton after the sides of the
head shield had broken off. The
marginal drawing shows a complete
specimen of *Zacanthoides*.

291 Albertella

Description: ⅞" (22 mm) long, ½" (13 mm) wide.
Narrowly oval; head shield wider than

thorax or tail. Glabella long, straight-sided, rounded in front and touching front border, with several pairs of partial transverse furrows. Lateral-rear corner spines long, extending well out from head shield; eyes large, crescent-shaped. Facial sutures intersect front margin to sides of glabella, curve in to front edge of ridge over eyes, follow ridge almost to rear margin, then bend and cross shield transversely to intersect rear margin below base of corner spines. Thorax has 7 segments; wide middle lobe usually has bumps or small spines in center; narrow pleura end in downward-curving spines; spines on 3rd or 4th segment particularly long. Tail shield narrow, with relatively wide, segmented middle lobe; pleural grooves bend down and end before smooth margin. 2 long spines extend from margin.

Age: Middle Cambrian.

Distribution: Western North America.

Comments: Many Cambrian trilobites, like *Albertella*, have enlarged pleura on one of the thoracic segments. These may have borne small holes that functioned in reproduction by releasing eggs or sperm.

Albertella is in the same family as *Zacanthoides*, and the 2 genera are very similar. *Zacanthoides*, however, has more spines on the tail shield margin, and lacks the large 3rd or 4th segment in the thorax. The marginal drawing shows a complete specimen of *Albertella*.

ORDER PTYCHOPARIIDA
(Lower Cambrian through Middle Permian)

This is the largest order of trilobites. Some experts now divide the group into several orders. Ptychopariids are variable in most characteristics. Most have facial sutures that cross the front margin of the head to the sides of the glabella, and the rear margin inside the bases of the lateral-rear corners. But some have facial sutures that cross the sides of the head shield above the lateral-rear corners, and in a few the facial sutures run along the margins and are not usually visible. The glabella is generally simple; it tapers toward the front, and most commonly ends well before the front margin. There are more than 3 segments in the thorax.

283 Elrathia

Description: 1″ (25 mm) long, ⅝″ (16 mm) wide. Broadly oval; head shield much larger than tail shield, but both semicircular. Glabella short, not reaching border, tapering in width toward front; transverse furrows curve downward. Short spines at lateral-rear corners of head shield. Facial sutures extend from front margin well to sides of glabella, curve above small eyes, then to rear margin just inside bases of corner spines. Thorax has 12–17 segments, sharply defined narrow middle lobe, and nearly flat pleura with distinct oblique furrows. Tail small, with few middle lobe segments; grooves on pleura indistinct. Middle lobe ends just before rear border.

Age: Middle Cambrian.

Distribution: Widespread in the western United States.

Comments: The specimen of *Elrathia* illustrated is a dorsal skeleton that has been molted. The marginal drawing shows a complete specimen of *Elrathia*. Other broadly oval trilobites with similar characteristics are *Elrathina*, *Ehmania*, and *Conocoryphe*. *Elrathina*, found in Middle Cambrian strata of western Canada, has a broader head

shield, a glabella extending closer to
the front border, and a much smaller
tail. *Ehmania,* found in Middle
Cambrian regions of the northwestern
United States, has a slimmer outline, a
glabella reaching almost to the front
border, and a slightly smaller tail.
Conocoryphe, from the Middle Cambrian
rocks of eastern North America, lacks
eyes, has furrows from the front corners
of the glabella to the border, and has
bumps on the head shield.

297 Tricrepicephalus

Description: 2⅜″ (60 mm) long, 1¾″ (44 mm)
wide. Broadly oval; head larger than
tail. Glabella tapering forward,
rounded on front, ending well before
front margin; glabellar lobes marked by
2–3 pairs of short transverse furrows. 3
evenly spaced, round or elliptical pits
in border furrow in front of glabella.
Small bump or spine in middle of rear
glabellar lobe. Lateral-rear corners of
head shield have long, backward-
directed spines. Eyes extend to front
edge of glabella. Facial sutures cross
front margin to sides of glabella, follow
ridges over eyes, and cross rear margin
at inside of bases of spines. Thorax has
12 segments. Tail shield small, oval,
broader than long; 3 rings on middle
lobe, which is wider than pleura.
Pleura extend down to form 2 long,
hollow, rounded spines at rear corners
of tail. Border below middle lobe
continues under spines.

Age: Upper Cambrian.

Distribution: Widespread in North America.

Comments: The specimen illustrated in Plate 297
is the underside of the tail and part of
the thorax. The marginal drawing
shows a complete specimen of
Tricrepicephalus.
Crepicephalus is in the same family as
Tricrepicephalus and appears similar. The

eyes of *Crepicephalus* are set farther forward relative to the glabella than those of *Tricrepicephalus,* and the animal lacks the pits in the front border just above the glabella and the bump or spine on the last glabellar lobe. It has 4–5 rings on the middle lobe of the tail shield.

302 Dikelocephalus

Description: ¾" (19 mm) long, 1¼" (32 mm) wide. Large, with semicircular head shield and finlike tail shield. Glabella low, 4-sided, its front almost straight across, with 2 pairs of transverse furrows. Last furrows cross glabella, bending slightly down near middle. Area above glabella about ⅓ length of head. Eyes small, set close to glabella. Facial sutures join at midpoint of front margin, curve to sides, then down to ridges over eyes, and intersect rear margin midway between short corner spine and base of glabella. Thorax probably has 12 segments. Tail shield broadly oval, with 4 distinct rings on middle lobe, which ends well before rear margin. 10–12 pleural ribs merge into broad, flat, semicircular border. Lateral-rear corners of tail have short, flat spines.

Age: Upper Cambrian.

Distribution: Central and western North America.

Comments: *Dikelocephalus* is widely distributed but never common, and never found in a state of complete preservation. The specimen illustrated in Plate 302 is only a tail, but the marginal drawing shows a reconstructed specimen of *Dikelocephalus.*
Briscoia, found in the Upper Cambrian throughout North America, has a very similar tail, although the small spines at the outer lower corners of the tail are absent. *Briscoia* shows an extra, faint ring near the end of the middle lobe.

287 Olenus

Description: ⅜" (10 mm) long, ¼" (6 mm) wide.
Flat; almost triangular in outline; with
broad head shield. Glabella straight-
sided, rounded at front, ending well
before front border; 2–3 pairs short
glabellar furrows bend backward. Eyes
small, set about midway between sides
of glabella and lateral margin. Spines
extend from lateral-rear corners of head
shield. Facial sutures cross front margin
above eyes, follow ridges over eyes,
then extend to rear margin, crossing
about ⅓ distance between spines and
base of glabella. Thorax has 13–15
segments, middle lobe that tapers
evenly toward tail, and short,
downward-curving spines on pleural
extremities. Tail shield small,
triangular to semicircular, with narrow,
raised border, sometimes with minute
marginal spines.

Age: Upper Cambrian.

Distribution: Widespread in North America.

Comments: Very tiny *Olenus* specimens have been
found, some less than ½₃₂" (1 mm)
long. These little *Olenus* are immature,
and quite different in shape from the
adults. They are rounded, consisting
mostly of a head with a tiny tail, and
are more spiny than the adults. They
were probably planktonic—that is,
they floated near the surface of the
ocean. The spines would have helped to
keep them from sinking.

The specimen illustrated in Plate 287
lacks the sides of the head shield
because it is a molted skeleton.
The marginal drawing shows a
complete specimen of *Olenus*.

286 Triarthrus

Description: ¾" (19 mm) long, ⅜" (10 mm) wide.
Elongate-oval; head shield large,
semicircular; tail shield small. Glabella

almost rectangular, slightly longer than wide, with 2–4 pairs transverse furrows, and 1–2 pairs shorter furrows near front end in some species. Spines at lateral-rear corners absent. Eyes small. Facial sutures intersect front margin near sides of glabella, follow curve of ridges over eyes, and intersect rear margin just inside lateral-rear corners. Area of head shield outside sutures very narrow. Thorax with 13–16 segments; middle lobe wider than pleura, pleural extremities obliquely truncated or rounded; spines and bumps on middle lobe absent. Tail shield small, semicircular, with 3–5 rings on middle lobe and evenly rounded margin.

Age: Ordovician.

Distribution: Widespread in North America.

Comments: *Triarthrus* specimens from the Utica shale in New York are sometimes composed of pyrite, an iron sulfide mineral. The pyrite has replaced the entire *Triarthrus* skeleton, both the mineralized and the unmineralized parts. In such specimens, the small jointed legs and antennae attached to the ventral side of the animal are preserved. Because pyrite is opaque to X rays, these fossils can be X-rayed while still enclosed in the shale, and the X-ray film will show fine details of the anatomy.

---

281 Cedaria

Description: ⅜″ (10 mm) long, ¼″ (6 mm) wide. Oval; head and tail shields both semicircular, about equal in size. Glabella smooth, tapering to rounded front, ending well before front margin. Border on head shield well defined and wide. Lateral-rear corner spines of medium length. Medium-size eyes located slightly forward of mid-glabella. Facial sutures follow front margin toward side, then bend to follow eye ridges. At lower end of ridges, sutures turn 90° toward sides, cross side border, then turn back to intersect rear margin just inside bases of spines. Thorax has 7 segments; middle lobe narrower than blunt-ended pleura. Tail shield has long, low, tapered middle lobe with 5–6 rings that reaches almost to wide rear border; 4–5 tail pleura have well-defined grooves.

Age: Upper Cambrian.

Distribution: Widespread in North America.

Comments: Isolated head shields of *Cedaria* and other trilobites are commonly found, and it is usually possible to identify a genus from the head alone. Often the head shields are not complete, lacking the side areas outside the facial sutures that would have split off when the trilobite molted, shedding its skeleton. These head shields may seem like strange fossils until one realizes what parts are missing.

The marginal drawing shows a reconstructed specimen of *Cedaria*.

269, 271, 307 Homotelus

Description: 5″ (127 mm) long, 2¾″ (70 mm) wide (pl. 269); ¾″ (19 mm) long, ⅜″ (10 mm) wide (pl. 271); 2¾″ (70 mm) long, 2⅜″ (60 mm) wide (pl. 307). Oval; middle lobe poorly defined; head and tail shields semicircular, equal in size and shape. Glabella indicated by weakly raised, smooth area near rear of head shield that fades out at front margin; border and spines absent. Eye moderate in size. Facial sutures join at center of front margin, then follow margin to just above eyes, bend to follow ridges over eyes, then bend outward to intersect rear margin ⅓ distance from side to midline of head shield. Thorax has 8 segments and

slightly raised middle lobe twice width
of pleura. Tail shield smooth; poorly
defined middle lobe tapers rapidly, has
no rings, and ends slightly before faint,
flattened border.

Age: Upper Ordovician.
Distribution: Ohio.
Comments: *Homotelus* belongs to a subfamily in
which the loss of the middle lobe has
progressed substantially. Plate 269
shows a large, complete skeleton of
Homotelus; Plate 271 shows a group of
skeletons. These latter, beautifully
preserved specimens are banded
together, probably pushed into a
cluster by water currents. The skeletons
do not represent molts, since their head
shields are intact. They form part of a
large assemblage that covers an
extensive area of a single rock layer.
Some catastrophe apparently killed all
these *Homotelus* at once. The undersides
of several of the specimens are turned
up; in 2 of these, the hypostoma can be
seen lying on the underside of the head
shield. The hypostoma is a calcified
plate, usually rounded, that covered the
mouth area and, in *Homotelus,* has 2
rear prongs. Plate 307 shows an
isolated hypostoma with a bit of the
front end broken away. Hypostomas,
known from about ⅓ of all trilobite
species, are usually preserved this way,
separated from the rest of the skeleton.
2 other trilobites are very similar to
Homotelus. Isotelus, found throughout
North America in the Middle and
Upper Ordovician, has a concave border
around the head shield. 1 specimen,
found in Ohio, shows traces of a color
pattern, with dark areas along the
middle lobe of the trunk. *Isoteloides,*
found throughout North America in
the Lower Ordovician, has a well-
defined, flattened border around the
head shield, a better-defined glabella
than that of *Homotelus,* which ends at
the wide front border, small spines at
the lateral-rear corners, a relatively

narrow middle lobe on the thorax, and some faint traces of rings and ribs on the tail shield.

301 Pseudogygites

Description: ⅞" (22 mm) long, 1¼" (32 mm) wide. Oval; head and tail shields about equal in size and shape. Glabella expanded forward, reaching to wide front border. Short spines extend back from lateral-rear corners of head shield. Eyes small, located opposite middle of glabella. Facial sutures meet at center of front margin, curve around front of glabella, follow ridges over eyes, and intersect rear margin about ⅓ distance from side of head shield to midline. Middle lobe of thorax same width as pleura. Tail shield has prominent, faintly ringed middle lobe that extends almost to smooth rear border; pleural ribs of tail strong, reaching almost to border.

Age: Middle and Upper Ordovician.

Distribution: Widespread in North America.

Comments: Trilobite fossils are occasionally found aggregated. The *Pseudogygites* fossils illustrated in Plate 301 are mostly tail shields. What event would have produced such an aggregation? What happened to the thoraxes and heads? If the tails did, indeed, come from skeletons that had been molted, the head shields would have been split into 3 sections during that molt. A small disturbance may have caused the thoraxes to separate from the heads and tails. With each skeleton in 5 pieces, even a slight current could winnow out certain pieces, sorting them by their resistance to the water movement. All the tail shields may have been moved a small depression that was subsequently filled in with mud, preserving the tails. Other depression may have contained the head shields and the thoraxes.

270 Illaenus

Description: 1¼" (32 mm) long, ¾" (19 mm) wide. Oval; head and tail shields smooth, wide, short, about equal in size and shape. Head shield a rounded triangle; glabella indistinct, hourglass-shaped, lacking front boundary. Border absent. Eyes large, near side margins. Facial sutures cut off corners of shield, following curve of eyes. Thorax has 10 segments; indistinct middle lobe equal in width to pleural lobes, which are not furrowed. Tail shield contains short, indistinct middle lobe that narrows toward rear; rings, furrows, and borders absent.

Age: Ordovician.

Distribution: Widespread in North America.

Comments: Some species of *Illaenus* grew to 16" (41 cm) long, and were giants among trilobites. *Illaenus* is called a smooth trilobite because its middle lobe has almost disappeared.
Plate 270 shows an internal mold of *Illaenus*. The marginal drawing shows a reconstructed specimen.
Illaenus and other trilobites with large, smooth heads and tails, like *Bumastus* and *Homotelus*, may have lived with their tails and thoraxes buried vertically in the sediment and their heads resting horizontally on the surface. This position would have protected the animal's body and stabilized it in strong currents. Such animals may have filter-fed in this position, causing water to run under their head shields and seizing small food particles with specially modified legs.

272 Bumastus

Description: 1½" (38 mm) long, 1" (25 mm) wide. Oval; head and tail shields smooth, semicircular, about equal in size. Middle lobe of head shield very wide, slightly raised, indicated by faint furrows to sides. Furrows curve out slightly near rear margin of head shield, then in toward center, then out again, ending well before front margin. Corner and border spines absent. Eyes large, set near lateral-rear corners. Facial sutures cut off corners of shield that contain eyes. Thorax has 8–10 segments; middle lobe may be absent or indistinctly indicated by longitudinal furrows. Tail shield perfectly smooth, with middle lobe, furrows, and border absent.

Age: Middle Ordovician through Silurian.

Distribution: Widespread in North America.

Comments: James Hall, one of the first and most famous of all American geologists, found *Bumastus* for the first time in 1839, in a gorge near Lockport, New York, that had been cut for the Erie Canal. This discovery was made only months after *Bumastus* was first described from British rocks. Hall was delighted to find British fossil genera in America, since it meant that North American strata could be correlated in age with rocks across the Atlantic Ocean. Hall took the famous English geologist Charles Lyell to the same gorge during Lyell's first visit to North America in 1841, specifically for the purpose of finding *Bumastus.*

The specimen illustrated in Plate 272 is an internal mold of *Bumastus.* The marginal drawing shows a reconstructed specimen.

277 Griffithides

Description: 1⅝″ (41 mm) long, ¾″ (19 mm) wide. Oval; head shield slightly larger than tail shield. Head strongly convex, semicircular in outline. Glabella large, expanded in front, reaching to margin. Transverse furrow at base of glabella; other furrows cut off basal corners to form long, triangular lobes. Border on head shield wide, set off by sharp furrow. Lateral-rear corners rounded or with short spines. Facial sutures cross front margin to either side of glabella, curve out to follow eye ridges, and bend out slightly at rear border to intersect rear margin midway between corner and beginning of glabella. Thorax has 8 segments, with middle lobe as wide as pleural lobes, and pleura with grooves and truncated ends. Tail shield convex, semicircular; middle lobe broad, with more than 6 rings. Incised lines in pleural area of tail show pleural and interpleural furrows. Middle lobe does not extend to margin; well-defined border absent. Surface texture grainy.

Age: Lower and Middle Mississippian.

Distribution: Widespread in North America.

Comments: *Phillipsia,* a trilobite of the same age and with the same distribution as *Griffithides,* is closely related to it. *Phillipsia* has a glabella with almost parallel sides that extend only to the front border. The small, triangular lobes are shorter than those of *Griffithides,* and resemble quarter-circles.

The specimen of *Griffithides* illustrated in Plate 277 has a crushed head, and the fossil is surrounded by the round stem plates of crinoids. The marginal drawing shows a reconstructed specimen of *Griffithides.*

298 Cryptolithus

Description: ½″ (13 mm) long and wide. Head shield wide, thorax short, tail shield small, triangular. Head shield crescent-shaped, with wide border, sloping outward. Band around border has radially arranged, large, round pits. Outer rows have small radial ribs between pits; inner rows have concentric ridges separating pits. Glabella large, raised, bulbous, extending forward to border, with raised lobes to each side. Glabella and raised lobes to each side smooth, except for slightly indented furrows on glabella and, in some species, very small eye-bumps in centers of side lobes. Spines twice length of exoskeleton, extending backward from lateral-rear corners of head shield. Single spine projects backward over thorax from rear of glabella. Facial sutures follow margin of head shield, cutting across bases of corner spines. Thorax contains 6 segments; middle lobe straight, with furrows to side broad and shallow. Tail shield short, wider than long, with many fused segments and narrow, bent border.

Age: Ordovician.

Distribution: Quebec, New York, New Jersey, Pennsylvania, Virginia, Ontario, Kentucky, Oklahoma.

Comments: *Cryptolithus* was blind and may have had a life similar to that of living horseshoe crabs, which it resembles in shape. Horseshoe crabs plow through sand and mud, finding small bits of food, living and dead, in the sediment. The pits around the head shield of *Cryptolithus* are in fact perforations, and may have functioned as a sieve. As *Cryptolithus* slapped its head shield against the sediment, water would rush through the perforations, and tiny food particles would be strained out.
2 genera closely related to *Cryptolithus* and very similar to it are *Cryptolithoides* from the Middle Ordovician of Oklahoma and Texas, and *Tretaspis,* from the Middle and Upper Ordovician of eastern North America. *Cryptolithoides* can be distinguished by the more rectangular head shield and the irregular arrangement of all but the 2 outer rows of pits. *Tretaspis* has a glabella with a very round, swollen frontal lobe clearly set off from the

glabellar base, and lacks the median
spine extending from the glabellar
base.

ORDER PHACOPIDA
(Lower Ordovician through Upper
Devonian)

Many of the most common
Postcambrian trilobites are in this
order. Most phacopids have facial
sutures that begin on the front margin
of the head shield and, after following
the eye ridges, end above the lateral-
rear corners. In some, the sutures cross
the rear margin inside the lateral-rear
corners. The glabella commonly
becomes wider in front, but may be
tapered. There may be a narrow area in
front of the glabella. The thorax has
8–19 segments. The tail is typically
medium to large in size, but is small in
a few early genera.

305 Cheirurus

Description: ¾″ (19 mm) long, 1¼″ (32 mm) wide.
Oval; head shield semicircular, larger
than tail shield; surface covered with
small bumps. Glabella expanded
forward, overhanging front margin,
marked by 4 pairs of transverse furrows;
2nd-to-last furrow extends backward
and intersects last furrow, forming X
over rear of glabella. Head shield
has border on sides and rear; lateral-rear
corners have short spines. Facial sutures
intersect front margin at sides of
glabella, extend along ridges over eyes,
and bend at right angles just below
eyes to intersect side margins just
below eye level. Thorax has 11
segments; pleura have oblique furrows
near fairly wide middle lobe, and
pointed extremities that do not bend

backward. Tail shield small, with 3 pairs of heavy radiating pleural spines, almost equal in size, and a smaller middle spine.

Age: Upper Ordovician through Silurian.

Distribution: Widespread in North America.

Comments: The specimen shown in Plate 305 is a tail. Whole skeletons and even entire head shields of *Cheirurus* are very rare in North America. The tail, however, is a common fossil and so distinctive, with its bumpy surface and stout spines, that generic identification can usually be made from this body part alone. The hypostoma, or ventral shield, of *Cheirurus*—a flat, oval to triangular plate about the length of the glabella —is also occasionally found as a fossil. The marginal drawing shows a reconstructed specimen of *Cheirurus*.

295 Ceraurus

Description: 1¼″ (32 mm) long, ⅝″ (16 mm) wide. Oval to somewhat triangular; head shield much wider than thorax; tail very small except for very long, stout spines. Surface covered with coarse bumps. Head shield trapezoidal, with front and rear margins almost parallel. Glabella large, expanding forward and overhanging front margin, marked by 3 pairs of short, transverse furrows that do not meet. Lateral-rear corners have long spines that extend out and back, making 45° angle with midline of body. Thorax usually has 11 segments; pleura end in short spines that point outward on 1st segments but recurve toward rear in lower segments. Tail shield distinctive, consisting of only a few middle lobe rings and 2 very heavy, sharply pointed spines to either side of middle lobe that extend out and back at 45° angle.

Age: Middle and Upper Ordovician.

Distribution: Widespread in North America.

Comments: The specimen in Plate 295 is complete.
As in the case of *Cheirurus,* however,
most recognizable fossils of *Ceraurus* are
only tail shields. *Cheirurus* and *Ceraurus*
are very closely related genera, but
fortunately their tails are distinctively
different. That of *Cheirurus* has 7
marginal spines; that of *Ceraurus* has
only 2 very long spines.

276, 300 Encrinurus

Description: ¾" (19 mm) long, 1" (25 mm) wide
(pl. 276); ⅝" (16 mm) long and wide
(pl. 300). Teardrop-shaped; head shield
broad, semicircular, covered with
coarse bumps; tail shield long and
tapered. Glabella large, raised,
expanded forward, overhanging front
margin; 3 pairs of short, shallow,
transverse furrows mark glabella. Eyes
small, stalked, located near front
margin. Thorax tapers evenly to tail,
has 11–12 segments; pleura end
bluntly. Tail shield long, narrow,
triangular, with numerous rings,
usually incomplete, on middle lobe,
and 5–10 pairs of backward-curving
ribs on pleural lobes. Small bumps
usually present on middle lobe, and
sometimes on ribs.

Age: Middle Ordovician through Silurian.
Distribution: Widespread in North America.
Comments: The specimens of *Encrinurus* illustrated
show part of an isolated head shield
(pl. 276) and a tail shield (pl. 300). In
Plate 276, a second trilobite head is
shown emerging below the complete
head at the top, separated by a double
line. Trilobite heads are often preserved
in this fashion. *Encrinurus* tail shields
are often found isolated from the rest of
the skeleton, and are distinctive
enough that they can usually be
identified. The bumps on the head
shield of *Encrinurus* are very prominent,
and may have served a sensory function

since bumps on the skeletons of living arthropods often have sensory hairs growing out of the center.
The marginal drawing shows a reconstructed specimen of *Encrinurus*.

280 Calymene

Description: 1⅛″ (28 mm) long, ⅝″ (16 mm) wide. Broadest across semicircular head shield; thorax tapering to small, semicircular tail shield. Glabella convex, raised well above rest of head shield, tapering to front border. Furrows on glabella pinch off 3 pairs of rounded lobes on sides; largest pair of lobes near base of glabella, 2 smaller pairs forward. Border convex, separated from rest of shield by distinct furrow, and extended completely around head shield. Lateral-rear corners of head shield rounded. Eyes small, located at level of middle glabellar lobes. Facial sutures cross front margin and border at sides of glabella, extend along ridges over eyes, and cross diagonally to lateral-rear corners. Thorax usually has 13 segments; thorax and middle lobe taper evenly; pleural extremities rounded. Tail shield has 6 complete rings in middle lobe and 6 deep grooves on each side that extend to borderless margin. Additional shallow grooves present near margin.

Age: Lower Silurian through Middle Devonian.

Distribution: Widespread in North America.

Comments: *Calymene* is often found enrolled—that is, with the tail shield turned down and around so that the lower surface is pressed against the underside of the head shield. This posture, common in many trilobites, must have aided the animal in self-defense. Only a small part of the underside of a trilobite was mineralized, and thus it is on the underside that it was most vulnerable to predators. A typical predator probably tried to turn a trilobite over, exposing the "soft underbelly," in order to begin its meal. Quick enrollment by the trilobite must have been effective in deterring at least those predators not large enough to gulp the trilobite down whole.

The marginal drawing is a side view of *Calymene* that shows the shape of the border of its head shield.

275 Flexicalymene

Description: ¾" (19 mm) long and wide. Head shield semicircular, broad; thorax tapering; tail shield small, semicircular. Glabella convex, tapering to front border, with 3 sets of lateral lobes, largest at base and decreasing in size toward front. Distinctive front boundary on glabella absent; deep but broad furrow separates glabella from border, which is sharply raised and liplike. Facial sutures extend from front margin well to sides of glabella, along eye ridges, and across to rounded lateral-rear corners. Thorax has 12–13 segments. Tail shield has middle lobe extending to margin, and prominent pleural grooves. Tail shield border absent.

Age: Middle and Upper Ordovician.
Distribution: Widespread in North America.
Comments: Plate 275 shows an enrolled *Flexicalymene* specimen. *Flexicalymene* can easily be confused with *Calymene*, especially since both genera are often found enrolled. The borders around the animals' head shields are distinctive, however. The furrow setting off the border in *Calymene* is sharper than that of *Flexicalymene*, and the border is not raised as much as it is in *Flexicalymene*. The marginal drawing is a side view of the head of *Flexicalymene*, and illustrates its raised border.

273 Dipleura

Description: 1¾" (44 mm) long, 1⅜" (35 mm) wide. Coffin-shaped; head and tail shields triangular, nearly equal in shape, but head shield somewhat larger than tail shield. Head shield more than twice as wide as long. Glabella smooth, almost square, only slightly longer than wide. Glabella ends well before margin. Border absent. Eyes small. Facial sutures extend from front margin above glabella, along eye ridges, and cut side margin just above rounded lateral-rear corners. Thorax has very wide, indistinct middle lobe, raised only slightly above pleura. Tail shield smooth, longer than wide, with many rings in broad, tapering, faint middle lobe. Middle lobe does not extend to rear margin. Very faint interpleural grooves cut sides of tail shield.

Age: Middle Devonian.

Distribution: Eastern and central North America.

Comments: The tail of the *Dipleura* specimen in Plate 273 is rolled under the body. The marginal drawing shows a reconstruction of *Dipleura*. *Dipleura* and *Trimerus* are very similar. Some trilobite experts now put them in the same genus, with *Dipleura* considered a subgenus of *Trimerus,* and called *Trimerus (Dipleura)*. *Dipleura* can be distinguished from *Trimerus* most easily by its tail shield, which is smoother, with only a faint indication of a middle lobe and segmentation. The shape of the heads of the 2 animals is also somewhat different, with the head of *Dipleura* shorter and broader, and the sides of the glabella more nearly parallel.

278 Phacops

Description: 1⅝" (41 mm) long, ¾" (19 mm) wide. Oval; head and tail shields bluntly rounded; sides straight, parallel. Head shield almost completely covered by glabella and eyes. Glabella large, bulbous, broadest at front, reaching forward almost to margin. Border and distinct marginal furrow present, sometimes obscured by glabella in front. Eyes relatively large, crescent-shaped, located close to glabella, slightly back of glabellar midline. Lateral-rear corners curve backward but are broadly rounded. Glabella covered with coarse, flattened bumps. Thorax has 11 segments, all about equal in size; middle lobe arched, bordered by distinct furrows. Sides of segments with nearly straight margins, rounded at ends. Tail shield similar in shape to head shield but smaller; transverse grooves indicate 10–11 segments; border and spines commonly absent.

Age: Devonian.

Distribution: Widespread in North America.

Comments: *Phacops* is the best known of all North American trilobites and is often beautifully preserved. The eyes are very prominent and positioned so that *Phacops* could see in nearly all horizontal directions at once. The eyes are different from those of most trilobites, with rounded instead of hexagonal lenses. Many *Phacops* fossils are found enrolled, with the undersides of the head and tail shields pressed together.
Reedops is a very similar trilobite of the same age found in Tennessee and Oklahoma. *Reedops,* however, has a more bulbous glabella that extends far forward over the front margin, and also has somewhat smaller eyes.

290 Dalmanites

Description: 2¾" (70 mm) long, 1⅝" (41 mm) wide. Oval; head wider than tail. Head shield crescent-shaped, with lateral

spines curving down and extending to 5th trunk segment. Front margin may have single forward-pointing spine at center. Glabella bulbous, inverted-pear-shaped, strongly indented by 2 or more pairs of transverse furrows. Head shield border wide, flat, with distinct furrow. Eyes relatively large, crescent-shaped, located close to glabella and slightly back of mid-length. Facial sutures join at center of head shield close to glabella, follow front margin of glabella, then along ridges over eyes before curving out to side margins, well above posterior margin of head shield. Thorax has 11 segments of nearly equal size; middle lobe arched, with well-incised grooves to either side. Pleura have nearly straight margins and rounded or bluntly angled extremities. Tail shield long, semicircular, with transverse grooves indicating 11–16 segments. Middle lobe of tail slender, raised; border absent; long or short median spine points backward. 6–7 pairs of ribs on pleural areas.

Age: Silurian.

Distribution: Eastern and central North America.

Comments: The eyes of *Dalmanites* are large, with the long axis vertical, and set high on the head shield. *Dalmanites* could thus see in nearly all horizontal directions at once. The advantage to seeing well probably was one of defense rather than offense. Trilobites were not predators that saw prey and pursued it; they did not have jaws, and their mouths were very small. If they ate live food, it must have been tiny—a creature like a worm that they would find in the sediment as they plowed along. In fact, its high-set eyes probably allowed *Dalmanites* to see above the sediment while it hunted for bits of food with its legs. Some trilobites, perhaps including *Dalmanites,* may actually have ingested quantities of sediment, and then digested out the organic material that it contained.

3 genera resemble *Dalmanites* and might be confused with it. *Odontochile,* from the Lower and Middle Devonian throughout North America, has facial sutures set forward slightly from the glabella, and a tail shield with more segments, 16–22 rings and 12–15 pairs of ribs. *Synphoria,* from the Lower Devonian of eastern North America, has 2 pairs of deep pits instead of furrows on the glabella, and a head shield margin that may be smooth or scalloped, or have a single median projection like *Dalmanites.* Its tail has about 15 rings. *Odontocephalus,* from the Middle Devonian of eastern and central North America, has a glabella like *Synphoria,* but with a wide front margin perforated with oval holes. Its tail shield has 2 backward-directed spines.

289 Greenops

Description: 1″ (25 mm) long, ⅝″ (16 mm) wide. Oval; head and tail shield semicircular, moderately convex, almost equal in size and shape. Glabella mushroom-shaped, greatly expanded toward front and extending almost to margin; 4 pairs of glabellar furrows; 2nd pair from front reduced to small, elongate pits. Border indistinct. Lateral-rear corner spines broad, blunt, long. Eyes large. Facial sutures follow front margin of glabella, trend along ridges over eyes, and intersect lateral margins above bases of spines. Thorax has gently tapering middle lobe, approximately equal in width to pleura. Distinct furrows between and within pleura. Tail shield broad; 7–10 rings in middle lobe that bend forward near front of tail. 5 pairs of ribs and 5 pairs of short, gently curving, blunt spines and single median spine, all approximately equal in length.

Age: Middle Devonian.

Distribution: Widespread in North America.
Comments: The tail shield of *Greenops* is often
found independent of the rest of the
skeleton. While many other trilobite
genera have a fringe of spines along the
tail margin, the tail of *Greenops* is
distinctive because there is a median
spine and always 5 additional pairs of
spines of equal length.

The marginal drawing shows a
reconstructed specimen of *Greenops*.

SUPERCLASS CRUSTACEA
(Cambrian to Recent)

The Crustacea are mainly a marine
group, although some are common in
fresh water and some are partially
terrestrial. The legs of crustaceans are
highly differentiated, and typically
have 2 branches. The jaws are formed
from interior segments of legs, not
from the tips as in Uniramia. The
common marine crustaceans are the
ostracodes, amphipods, barnacles,
shrimps, lobsters, and crabs.
There are about 2000 genera of fossil
crustaceans, divided among 7 classes.
Only the classes Ostracoda,
Branchiopoda, Cirripedia, and
Malacostraca will be considered here.
The most abundant crustacean fossils,
with more than 1000 genera, are the
ostracodes—usually microscopic
animals with a bivalved shell. There are
about 100 fossil genera of branchiopods
—small freshwater forms that may also
have bivalved shells. About 100 fossil
genera of cirripeds, the barnacles, are
included in this superclass as well. The
malacostracans—usually large, shrimplike crustaceans, many of which have
calcified skeletons—are represented by
about 600 fossil genera.

CLASS OSTRACODA
(Cambrian to Recent)

Ostracodes, sometimes called mussel or seed shrimp, are tiny, mostly microscopic crustaceans enclosed by bivalved shells that are usually calcified but sometimes composed only of chitin. The head is undifferentiated and bears 2 pairs of *antennae* and 2 additional pairs of appendages. The thorax has 1–3 pairs of appendages, and a last, well-developed pair that projects down and forward. Most ostracodes are aquatic, but a few forms are terrestrial.

116 Leperditia

Description: ½" (13 mm) long, ¼" (6 mm) high. Oval, elongate, bivalved, with long, straight hinge; margins rounded. Shell slightly higher at 1 end of hinge than other. Valves only slightly inflated. Surface usually smooth, but may be grainy or covered with tiny pits.

Age: Ordovician through Pennsylvanian.

Distribution: Widespread in North America.

Comments: While most ostracodes are microscopic, *Leperditia* is one of the largest, commonly exceeding ⅜" (10 mm) in length. The shell of *Leperditia* looks just like that of a small clam, but the animal inside was very different from a clam. It was shrimplike, with long antennae and several small legs. Ostracodes are crustaceans, distantly related to lobsters, crabs, and barnacles. They have the best fossil record of any crustacean group, with more than 10,000 fossil species described, and they have been an important part of marine life since the Cambrian. Living ostracodes feed in many different ways. Some are predators, some eat plants, and others filter food from the water.

CLASS BRANCHIOPODA
(Lower Devonian to Recent)

Branchiopods are a diverse group of
small crustaceans. Many have a dorsal
shield, others a bivalved shell, and still
others have no shield or shell over the
body. All have flattened, leaflike trunk
appendages. The group with the best
fossil record is the order Conchostraca,
the tiny clam shrimps, found only in
fresh water. Their bivalved shells,
composed of chitin sometimes
impregnated with calcite, are not
hinged dorsally, but folded. There are
10–32 trunk segments, each with a
pair of appendages.

117 Cyzicus

Description: ⅛″ (3 mm) long and high. Small,
bivalved, compressed, roughly oval.
Ornamented with concentric ridges
with irregular, branching radial lines
between. Hinge long and straight;
teeth absent; high, small umbones
present near 1 end.
Age: Lower Devonian to Recent.
Distribution: Widespread in North America.
Comments: Branchiopods are called clam shrimp
because the shell looks like a tiny clam
shell, but the animal inside is a
crustacean. Clam shrimp feed by
straining food out of the water with
their feathery appendages.

CLASS CIRRIPEDIA
(Silurian to Recent; possibly also in the
Ordovician)

There are 5 orders of barnacles, all
marine, but only 2 have a fossil record:
the thoracicans, which include the
gooseneck barnacles and the acorn
barnacles, and the acrothoracicans,

which live in burrows. Adult barnacles are typically cemented to the substrate; a mantle encloses the body and secretes a calcareous shell. There are usually 6 pairs of trunk appendages, which have 2 branches covered with bristles. Some barnacles have become parasitic and are highly modified.

316, 317 **Balanus**

Description: 6¼" (16 cm) high, 4" (10 cm) wide (pl. 316); ⅝" (16 mm) high, ⅜" (10 mm) wide (pl. 317). Cup-shaped. Walls formed by 6 jointed, calcareous plates resting on roughly circular base; 4 more plates, forming conical cover, usually separated from cup after death. Plates of wall roughly triangular, widest at base, with faint or pronounced horizontal growth lines, frequently with faint or heavy vertical ribs. Base sometimes found alone, cemented to hard substrate; rim thickened, with small radial ribs becoming fainter toward center.

Age: Middle Eocene to Recent.

Distribution: Widespread in North America.

Comments: *Balanus* is one of the acorn barnacles still common on all coasts of North America. It usually lives between the high- and low-tide lines, wherever there is a firm substrate on which to cement the base. Louis Agassiz once described a barnacle as "nothing more than a little shrimplike animal, standing on its head in a limestone house and kicking food into its mouth." The animal in the house is really not so different from a shrimp, except that its head has been modified to attach to the substrate and secrete calcium carbonate for the shell. 4 movable plates make a cover that can be pulled down to close the opening for protection when danger or low tide threatens. *Balanus* feeds by extending its bristly legs out of the top of the shell and sweeping them through the water, straining out small particles of food. Plate 316 shows one of the largest barnacles that ever lived; Plate 317 shows barnacles of a more usual size.

471 Acrothoracican barnacle burrows

Description: Burrows ⅛" (3 mm) long, 1/16" (2 mm) wide. Oval to teardrop-shaped, slitlike holes in shells or calcareous rock. Depths of holes vary from ½ length to greater than full length of hole.

Age: Pennsylvanian to Recent.

Distribution: Widespread in North America.

Comments: Instead of secreting a house around themselves like many other barnacles, acrothoracican barnacles burrow into shell or rock, and pull their bodies down into the burrows for protection. There are about 30 living species of acrothoracicans; only 2 of these can be identified to genus from the burrow alone. Therefore it is difficult to reliably name these on the basis of burrows alone. Furthermore, barnacle burrows are trace fossils, and it is recognized that different animals may leave the same traces.

CLASS MALACOSTRACA
(Cambrian to Recent)

This large order of living crustaceans includes the isopods, amphipods, shrimps, lobsters, and crabs. Most are marine, but some live in fresh water. A few forms are terrestrial. All malacostracans have 8 segments in the thorax and 6–7 in the abdomen. The head and thorax are typically covered with a single plate, called the *carapace*. Most malacostracans have compound

eyes on stalks. There are 5 pairs of
appendages on the head: 2 pairs of
antennae and 3 pairs modified for
chewing food. Most malacostracans use
the flexible abdomen for swimming,
but many crabs have it tucked under
the thorax.

308 Callianassa

Description: 1⅛" (41 mm) long, ¾" (19 mm) wide.
Lobsterlike; head shield almost
triangular. Abdomen has 6 segments,
the 6th elongated, ending in 5 rounded
lobes of the tailfan. 1st pair of
appendages have very large claw on 1
side, smaller claw on the other. Claws
compressed. 2nd pair of appendages
have much smaller claws.
Age: Upper Cretaceous to Recent.
Distribution: Widespread in North America.
Comments: Living *Callianassa,* called ghost
shrimps, dig deep, extensive burrows
near shore in marine areas along North
American coasts. Most of the skeleton
is only lightly mineralized, but the
claws are large and heavily calcified.
Most *Callianassa* fossils thus consist
only of claws, and consequently most
species have been described on the basis
of the claws alone.

309 Acanthotelson

Description: 2" (51 mm) long, ½" (13 mm) wide.
Small, shrimplike; head small,
about length of 2 trunk segments. Trunk
long, narrow, with 14 well-defined,
equal segments. 4 heavy antennae
extend forward from head; many
walking legs on trunk, 1st pair longer
than others. Tail has 5 spines radiating
from last segment.
Age: Pennsylvanian.
Distribution: Central North America.

Comments: Shrimplike animals are rarely preserved
in their entirety as fossils because their
exoskeletons are either uncalcified or
only lightly calcified. Most or all of the
skeleton is organic, and is usually
consumed by various organisms after
the death of the animal. *Acanthotelson* is
well known only because it lived in a
place and at a time when ironstone
nodules were forming quickly around
buried organisms. The nodules sealed
the organisms off from oxygen, halting
decay and preserving even the soft
parts. This happened in central North
America in the Middle Pennsylvanian.
Because the nodules were first found in
the streambed of Mazon Creek in
northern Illinois, the fossils are called
the Mazon Creek flora and fauna, but
such nodules have subsequently been
found in coal strip-mining areas over
much of the Midwest. Shrimp are
among the most common fossils in
these areas, along with beautifully
preserved plants.

SUPERCLASS CHELICERATA
(Cambrian to Recent)

Most living chelicerates are spiders, a
large and successful terrestrial group,
and scorpions, closely related to
spiders. In the fossil record, horseshoe
crabs and eurypterids, both members of
the class Merostomata, are important.
The chelicerate body is divided into a
cephalothorax (the combined head and
thorax) and an abdomen. There are no
antennae, and the first pair of
appendages, called *chelicerae,* end in
pincers. Chelicerae are used in feeding,
and in the spiders are modified as
poison-injecting fangs.

CLASS MEROSTOMATA
(Cambrian to Recent)

Merostomes are a largely extinct group characterized by paired appendages on the anterior cephalothorax and 5–6 pairs of abdomenal appendages modified as gills. The first 2 appendages on the cephalothorax are pincers. There is a long, spikelike tail.

SUBCLASS XIPHOSURA
(Ordovician to Recent)

Living xiphosurans, of which there are only 3 genera, are called horseshoe crabs. Fossil xiphosurans, even in the Paleozoic, are scarcely different from the living genus *Limulus*.

Xiphosurans have a large, laterally expanded cephalothorax, generally with lateral-rear corner spines, an abdomen with a varying number of segments that may be fused or jointed, and a long, spikelike tail.

There are 64 genera of fossil xiphosurans. Most occur in fresh water or brackish sediments, which are rare, and this fact, combined with the fact that the skeleton is only lightly, if at all, calcified, may explain the fossils' relative rarity.

304 Euproops

Description: 1″ (25 mm) long and wide. Almost circular; divided into semicircular head shield, rounded abdomen, and long, straight tail spine. Head shield drawn down at lateral-rear corners into long or short, sharply pointed spines. Eyes small, located on outside edge of ridges that curve inward below eyes.
Abdomen has raised middle lobe and 6 rings; 1st and 3rd rings each have small

bump; last ring longer, with larger bump or spine. Ribs extend from between middle lobe rings across flattened rim, and are prolonged as marginal spines. Large, long spine extends down from last middle lobe ring.

Age: Pennsylvanian through Permian.

Distribution: Widespread in North America.

Comments: *Euproops* is a horseshoe crab and remarkably similar to the living horseshoe crab, *Limulus*. The major difference is in the abdomen, which shows more evidence of segmentation in *Euproops* than in *Limulus*. The mode of life was probably the same for *Euproops* as it is today for *Limulus*, which lives on the sea floor of the continental shelf and plows along in mud and sand, picking up small bits of food—usually small clams, snails and worms.

Paleolimulus, found from the Mississippian through the Permian in Illinois and Kansas, is another fossil horseshoe crab. *Paleolimulus* has straight ridges below the eyes and joined spines projecting from the abdomen margins. The specimen of *Euproops* in Plate 304 was preserved in an ironstone nodule. Soon after the animal died, the iron in the sediment in which it was soon buried was attracted to the rotting animal and caused the sediment to gel around it, sealing it off from oxygen and slowing decay. The iron-rich sediment around the animal turned to ironstone, forming the nodule, and the mud that surrounded the nodule turned to shale. At the Mazon Creek collecting localities in northern Illinois, nodules like this contain one of the most spectacular fossil faunas in the world.

SUBCLASS EURYPTERIDA
(Cambrian through Permian)

Eurypterids are similar to xiphosurans,
but have a smaller cephalothorax
without lateral extensions. The
abdomen has an anterior part with 7
segments and a narrower, posterior part
with 5 segments. The abdomen lacks
appendages except for gills on the first
6 segments. The appendages on the
cephalothorax consist of a pair of
pincers, 4 pairs of walking legs, and 1
pair of large, elongated, paddlelike
appendages. The 4th pair of walking
legs is also paddlelike in some genera.
Eurypterids lived in both salt and fresh
water. Some fossil specimens are 9'
(3 m) long, the largest of all known
arthropods.

306 Eurypterus

Description: 5½" (140 mm) long, 1½" (38 mm)
wide. Large, scorpionlike; head shield
fairly small, body elongate. Head
almost square, but slightly rounded at
corners, somewhat convex, smooth
except for elevated kidney-shaped eyes
and 2 tiny bumps at center of head
between eyes. Narrow marginal furrow
around head. Head ⅐ to ⅕ length of
entire body, including tail. Abdomen
has 12 segments; 1st segment same
width as head; following segments
increase slightly in width to 4th, then
decrease in width while increasing
slightly in length. Last segment
double-pronged, with long, sharply
pointed spine extending down from
cleft. 5 pairs of legs restricted to head;
each pair longer than the pair in front.
5th pair long, ending in large, oval
paddles.

Age: Ordovician through Pennsylvanian.
Distribution: Widespread in North America.
Comments: *Eurypterus* is typical of the subclass

Eurypterida. Whole eurypterids are rare as fossils, and eurypterid fragments are more frequently found. They occur in limestones, dolomites, sandstones, and siltstones, and probably lived in marine, brackish, and freshwater conditions. *Eurypterus* might have had a life similar to that of living horseshoe crabs, which are restricted, by their small mouths and lack of strong claws, to eating animals no larger than small clams. Other eurypterids might have been more ferocious. Some paleontologists have speculated that it was for protection against eurypterid predators that the ancestors of fish evolved skeletons, thus becoming vertebrates. *Pterygotus,* found throughout North America from the Ordovician through the Devonian, is a relative of *Eurypterus.* The 1st pair of legs in *Pterygotus* are modified into large claws that may have been used to seize and tear apart prey. The tail of *Pterygotus,* unlike that of *Eurypterus,* ends in a flat, rounded plate. *Hughmilleria* is an eastern North American eurypterid, found in Ordovician through Permian rocks, that also had claws on the first appendages. Its tail, like that of *Eurypterus,* also ends in a spine, but its head is more rounded in front than that of *Eurypterus,* and the segment before the tail spine is not cleft. *Stylonurus,* found from the Silurian through the Devonian in eastern North America, has a head and body shaped like those of *Hughmilleria*—that is, the head is more rounded in front than that of *Eurypterus* and the segment before the tail spine is not cleft—but the appendages are very different from those of the other genera. Except for the very short, antennalike 1st pair, they are long and narrow, and the 2nd and 3rd pairs have many short spines.

Phylum Bryozoa
(Ordovician to Recent; possibly also in the Upper Cambrian)

Bryozoans are sometimes called moss animals. All are colonial, but their colonies, whether living or fossil, are usually small and easily overlooked. Bryozoans may be encrusting, massive, branching, or sheetlike. They have a planktonic larval stage; after a larva settles on a firm substrate, it buds off new individuals to form a colony. Each animal in a colony has an external membranous or calcareous sac, called a *zooecium,* surrounding the soft parts. Zooecia may be connected to one another by a threadlike stolon. A crown of *tentacles* used for food-gathering can be extended from an opening in the sac, and then retracted by muscles. The tentacles are hollow and are extended by a rise in fluid pressure within the zooecium; the way in which the pressure is raised varies and is important in classification. The *digestive tract* is U-shaped, and the *anus* lies near the mouth, outside the ring of tentacles.

Bryozoans are divided into 3 classes, the Phylactolaemata, the Stenolaemata, and the Gymnolaemata. The Phylactolaemata are found only in fresh water and never have calcified zooecia. Fewer than 5 genera are found fossilized, and so they are not treated here.

2 sets of measurements are given for the bryozoans illustrated. First, length and width are given for an individual zooecium as it appears on the colony surface. Then height and width are given for the colony, and length and width for the branches.

CLASS STENOLAEMATA
(Ordovician to Recent)

The class Stenolaemata has a rich fossil
record and includes the orders
Cyclostomata, Cystoporata,
Trepostomata, and Cryptostomata.
Only the Cyclostomata have living
members. All are marine; the zooecia
are tubes with calcified walls fused to
the walls of adjacent zooecia. The
openings of the zooecia are circular and
terminal. Protrusion of the tentacles
does not depend upon deformation of
the zooecial walls, but results from
changes in pressure caused by
contraction of muscles in the tops of
the tubes.

ORDER CYCLOSTOMATA
(Ordovician to Recent)

Cyclostome zooecia are simple
calcareous tubes with round,
unconstricted openings. The walls are
thin, with very fine pores, and the
openings have no lids. The colony form
varies from delicate, threadlike
branches to growths 4″ (10 cm) or more
in diameter. The tubes generally do not
contain transverse partitions. Accessory
tubes are present in many genera.
These may be open or closed, and are a
different size and shape from the
zooecia. The function of these tubes is
not known. Some zooecia may be much
enlarged as *brood chambers*.

448, 450 **Stomatopora**

Description: Zooecium ⅟₅₀″ (0.5 mm) long, ⅟₁₇₀″
(0.15 mm) wide. Colony 1¼″ (32 mm)
high, ½″ (13 mm) wide (pl. 448); ⅝″
(16 mm) high, ⅜″ (10 mm) wide (pl.
450). Encrusting colony of tiny,

branching, curving tubes, with round openings fairly evenly spaced along tubes. Tube somewhat swollen at position of opening. Openings have raised edges.

Age: Ordovician to Recent.

Distribution: Widespread in North America.

Comments: *Stomatopora* is a very simple, primitive bryozoan that still lives in the ocean today. Its tubes hold tiny animals that extend their crowns of tentacles out of the round pores. The beating of small hairs on the tentacles sets up a current that carries small particles of food, such as single-celled algae, into the mouth at the base of the tentacles.

In Plate 448, *Stomatopora* is shown growing over another bryozoan. In Plate 450, it is shown growing on the inside of a scallop shell. The drawing in the margin shows part of a colony, much enlarged.

---

395 Idmidronea

Description: Zooecium ⅟₂₅₅″ (0.1 mm) long and wide. Branch ⅝″ (16 mm) long, ⅟₃₂″ (1 mm) wide. Colony of erect branches; 1 side only has zooecia, indicated by round openings with raised, thickened rims, not covering whole surface. Zooecia may be arranged in transverse rows extending out from longitudinal central keel. Opposite side has shallow, inconspicuous openings.

Longitudinal section through center of side with zooecia shows zooecial tubes extending obliquely downward from surface, without partitions but with walls much thickened near surface. Opposite side shows narrower tubes extending obliquely upward from surface.

Age: Eocene.

Distribution: Atlantic and Gulf coastal plains.

Comments: *Entalophora* is another common cyclostome, found in North America from the Jurassic to the Recent, and similar to *Idmidronea*. *Entalophora* has zooecia completely around the branches, and large brood chambers that appear as smooth swellings with only 1 aperture.

396 Hornera

Description: Zooecium 1/50" (0.5 mm) long and wide. Branch 3/4" (19 mm) long, 1/16" (2 mm) wide. Colony of erect branches, with irregular longitudinal grooves. Zooecia, restricted to front of branches, project as small, rounded tubes above branch surface. Branch covered with small pores. Brood chambers large, sac-shaped, located on back of branches. Longitudinal section shows long, oblique tubes of zooecia without partitions; zooecia separated by closely layered shell cut by narrow, shallow tubes that open on surface as pores.

Age: Eocene to Recent.

Distribution: Atlantic and Gulf coastal plains.

Comments: Brood chambers are common in cyclostomes. They are specialized zooecia where eggs form, are fertilized, and develop into tiny, saclike larvae before being released to found new colonies. Many specimens of *Hornera* are twiglike or branching, but some form latticelike or trellislike sheets, and are especially common in the Castle Hayne limestone of North Carolina.

The marginal drawing shows an enlarged section of a *Hornera* colony. The tubes have been worn down somewhat by abrasion.

403 Ceriopora

Description: Zooecium 1/125" (0.2 mm) long and wide. Colony 1/2" (13 mm) high, 5/8"

(16 mm) wide. Colony massive, encrusting, or branching. Zooecia tiny, closely packed, with angular to rounded openings. Small pores absent. Longitudinal section shows tubes with few to many straight, transverse partitions; wall structure porous.

Age: Triassic through Pliocene.

Distribution: Widespread in North America.

Comments: The top marginal drawing shows an enlargement of the surface of a colony. The lower picture is of a longitudinal section showing the transverse partitions in the tubes. *Heteropora* is a genus similar to *Ceriopora*, but it has numerous small pores opening between the zooecia. It ranges from the Triassic to the Recent in North America, and is still found living in coastal waters near Seattle.

451 Lichenopora

Description: Zooecium ½₅₅″ (0.1 mm) long and wide. Colony ⅛″ (3 mm) high and wide. Colony tiny, disc-shaped, wartlike, often attached to firm object, with zooecia opening on upper surface, arranged in rows radiating from center, separated by small pits. Center depressed, containing brood chambers indicated by a few large openings.

Age: Cretaceous to Recent.

Distribution: Widespread in North America.

Comments: *Lichenopora* is a distinctive bryozoan, and quite common in some Cenozoic marine sediments along the Atlantic and Gulf coastal plains. It is often found growing on other bryozoan colonies, as it appears in Plate 451. It is very small, however, and barely visible to the naked eye. To appreciate the details of its morphology, a hand lens or, even better, a dissection microscope is useful.

449 Hederella

Description: Zooecium ⅛″ (3 mm) long, ¹⁄₃₂″
(1 mm) wide. Colony 1¼″
(32 mm) high, ¾″ (19 mm) wide.
Colony encrusting; zooecia are short,
cylindrical tubes, bending alternately
right and left from tubular axis.
Simple, oval, unconstricted openings at
ends of tubes. Tubes covered with fine
circular rings and longitudinal lines.

Age: Silurian through Pennsylvanian.

Distribution: Widespread in North America.

Comments: The genus *Hederella* includes at least 60
described species and is a common
fossil in the middle Paleozoic. The
colony appears very plantlike, with the
small tubes that branch off from the
main tube resembling leaves. *Reptaria,*
from the Silurian and Devonian in
North America, has a similar
appearance but lacks the continuous
axial tube. Its zooecia arise out of the
base of each preceding zooecium, and
the sides of the colony are parallel. The
Hederella specimen illustrated is shown
growing on a rugose coral.

ORDER CYSTOPORATA
(Lower Ordovician through Upper
Permian; possibly also in the Upper
Cambrian)

Cystoporate colonies are encrusting,
massive, or branching. The branches
may be cylindrical or flattened, with
zooecia on both sides. The zooecial
openings are rounded, frequently with
a raised, hoodlike rim on 1 side.
Zooecia are isolated from one another or
grouped in clusters separated by
vesicular or solid skeletal material.
Most colonies have depressed or flat
areas on the surface that may contain
vesicular or solid skeletal material,
small pores or tubes whose function is
unknown, or a combination of these.

406 Ceramopora

Description: Zooecium 1/50" (0.5 mm) long and
wide. Colony 1¾" (44 mm) high, 2⅛"
(54 mm) wide. Colony disc-shaped or
massive. Zooecia relatively large,
irregular, with oblique openings
radiating from regularly spaced
depressions. 1 side of opening often
modified into overarching hood. Small
pores between zooecia. Zooecia tubular
in longitudinal section; few or no
platforms across tubes; walls covered
with pores.

Age: Ordovician through Devonian.

Distribution: Widespread in North America.

Comments: Illustrated is a massive, solidified
colony of *Ceramopora* in which the
original skeletal material was
calcite. After death and burial, the
calcite skeleton was replaced by the
mineral silica, as often occurs with
calcite fossils. Since silica is more
resistant to dissolution than calcite, the
fossil is often better preserved. If the
fossil is embedded in limestone, the
limestone can be dissolved with
hydrochloric acid, which does not
dissolve the silica. Silicified fossils often
weather out naturally from outcrops
when the rock in which they are

embedded is eroded away. The drawing
in the margin shows an enlargement of
the surface of a *Ceramopora* colony.
Ceramoporella is a similar and closely
related genus that is distributed
throughout North America in the
Ordovician and Silurian. *Ceramoporella*
can be distinguished from *Ceramopora*
by its encrusting growth form, which
creates thin layers of short tubes, often
over other bryozoans.

413, 416 Fistulipora

Description: Zooecium 1/50" (0.5 mm) long and
wide. Colony 2⅜" (60 mm) high, 1¾"

(44 mm) wide (pl. 413); ⅜″ (10 mm) high and wide (pl. 416). Colony massive or branching; surface covered with tiny openings of zooecia. Zooecia tubes cylindrical or compressed, with hoodlike extension of rim over opening but not completely covering it. Surface usually covered with small depression bumps, or flat areas that contain openings smaller than zooecia; very large zooecia radiate from each bump depression, becoming smaller toward edge of colony. Underside of massive colony has wrinkled base. Longitudinal section shows tubes with few partitions areas between tubes occupied by cells with walls convex toward colony surface.

Age: Silurian through Permian.

Distribution: Widespread in North America.

Comments: *Fistulipora* is similar to many other cystoporates that are massive, branching, or encrusting and have tiny zooecia. Sometimes the use of a hand lens can help to distinguish one genus from another; at other times the only way to identify a specimen is to make thin sections in a plane parallel to the colony surface as well as sections perpendicular to the surface.

 The marginal drawing shows an enlargement of a *Fistulipora* colony surface, with the zooecia radiating from a depression.

423 Constellaria

Description: Zooecium ¹⁄₂₅₅″ (0.1 mm) long and wide. Branch 1¾″ (44 mm) long, ⅜″ (10 mm) wide. Colony of erect frond or flattened branches; surface covered with distinctive star pattern formed by clusters of tiny pores. Zooecial openings irregularly rounded. Some small pores present between zooecia. Longitudinal section shows straight, complete partitions across tubes.

Age: Ordovician.
Distribution: Widespread in North America.
Comments: *Constellaria* is one of the easiest
bryozoan genera to identify because the
star pattern can be seen readily without
a hand lens or microscope. There is
only one genus in the same family that
might be confused with *Constellaria*,
and this is *Stellipora*, found in the
Ordovician of North America. The
growth form of *Stellipora* is different
from that of *Constellaria* in that
Stellipora is always encrusting, and all
of its small tubes are restricted to the
star-shaped clusters.

ORDER TREPOSTOMATA
(Lower Ordovician through Lower
Triassic; possibly also in the Upper
Triassic)

Trepostomes formed stony, massive, or
stemlike colonies that were important
in sediment-forming. Some colonies
were more than 20″ (51 cm) across. The
zooecia are long, calcareous tubes,
usually with many transverse
partitions. If a colony is cut open
longitudinally, it can be seen to have 2
regions. Toward the center is the
immature region, where the walls
between zooecia are thin and the
partitions widely spaced. The outer,
mature region has thickened walls
between tubes and more closely spaced
transverse partitions. Most genera have
bumps or depressions on the colony
surface where zooecia are smaller or
larger than average. There may be
skeletal rods that parallel the tubes and
terminate on the surface as small *spines*.

392, 412   Monticulipora

Description: Zooecium $\frac{1}{100}''$ (0.25 mm) long and wide. Branch 4″ (102 mm) long, 3⅛″ (79 mm) wide (pl. 392). Colony 1¼″ (32 mm) high, 2⅜″ (60 mm) wide (p 412). Colony usually massive, but m: be encrusting or frondlike; surface m: be covered by prominent, evenly spac bumps. Zooecia very tiny, closely packed, with polygonal tubes and openings. Smaller tubes between larg zooecia sometimes present, but uncommon. Tiny spines between zooecia present on surface. In longitudinal section, no boundaries visible between walls of individual tubes; tubes have straight, transverse partitions as well as curved, incomple partitions.

Age: Ordovician.

Distribution: Widespread in North America.

Comments: *Monticulipora* is a characteristic trepostome. "Trepostome" means "changing mouth" and refers to the change in the character of the tubes a: they pass from the immature (inner) region of a colony to the mature (oute region. A thin section must be made perpendicular to the surface of the colony in order to see this change. In the immature region, the tubes have thin walls and few transverse partition In the mature region, the tubes have many complete, straight partitions, many incomplete, curved partitions, and thick walls. Further, the mature region in many specimens has additional kinds of tubes, some small: and with tiny spines, between the zooecia. The photographs show 2 different colony forms. Plate 412 sho a massive *Monticulipora* specimen with prominent bumps on the surface. Plat 392 shows a branching colony that lacks the prominent bumps. *Monticulipora* was formerly known as *Monticuloporella*.

418 Prasopora

Description: Zooecium ⅛₀″ (0.3 mm) long and
wide. Colony ⅝″ (16 mm) high and
wide. Colony massive, disc-shaped or
hemispherical; not attached. Base
covered with concentric wrinkles.
Zooecia prismatic or cylindrical, with
adjacent walls fused together; zooecial
tubes lined by curved, incomplete
partitions that overlap, and crossed by
complete, straight partitions. Small
tubes between zooecia, prismatic in
cross section, with crowded partitions.
Spines few, inconspicuous.

Age: Ordovician.

Distribution: Widespread in North America.

Comments: The colonies of *Prasopora* have no
indication of attachment at their bases.
In order to avoid being overturned,
Prasopora must have lived in relatively
quiet waters, but since most sediment
settles in quiet water, what prevented
the colonies from being swamped by
sediment? Perhaps specialized
individuals could "sweep" sediment off.
But *Prasopora* could not have prevented
sediment from accumulating around its
base and, if the rate was great enough,
smothering the animals. In fact, well-
preserved *Prasopora* colonies may have
resulted from just such a situation—a
colony being quickly buried while still
alive. *Prasoporina* is a very similar
Ordovician genus in North America
that differs by having fewer curved and
straight cross-partitions in the zooecial
tubes.

Batostomella

Description: Zooecium ⅟₂₅₅″ (0.1 mm) long and
wide. Branch 1⅜″ (35 mm) long, ⅛″
(3 mm) wide. Colony of slender
branches; surface with rounded to oval
zooecial openings, well separated, with
occasional tiny openings and many

tiny, rounded spines between zooecia
Longitudinal section shows thick wall
between zooecia; tubes with few
straight, complete partitions.

Age: Ordovician.

Distribution: Widespread in North America.

Comments: Branching colonies similar to
Batostomella are *Hemiphragma* and
Batostoma, both the same age and with
the same distribution as *Batostomella.*
Longitudinal sections show both genera
to have many more partitions than
Batostomella; the partitions of
Hemiphragma do not completely cross
the tubes in the outer areas. The top
marginal drawing shows some branches
of a *Batostomella* colony, and the other
an enlargement of the colony surface.
Many specimens of *Batostomella* were
formerly called *Bythopora.*

399 Tabulipora

Description: Zooecium ¹⁄₂₅₅″ (0.1 mm) long and
wide. Branch 1″ (25 mm) long, ¼″
(6 mm) wide. Colony branching or
massive; zooecial openings placed well
apart, rounded or oval; area between
openings has many small, rounded
spines. Small pores absent.
Longitudinal section shows walls of
tubes greatly thickened in places;
transverse partitions within tubes
straight, but often with hole in center.

Age: Mississippian through Permian.

Distribution: Widespread in North America.

Comments: Many bryozoan genera, like *Tabulipora*
have both massive and branching
forms. The sturdy, massive forms were
best adapted to live where water
currents were strong and could sweep
sediment off the colony. The more
delicate, branching forms tended to
live in quieter areas where their
fragility would be less of a hazard, and
where the sediment in the water could
fall between the branches instead of

piling up, as it would on a massive colony. Fossil groupings tend to show this division, with delicate, branching, lacy bryozoans preserved together, and massive bryozoans preserved together elsewhere.

400 Rhombotrypa

Description: Zooecium ⅛₅″ (0.3 mm) long and wide. Branch 1⅝″ (41 mm) long, ⅜″ (10 mm) wide. Colony branching; cross section of branch shows square zooecial tubes. Surface of colony shows zooecia closely packed, polygonal in outline. Longitudinal section shows few straight partitions across tubes; walls of tubes are double.

Age: Ordovician.

Distribution: Widespread in North America.

Comments: Although there is usually nothing distinctive about the surface of *Rhombotrypa,* if 1 of the branches is broken, the typical quadrate pattern of the square tubes is seen and identification easily made. In a few specimens, the branch surface shows zooecial openings arranged in a fanlike splay pattern, which is a geometric result of the tubes growing as square prisms. At least twice, later in the Paleozoic, other trepostomes evolved the same square-grid appearance in cross section, and so illustrate the phenomenon of convergent evolution.

393 Parvohallopora

Description: Zooecium ⅛₅″ (0.3 mm) long and wide. Branch 2⅛″ (54 mm) long, ¼″ (6 mm) wide. Colony branching; branches may intertwine to form clumps 12″ (30 cm) wide. Zooecia rounded or cylindrical; many small, prismatic pores present between zooecia. Some species have regularly spaced bumps; others have pronounced rings around branches. Longitudinal section shows straight, complete, transverse partitions in tubes, sometimes sparse near center of branches. Small, prismatic tubes have dense partitions.

Age: Ordovician.

Distribution: Widespread in North America.

Comments: An easily identifiable *Parvohallopora* species is *P. rugosa,* illustrated in Plate 393. This species has pronounced rings around its branches that are, in fact, unusually elongated monticules—the surface bumps that many bryozoans show. In some genera these bumps have a special structure and, for example, may contain only the small pores commonly seen between zooecia on other genera. But in *P. rugosa,* the monticules are no different, except in elevation, from the other areas. *Hallopora,* sometimes known as *Calopora,* is a very similar branching form with somewhat larger zooecia, found especially in Silurian rocks in eastern North America.

ORDER CRYPTOSTOMATA
(Middle Ordovician through Upper Permian; possibly also in the Lower Triassic)

Most cryptostome colonies are delicate, latticelike fronds or slender, branching stems, but some consist of a sheetlike expansion with a single layer of zooecia on each side. All are calcareous. The zooecia are simple tubes, with immature and mature regions as in the trepostomes, but the tubes are much shorter. The outer part of the tube is a *vestibule,* marked off from the inner part by a shelflike partition. The area between adjacent vestibules is filled with either vesicular material or solid shell. Skeletal rods are common, as in the trepostomes, and end in small spines on the surface of the colony.

378, 385 Fenestella

Description: Zooecium ⅟₈₅″ (0.3 mm) long, ⅟₁₂₅″ (0.2 mm) wide. Colony 1⅛″ (28 mm) high, 1¼″ (32 mm) wide (pl. 378); 3⅛″ (79 mm) high, 2″ (51 mm) wide (pl. 385). Colony lacy, funnel- or fan-shaped; narrow, rigid, straight or sinuous branches connected at fairly regular intervals by crossbars. Branches have 2 longitudinal rows of openings on 1 side only, usually separated by plain or bumpy keel. Openings absent on crossbars. "Windows" between crossbars with 2–8 openings in single row opposite "windows." Spines may be present.

Age: Silurian through Permian.

Distribution: Widespread in North America.

Comments: *Fenestella* is one of the most common and distinctive of all bryozoan genera; its beautiful lacy colonies are found in many Paleozoic rocks. The colonies attached to rocks or shells and grew upright, their open, latticelike structure offering minimal resistance to currents.

Colonies of *Fenestella* are often preserved with the back side up and the openings hidden. In this case, it can be difficult to identify the genus, since one of its most important characteristics is the presence of only 2 rows of openings on the branches. *Fenestella,* like other bryozoan genera, may also be preserved as molds in rock. Again, it may be very difficult to determine the position of the openings in that state of preservation, and identification may have to be tentative. Both plates show nearly complete fronds. The marginal drawing shows a small section, much enlarged, of a *Fenestella* colony.

Several related genera are very similar to *Fenestella,* but can be distinguished in well-preserved specimens. *Fenestralia,* found in the Mississippian of central North America, has 4 rows of zooecia, 2 on each side of a median keel. *Fenestrapora,* from the Devonian in North America, bears pores or pits on the rear of the colony and on the wide summits of the keels. *Semicoscinium,* from the Silurian through the Devonian in North America, has expanded crossbars between its upright branches, so the "windows" are small, oval holes, and the keel on the front between the zooecia is very high and expanded at the top. In *Loculipora,* found in the Silurian and Devonian of North America, the branches and crossbars are arranged so that the colony appears to be built of tiny hexagonal blocks, each block with an oval hole in the center that is surrounded by zooecial openings. *Fenestrellina,* ranging from the Mississippian through the Permian in North America, has widely separated crossbars and very long upright branches, but is otherwise like *Fenestella,* with only 2 rows of zooecia. For many years there has been confusion about the proper name for *Fenestella.* In many books, the name *Fenestrellina* has been used to include both *Fenestella* and *Fenestrellina,* but most bryozoan experts now agree that these should be considered separate genera.

455, 456   Archimedes

Description: Zooecium $\frac{1}{50}''$ (0.5 mm) long, $\frac{1}{85}''$ (0.3 mm) wide. Axis 4″ (102 mm) long, $\frac{1}{8}''$ (3 mm) wide (pl. 455); 3$\frac{1}{8}''$ (79 mm) long, $\frac{3}{4}''$ (19 mm) wide (pl. 456). Lacy colony of latticelike branches and crossbars, with 2 longitudinal rows of zooecia on branches; zooecia on crossbars absent. Colony supported by screwlike axis of layered, calcified tissue that held edge of lacy colony. Axis usually preserved alone.

Age: Mississippian.

Distribution: Widespread in North America.

Comments: This is one of the most distinctive Upper Paleozoic fossils, seldom confused with any other. Its axis was a solid structure, resistant to destruction, and thus more commonly found than the lacy part of the colony that bore the zooecia. The axis held the delicate lacework upright in a spiral funnel, with the zooecia facing outward. This arrangement could fit more zooecia into a limited space than was possible in the more common fan-shaped colonies of *Fenestella* and similar genera.

Plates 455 and 456 both show isolated axes, 1 small and 1 large, that have been preserved without the fronds. The marginal drawing shows part of a complete *Archimedes* colony with the fronds still attached to the axis. The fronds in front have been omitted so that the axis can be seen.

380, 381 Polypora

Description: Zooecium ¹⁄₄₀″ (0.6 mm) long, ¹⁄₁₂₅″ (0.2 mm) wide. Colony 3¾″ (95 mm) high, 2¾″ (70 mm) wide (pl. 380); ⅝″ (16 mm) high, ½″ (13 mm) wide (pl. 381). Colony flat, lacy, consisting of upright branches bearing 3–8 longitudinal rows of zooecia, each represented by a small, rounded hole. Branches connected by crossbars to form fanlike colonies. Zooecia on crossbars absent. Keel usually absent on front of branches, but central bumps may be present.

Age: Silurian through Permian.

Distribution: Widespread in North America.

Comments: Illustrated in Plate 380 is an almost

complete frond of *Polypora;* Plate 381
shows a small piece of another frond.
Polypora is often found in fossil
assemblages with both *Fenestella* and
Archimedes. All 3 genera are very similar
in structure and probably inhabited
similar environments.

Anastomopora, found in the Devonian in
North America, resembles *Polypora,*
with 3–7 rows of zooecia across the
branches, but the branches are wavy
and join together instead of being
connected by crossbars. *Thamniscus,*
ranging from the Silurian through the
Permian in North America, has
branches and zooecia like *Polypora,* but
its crossbars are much farther apart, or
nearly absent.

---

394 Acanthocladia

Description: Zooecium ⅟₈₅" (0.3 mm) long and
wide. Branch ⅝" (16 mm) long, ⅛"
(3 mm) wide. Colony composed of
strong branches with short, oblique,
closely and regularly spaced side
branches; crossbars usually absent. 3 or
more longitudinal rows of zooecia along
1 side of branches only. Longitudinal
section shows tubes of zooecia
constricted before opening and without
partitions.

Age: Pennsylvanian through Permian.
Distribution: Widespread in North America.
Comments: *Acanthocladia* may look very much like
a fossil plant, but a hand lens should
reveal the zooecial apertures. In the
Permian, it is sometimes found as small
tangled mats, or thickets.

---

453 Penniretepora

Description: Zooecium ⅟₃₂" (1 mm) long, ⅟₈₅"
(0.3 mm) wide. Branch ⅝" (16 mm)
long, ⅟₅₀" (0.5 mm) wide. Branching;

slender main branch bearing short, regularly spaced, oblique side branches. Crossbars absent. 2 longitudinal rows of zooecia on 1 side of main and side branches only. Keel between zooecial rows may be inconspicuous or strong.

Age: Devonian through Permian.

Distribution: Widespread in North America.

Comments: *Ptylopora* (sometimes incorrectly spelled *Ptilopora*), which is the same age and has the same distribution as *Penniretepora*, has a very similar growth form, but its side branches are connected by crossbars, and its keel bears small bumps. *Diploporaria*, from the Devonian through the Permian in North America, is also similar, but has very few or no side branches.

397 Rhombopora

Description: Zooecium $\frac{1}{50}$" (0.5 mm) long and wide. Branch $\frac{5}{8}$" (16 mm) long, $\frac{1}{16}$" (2 mm) wide. Colony branching, with solid, slender stems. Oval openings of zooecia aligned in regular, oblique rows. Openings may be separated by longitudinal ridges that bear large and small spines. Small pores between openings absent. Longitudinal section shows tubes with thick walls in mature regions and few partitions, all complete.

Age: Devonian through Permian.

Distribution: Widespread in North America.

Comments: *Bactropora* is a similar genus that may be confused with *Rhombopora*. *Bactropora*, which occurs from the Devonian through the Lower Mississippian in eastern North America, has a pointed lower end on its main branch, and very few side branches, or none. *Streblotrypa*, which is the same age and has the same distribution as *Rhombopora*, is also similar, but has numerous tiny pores between the zooecia.

388 Ptilodictya

Description: Zooecium ¼₀″ (0.6 mm) long, ¹⁄₈₅″ (0.3 mm) wide. Frond ¾″ (19 mm) long, ¹⁄₁₆″ (2 mm) wide. Sheetlike fronds with zooecia on both sides, regularly arranged in longitudinal rows in young sections of colony. In older sections, arrangement may be irregular. Zooecia narrow, oval, or rectangular. Low bumps present, but small pores and spines absent.

Age: Ordovician through Devonian.

Distribution: Widespread in North America.

Comments: The *Ptilodictya* fossil illustrated shows a mode of preservation that is common for bryozoans, but in which identification is difficult because fine details of the surface are often lost. In this specimen, *Ptilodictya* has been preserved as a mold. The bryozoan itself dissolved away, leaving a hollow in the rock that was partly filled in with yellow limonite. The pronglike elevations represent the sediment fillings of the openings. The marginal drawing shows an enlargement of the surface of *Ptilodictya*.

384 Rhinidictya

Description: Zooecium ¹⁄₈₅″ (0.3 mm) long, ¹⁄₁₂₅″ (0.2 mm) wide. Branch ½″ (13 mm) long, ⅛″ (3 mm) wide. Colony branching; branches narrow, compressed, straight-edged, with zooecia on both flattened sides. Openings aligned between longitudinal, slightly elevated or sinuous ridges that are covered with crowded row of small, blunt spines.

Age: Ordovician.

Distribution: Widespread in North America.

Comments: The arrangement of the zooecia in *Rhinidictya* is very similar to that in *Ptilodictya,* but the overall shape of the colonies is different. If only fragments of a colony are preserved, it is often difficult or impossible to tell these 2 genera apart.

387 Sulcoretepora

Description: Zooecium $\frac{1}{40}''$ (0.6 mm) long, $\frac{1}{85}''$ (0.3 mm) wide. Branch $\frac{3}{4}''$ (19 mm) long, $\frac{1}{16}''$ (2 mm) wide. Branches narrow, ribbonlike, with sides almost parallel; zooecia on both sides of branches arranged in longitudinal rows, with small ribs commonly between rows. Zooecia have hoodlike projections on 1 side of openings.

Age: Devonian through Permian.

Distribution: Widespread in North America.

Comments: *Taeniopora* is a closely related bryozoan, found in the Devonian through the Mississippian in North America, that differs from *Sulcoretepora* by having a low keel in the center of its branches. *Dichotrypa*, with the same age and distribution as *Taeniopora*, is also very similar, but grows in broad sheets instead of branches; also, there are bumps on its surface. Recent studies suggest that many of these genera should be classified as cystoporates instead of cryptostomes.

CLASS GYMNOLAEMATA
(Ordovician to Recent)

The class Gymnolaemata contains 2 orders, the Ctenostomata and the Cheilostomata, both with living members. Almost all are marine. The zooecia may be calcified or uncalcified, cylindrical or boxlike, and the protrusion of the tentacles depends on deformation of the body wall.

ORDER CTENOSTOMATA
(Lower Ordovician to Recent)

Ctenostome zooecia bud off from a creeping, tubelike base, and are typically not connected. This base may sometimes be partially calcified, but never the whole zooecium. Some ctenostomes have the ability to excavate cavities in calcareous material to hold the zooecia. Ctenostome means "comb mouth" and refers to the comblike row of *bristles* on the flexible fold of the body wall that closes off the opening of the zooecium.

405 Ropalonaria

Description: Zooecium $\frac{1}{32}$" (1 mm) long, $\frac{1}{255}$" (0.1 mm) wide. Colony $\frac{3}{4}$" (19 mm) high, 1" (25 mm) wide. Colony branches irregularly, like a feather, or is netlike. Tiny, oval excavations connected by threadlike stolons, usually on a coral, crinoid stem, or shell. Stolons about as long as excavations.

Age: Ordovician through Cretaceous.

Distribution: Widespread in North America.

Comments: The excavations in *Ropalonaria,* often filled with clay, originally held zooecia, which are not preserved. *Vinella,* especially common in the Ordovician in North America, resembles *Ropalonaria,* and also forms a delicate, netlike pattern on shells. *Vinella* has stolons that are either radially arranged or more irregularly placed than those of *Ropalonaria.* There is a single row of pores along the tops of the stolons. Since only the threadlike stolon is calcified, the zooecia with the comb-mouth are almost never preserved in fossils. Ctenostome stolons or their excavations were formerly thought to be trilobite eggs, sponge borings, or foraminiferans. The specimen of *Ropalonaria* in Plate 405 resembles a group of thin, intersecting threads. It has grown over a calcareous nodule that also contains the slitlike holes of a boring barnacle. The drawing in the margin shows an enlarged *Ropalonaria* colony.

ORDER CHEILOSTOMATA
(Lower Cretaceous to Recent; possibly also in the Upper Jurassic)

A typical cheilostome zooecium is boxlike or egg-shaped, but it may also be a short tube. Most colony forms are delicate, either encrusting single layers or slender, branching stems or fronds. The zooecia are usually arranged side by side. The zooecial opening is constricted and may be extended out as a tube. A lid closes the opening in living cheilostomes, but this is seldom preserved in fossils. Many zooecia have an attached brood chamber—a globular swelling near the zooecial opening. Specialized individuals are common in the colonies, some utilized for attachment, some modified into tiny snapping *jaws*. Others are like tiny brooms, sweeping the surface of the colony with long bristles.

386 Membranipora

Description: Zooecium 1/50″ (0.5 mm) long, 1/170″ (0.15 mm) wide. Colony 1/2″ (13 mm) high, 1¼″ (32 mm) wide. Colony encrusting or erect; if erect, zooecia on both sides of fronds. Zooecia rectangular, boxlike, aligned in parallel rows. Openings large, oval, occupying almost entire front of zooecia. Spines absent, but small, blunt knobs present on corners of zooecia in some species.
Age: Cretaceous to Recent.
Distribution: Widespread in North America.

Comments: *Membranipora* is one of the most
common Tertiary bryozoans, with more
than 600 described species. The small
polyps that lived in the zooecia fed by
collecting small particles of food on the
hairs of tentacles that they extended up
from inside the zooecia. The fossil
zooecia show very large openings
because most of the front of the
zooecium was covered with a thin,
chitinous membrane not preserved in
the fossil. This membrane had muscles
attached to it that extended to the floor
of the zooecium. When the bryozoan
wanted to feed, it would contract the
muscles, pulling the membrane down
and transferring pressure through the
liquid-filled zooecium so as to squeeze
the tentacles up through the opening.

382 Conopeum

Description: Zooecium ⅟₅₀″ (0.5 mm) long, ⅟₈₅″
(0.3 mm) wide. Colony ½″ (13 mm)
high, 1″ (25 mm) wide. Colony
encrusting, lacy; zooecia side by side in
single layer consisting of large, oval
holes surrounded by slightly raised,
oval rims fused to rims of adjacent
zooecia. Small, triangular holes appear
in framework where gaps are left
between rims. Brood chambers and
spines absent.

Age: Cretaceous to Recent.
Distribution: Widespread in North America.
Comments: The oval holes in the *Conopeum* colony
are not the original openings, but
represent the entire upper wall of the
zooecium, which was covered over by a
chitinous membrane; the membrane, in
turn, had within it a smaller opening
through which the tentacles were
protruded at one end of the zooecium.
The flexible front wall served the same
purpose as it did in *Membranipora*.
Muscles that were attached to the wall
pulled it down, increasing liquid

pressure within the zooecium and
squeezing the tentacles so as to eject
them from the zooecium.

398 Coscinopleura

Description: Zooecium ⅟₅₀" (0.5 mm) long, ⅟₈₅"
(0.3 mm) wide. Branch ¼" (6 mm)
long, ⅛" (3 mm) wide. Branches erect,
narrow, with zooecia on both sides.
Zooecia hexagonal, flat, slightly
elongate, arranged in rows; openings
semicircular. Edges of branches
bordered by zooecia with porous walls
and restricted openings.

Age: Paleocene.

Distribution: Atlantic coastal plain.

Comments: The porous cells on the borders of the
colony are specialized, with their
opercula, or opening-covers, modified
into long bristles that were used to
sweep foreign matter off the colony.
This cleaning function was important,
since many kinds of encrusting
organisms, including other bryozoans,
use bryozoan colonies as solid, firm
bases to grow on.

415 Cribrilaria

Description: Zooecia ⅟₃₂" (1 mm) long, ⅟₆₄"
(0.5 mm) wide. Colony ⅝" (16 mm)
high and wide. Zooecia arranged
regularly in diagonal rows; zooecia
oval, with small, semicircular openings
surrounded by rim with small spines on
outer edge. Ribs radiate inward from
margins, fuse along edges, covering
front of cell; pores in center and in
furrows between ribs. Small teardrop-
shaped openings between zooecia in
some specimens.

Age: Eocene to Recent.

Distribution: Widespread in North America.

Comments: *Cribrilaria* probably evolved from a

bryozoan, perhaps similar to *Conopeum*, with only a chitinous membrane in front. *Cribrilaria* has added protection, however: structures that begin as spines across the membrane and then grow together to form an almost solid cover. This cover contains only tiny pores to let water in and out as the membrane underneath it rises and falls to control the extrusion of the tentacles.

Cribrilina is a similar bryozoan of the same age and with the same distribution as *Cribrilaria*. *Cribrilina* has radiating ribs indicated only by rather large pores that are arranged in regular rows. There are no spines above the opening.

404 Celleporaria

Description: Zooecium ¹⁄₃₂″ (1 mm) long, ¹⁄₅₀″ (0.5 mm) wide. Colony 2⅜″ (60 mm) wide, 1⅝″ (41 mm) high. Colony massive or layered, free or encrusting. Zooecia boxlike to egg-shaped, with irregularly arranged, semicircular openings randomly oriented on surface of colony.

Age: Miocene to Recent.

Distribution: Atlantic and Gulf coastal plains.

Comments: The *Celleporaria* colony illustrated comes from a fossil bryozoan reef in the Bahama Islands. *Celleporaria* is the main bryozoan component of these reefs, which formed during the last few thousand years, after the most recent major glaciation.

383, 414 Schizoporella

Description: Zooecium ¹⁄₈₅″ (0.3 mm) long and wide. Colony ¾″ (19 mm) high, 1⅜″ (35 mm) wide (pl. 383); ⅜″ (10 mm) high, ¾″ (19 mm) wide (pl. 414). Colony branching, encrusting, or becoming nodular by formation of many layers. Zooecia in parallel rows; when well preserved, some specimens have raised, circular brood chambers in midline of each zooecium, making a kind of hood over straight side of semicircular opening. Lower side of opening has tiny notch in middle and small opening beneath notch. Zooecium covered with tiny openings. 1 or 2 small, teardrop-shaped, raised areas on each zooecium.

Age: Eocene to Recent.
Distribution: Atlantic and Gulf coastal plains.
Comments: *Schizoporella* is an important rock-forming organism. Much of the extensive Miami Limestone in southern Florida is composed principally of fossilized *Schizoporella* branches. This bryozoan occurs all along the Atlantic coast, both as fossil and living colonies. When colonies wash up on the beach, they look very similar to the fossil illustrated in Plate 383, which is a specimen in a typical state of preservation. The surface of the colony is eroded, but the cross section, visible where the colony has been broken, shows layering with the characteristic cellular structure.

The front, or outer, wall of *Schizoporella* is calcified and rigid. This wall cannot be pulled down to evert the tentacles, as is possible in *Membranipora,* which has soft front walls. Instead, *Schizoporella* has a compensation sac located within the zooecial cavity. This sac connects to the outside through the pore just below the opening. Water enters the sac when the muscles that hold it to the wall contract. As the sac dilates, pressure is exerted on the tentacles so that they are pushed out for feeding. The tentacles are withdrawn by muscles attached at their bases.

Phylum Brachiopoda
(Lower Cambrian to Recent)

Brachiopods were perhaps the most conspicuous, abundant, and diverse fossil animals in the Paleozoic Era, but there are only 70 brachiopod genera living today. All are marine, and many inhabit cold, deep water.

Brachiopods are solitary marine animals with very little power of movement. Today, most spend their lives attached by a fleshy stalk to a firm substrate, but many extinct brachiopods lay free on the sea floor, and some had spines that were used to adhere to objects or to act as stabilizers in soft sediment.

All brachiopods have a shell with 2 valves. Superficially, they look very much like the Bivalvia, but the symmetry and arrangement of the soft parts within the shells are different in these 2 groups.

A brachiopod is oriented quite differently from a bivalve, with the plane of commissure horizontal. One valve is dorsal and one is ventral. The posterior part of the animal is near the hinge line. The plane of symmetry passes through both umbones and the middle of the anterior margin, dividing each valve into symmetrical halves, one the mirror image of the other. The 2 valves are always unequal. The ventral valve, or *pedicle valve*, has the stalk (or pedicle) attached inside, and there is usually a hole or notch through which the stalk emerges. The dorsal valve, or *brachial valve*, bears the feeding organ, called a *lophophore*, or *brachia*.

When the valves of a brachiopod are closed, its form is often egg-shaped, but may be spherical or flat. The ventral valve is usually larger than the dorsal valve, and extends farther beyond the hinge, which is where the 2 valves articulate. Near the hinge are the *umbones*, the first-formed parts of the shells. The areas on the valves between the umbones and the hinge, called the *interareas*, may be flat or concave. The interareas may have a notch in the center.

Most brachiopods are sculptured in some way. The most common sculpture is of *radial ribs*, which radiate from the umbones and may be very fine or coarse. Many brachiopods have a *radial fold* down the center of the dorsal or ventral valve, and a complementary *groove* in the opposite valve. The margins of the valves may be complexly folded by radial ribs that lock the valves together when they are closed. Concentric ornamentation also occurs, but is rarely dominant. *Growth lines* are common and vary in intensity. Resting periods in growth may leave deep concentric grooves on a shell.

The shells of brachiopods are composed either of a mixture of chitin and calcium phosphate along with small amounts of other minerals, or of calcium carbonate and a small amount of organic material. Many features important for the identification of brachiopods are contained on the interior of the valves. Shells are rarely obtained free of the matrix, with the interior fully exposed. Paleontologists often make transverse sections of the posterior parts of shells that are filled with rock, to determine the internal structure from details of the hinge, *muscle scars*, *ridges*, *septa*, and calcareous supports for the lophophore.

The measurements given in the plate captions and the descriptions are of length and width. Length is the distance from the middle of the posterior margin of the ventral hinge area to the anterior margin; width is the maximum distance taken at right angles to the length.

CLASS INARTICULATA
(Lower Cambrian to Recent)

The inarticulates are so called because they lack articulation: unlike the articulates, they have no *teeth* or *sockets*. Instead, they rely on muscles to hold the valves together. The shell is generally composed of calcium phosphate and layers of chitin, but in some genera is of calcium carbonate. Most inarticulates are small, some almost microscopic, but a few reach 3″ (8 cm) in length. Inarticulates are never as large as the larger articulates. The valves of some inarticulates are flat, with elongated, parallel sides—an adaptive shape for living in a vertical burrow. Other inarticulates are teardrop-shaped or round. One or both valves may be extended into small cones. Concentric growth lines and ribs are common, but radial and spiny ornamentation occurs in only a few genera.

152 Lingula

Description: ⅝″ (16 mm) long, ⅜″ (10 mm) wide. Elongate-oval to tongue-shaped. Valves thin, often lustrous black, almost identical, both slightly convex. Fine concentric growth lines surround marginal umbones; posteriors pointed. Ventral valve has poorly defined pseudointerarea with triangular groove in middle for passage of stalk; dorsal valve has small pseudointerarea but no groove.

Age: Silurian to Recent (possibly also in the Ordovician).

Distribution: Widespread in North America.

Comments: Fossil *Lingula* and related genera with parallel sides probably behaved like living *Lingula*: a long, flexible stalk emerges from between the valves at the posterior and attaches to the bottom of a vertical burrow. The shell moves up and down in the burrow as the stalk extends and contracts. By withdrawing into the burrow and closing the valves, the animal can protect itself from drying out during low tides and temporary low salinity after rain storms. Living *Lingula* are often found in intertidal areas, and fossil *Lingula* also colonized this environment. The genus *Lingula* has survived virtually unchanged for at least 400 million years, making it among the more venerable of all "living fossils." *Trigonoglossa* is very similar and closely related to *Lingula*. It is found in Devonian through Pennsylvanian strata of Kentucky, the Mississippi Valley, Nebraska, Kansas, Oklahoma, and Texas. *Trigonoglossa* differs from *Lingula* in its triangular outline and strongly raised growth lines. *Barroisella,* found in the Upper Devonian through Mississippian of eastern North America, is also similar to *Lingula,* but has more rounded anterior and posterior margins, and a broad median ridge on the interior of the dorsal valve. It is common in black shale.

118 Dicellomus

Description: $\frac{1}{16}$" (2 mm) long and wide. Small; rounded to oval. Valves thick, polished, slightly convex, with marginal umbones and concentric growth lines. Ventral valve has groove dividing pseudointerarea; dorsal valve has more flattened posterior margin. Elongated, narrow grooves on interior of dorsal valve at hinge area.

Age: Upper Cambrian.

Distribution: Appalachian Mountains, Wisconsin, Missouri, Black Hills of South Dakota, Rocky Mountains.

Comments: *Dicellomus* appears to have lived on hard, sandy bottoms not far offshore.

Great numbers of its broken and sorted shells are often found in deposits in Montana and Wyoming. The marginal drawings show the dorsal and ventral valves.

104 Lingulella

Description: ⅝" (16 mm) long, ½" (13 mm) wide. Thin-shelled; elongate-oval to nearly triangular; ventral valve pointed at its posterior, dorsal valve more rounded. Covered with fairly prominent concentric growth lines; outer shell layer often chipped to reveal fine radial lines on inner layer. Pseudointerareas on both valves divided by groove.

Age: Lower Cambrian through Middle Ordovician (possibly also in the Upper Ordovician).

Distribution: Widespread in North America.

Comments: *Lingulella* is abundant in some Cambrian shallow-water deposits where the only other fossils may be the tracks of trilobites, called *Cruziana*. There is no direct evidence that *Lingulella* burrowed like *Lingula,* but its structure suggests that its habits were similar, and it is present in rocks with cross-bedding and ripple marks, indications of currents and sediment movement. Without a burrow, *Lingulella* would have been continually dislodged and washed away. *Lingulella* also occurred in deeper-water communities, especially during the Ordovician Period, but never with the abundance found in the deposits containing *Cruziana*.

Lingulepis and *Leptobolus* are 2 similar and closely related genera that are fairly common. *Lingulepis,* found in the Upper Cambrian of New York, the Appalachian Mountains, Wisconsin, and Texas, is like *Lingulella* but with the tip of its umbo much more elongated and pointed. *Leptobolus* is

found in the Middle and Upper
Ordovician of Ohio, Indiana, and
Kentucky. It has tiny, nearly oval
shells with a median ridge on the
interior of the ventral valve that forks
near the center of the shell.

Conotreta

Description: ¹⁄₁₆″ (2 mm) long and wide. Small;
conical valves circular in outline, with
fine concentric lines. Ventral valve a
high cone with small opening for stalk
at or near apex of cone; dorsal valve
gently convex; umbo marginal; both
valves with pseudointerarea divided by
groove.

Age: Ordovician.

Distribution: Eastern North America.

Comments: The stalk of *Conotreta* must have been
attached to some firm substrate. There
would have been many sites to which
Conotreta might attach, even on a soft
bottom: dead shells of all kinds, algal
fronds, bryozoans, or other
brachiopods. Another tiny inarticulate
brachiopod, *Prototreta,* from the Middle
Cambrian (and possibly also the Lower
Cambrian) and found in the Rocky
Mountain region of Montana, differs
slightly from *Conotreta* externally. The
groove dividing the ventral
pseudointerarea is more pronounced,
and the stalk opening is posterior to the
apex. Inside the dorsal valve is a high
median ridge that splays out toward the
opposite valve like the fingers of a
hand. The marginal drawing of
Conotreta shows a side view of the
ventral valve.

113   Acrothele

Description: ¼″ (6 mm) long, ⅜″ (10 mm) wide.
Small; ventral valve conical, dorsal

valve gently convex. Ventral valve circular, with fine concentric growth lines; elliptical stalk opening on posterior slope of valve just below apex. Dorsal valve has marginal umbo. Both valves have small, rounded spines on umbones.

Age: Middle Cambrian (possibly also in the Lower Cambrian).

Distribution: Newfoundland, New Brunswick, Quebec, Appalachian Mountains, Great Basin area, Rocky Mountains.

Comments: *Conodiscus,* from the Cambrian of Alaska and Nevada, has a ventral valve like *Acrothele,* but its dorsal valve is also conical. Both of the specimens of *Acrothele* illustrated in Plate 113 are ventral valves.

120 Schizocrania

Description: ¼″ (6 mm) long and wide. Small; both valves nearly circular. Ventral valve flat to slightly concave, ornamented with concentric lines; dorsal valve convex, with fine ridges radiating from a marginal umbo. Deep, wide, triangular notch for stalk extending from tip of umbo to posterior margin; pseudointerareas absent.

Age: Ordovician through Lower Devonian.

Distribution: Ohio Valley, Appalachian Mountains, Mississippi Valley.

Comments: The flat to concave ventral valve and the wide notch suggest that a short, thick stalk held *Schizocrania* tightly against a firm surface. Not surprisingly, specimens such as those illustrated are commonly found still attached to the valves of flat articulate brachiopods. Only the dorsal valves of the specimens in Plate 120 are visible.

114 Orbiculoidea

Description: ¼" (6 mm) long and wide. Small, rounded, thin, often shiny; valves ornamented by fine to well-developed, concentric growth lines. Ventral valve may be gently conical, flat, or gently concave; narrow groove for pedicle runs from apex near center of shell to small hole just before posterior margin. Dorsal valve conical, with apex almost central.

Age: Ordovician through Permian.

Distribution: Widespread in North America.

Comments: *Orbiculoidea* often occurs in deep-water Silurian graptolite communities preserved in black shale. Since the sea floor probably had no oxygen at the time of deposition, the brachiopods must have been attached to floating objects such as seaweed. Other inarticulates, such as *Conotreta* and *Lingulella,* also may have attached to floating objects.
2 very similar genera are: *Roemerella,* from the Devonian and found in New York, Indiana, and Kentucky, which is usually larger than *Orbiculoidea* and has a concave ventral valve; and *Lindstroemella,* from the Middle Devonian through the Pennsylvanian throughout North America, whose pedicle groove reaches all the way to the posterior margin.

78 Petrocrania

Description: ½" (13 mm) long, ⅜" (10 mm) wide. Irregular but approximately circular in outline; ventral valve flat; dorsal valve conical. Ventral valve cemented directly to substrate. Dorsal valve ornamented with concentric growth lines that may mimic ornamentation of host to which it is attached; margins not thickened. Dorsal valve has 4 large, rounded muscle scars, 2 near posterior margin and 2 touching near valve center. Interior of valves have fine pits.

Age: Middle Ordovician through Devonian (possibly also in the Permian).

Distribution: Widespread in North America.

Comments: *Crania*, a genus related to *Petrocrania*, appeared in the Mesozoic and is still living. It is rare as a fossil in North America, where it occurs in Tertiary sediments. It attached like *Petrocrania* and has a similar size and shape except for the valve margins, which are noticeably thickened. *Crania* has concentric growth lines; radial ribs are either absent or very faint, and may be spiny.

Both *Petrocrania* and *Crania* cemented the ventral valve directly to some hard surface, often another brachiopod or a mollusk. The chemical nature of the cemented attachment is unknown. Cementation evolved independently at least 4 times among the brachiopods. Plate 78 shows *Petrocrania* specimens attached to a strophomenid brachiopod. The radial ornamentation of the strophomenid is mimicked by that of *Petrocrania*, providing camouflage for the small shells.

115 Obolella

Description: ³⁄₈″ (10 mm) long; ½″ (13 mm) wide. Small; valves convex, oval to round, nearly equal in size and shape, with weak to prominent concentric growth lines. Ventral valve has pseudointerarea divided by groove along which stalk emerged; dorsal valve has marginal umbo.

Age: Lower Cambrian (possibly also in the Middle Cambrian).

Distribution: Widespread in North America.

Comments: Most of the inarticulate brachiopods had shells partially made of calcium phosphate. *Obolella* (along with *Petrocrania* and *Crania*) is one of the few to secrete shells of calcium carbonate like those of the articulate brachiopods. Since the 2 minerals are probably functionally identical, and since secreting calcium carbonate is more difficult biochemically, why did almost all brachiopods after the Cambrian secrete calcareous shells? The answer must be that calcium carbonate is much more abundant in sea water than is calcium phosphate. Once the brachiopods evolved the ability to secrete calcium carbonate (this happened independently at least 4 times), they had an unlimited source of material. A ventral valve is illustrated in the marginal drawing.

112 Micromitra

Description: ¹⁄₁₆″ (2 mm) long, ⅛″ (3 mm) wide. Tiny; valves oval, ornamented with regular, concentric growth lines and fine, radiating ridges. Ventral valve slightly more convex than dorsal valve. Stalk opening absent.

Age: Cambrian.

Distribution: Widespread in North America.

Comments: *Micromitra* is an important member of some shallow-water Cambrian communities, along with certain trilobites and other brachiopods. 2 other tiny inarticulates closely related to *Micromitra* are *Dictyonema* and *Paterina*. *Dictyonema,* from the Lower and Middle Cambrian (and possibly also the Upper Cambrian), found in Newfoundland, the Appalachian Mountains, Montana, Idaho, Utah, and Nevada, is ornamented with small superficial pits. *Paterina,* from the Cambrian and widespread in North America, is thicker than *Micromitra* and lacks radiating ridges. The marginal drawings show the ventral and dorsal valves of *Micromitra.*

CLASS ARTICULATA
(Lower Cambrian to Recent)

The shells of articulates are always composed of calcium carbonate. Unless secondarily lost, as in some of the strophomenids, hinge teeth are present in the ventral valve, and sockets in the dorsal valve. In the interior of the dorsal valve, the ridges that bound the dental sockets on the inside may be extended toward the dorsal valve as supports for the lophophore. The supports may join to make a complete loop that may be folded back on itself. Most articulates have an extension, called the *cardinal process,* at the umbo of the dorsal valve, for attachment of the opening muscles. The dorsal valve often has an internal ridge or septum down the middle. Scars of the closing muscles are located to either side of the midline.

The ventral valve contains the teeth, which may be supported by dental plates, and the opening- and closing-muscle scars. The closing-muscle scars are usually located near the center line, with the opening-muscle scars to the outside and anterior of the closing-muscle scars.

ORDER ORTHIDA
(Lower Cambrian through Upper Permian)

Orthids are very abundant in the Early Paleozoic. The hinge line is straight, not curved. Both valves are convex, but are not equal in convexity. There are usually triangular notches in both of the typically large ventral and dorsal interareas, with a cardinal process in the dorsal valve. Many genera lack pits in the shell interior. The muscle scar area in the ventral valve is usually large. There are never any ribbonlike supports for the lophophore, but *brachiophores* bound the inner side of the dental sockets and are sometimes considerably extended. Almost all genera have radial ornamentation.

101 Billingsella

Description: ¼" (6 mm) long, ⅜" (10 mm) wide. Almost square to elongate-oval. Valves convex, ventral valve deeper; radiating ridges fine. Shallow fold on anterior of ventral valve, shallow groove on anterior of dorsal valve. Broad interareas on both valves, with median notches covered by convex hoods. Ventral valve with opening for stalk at apex; teeth large; dorsal valve lacks cardinal process or has only simple ridge.

Age: Middle Cambrian through Lower Ordovician.

Distribution: Widespread in North America.

Comments: Among the earliest articulate brachiopods, *Billingsella* is a specialized form. In the Upper Cambrian in Montana and Wyoming, enormous numbers of *Billingsella* and *Eoorthis* may have attached to calcareous algae which formed widespread "meadows." A closely related genus, *Xenorthis,* differs from *Billingsella* by having the groove on the ventral valve and the fold on the dorsal valve. *Xenorthis* is found in Lower Ordovician strata in Maryland and Nevada. The specimen illustrated in Plate 101 is a dorsal valve. The marginal drawings show the interiors of the dorsal and ventral valves.

79 Eoorthis

Description: ½" (13 mm) long, ¾" (19 mm) wide. Almost square to round. Both valves convex, ventral valve deeper; radial ribs broad, with fine ribs superimposed. Hinge line wide, with large interarea and open notch on ventral valve; strong teeth on well-developed dental plates. Cardinal process small, inconspicuous; small brachiophores near hinge line on dorsal valve.

Age: Upper Cambrian.

Distribution: Wisconsin, Minnesota, Missouri, Oklahoma, Texas, Montana, Wyoming.

Comments: *Eoorthis* is a common brachiopod in the Upper Cambrian, sometimes occurring in such numbers that it formed a kind of "reef," leaving beds several inches thick that consisted mostly of *Eoorthis* shells. Like most other Cambrian articulates, *Eoorthis* favored well-oxygenated, shallow water and a hard sea floor.

Apheoorthis, in the same family as *Eoorthis* and very similar to it, ranges from the Upper Cambrian through the Lower Ordovician in the Appalachian Mountains, Oklahoma, the Great Basin, and the Rocky Mountains. It has ribs in groups of graduated sizes.

Nisusia, found in the Lower and Middle Cambrian in North America, is also similar to *Eoorthis.* It differs by having bumps along the radial ribs where strong concentric growth lines intersect.

Plate 79 shows both dorsal and ventral valves embedded in matrix. The marginal drawing shows the dorsal valve.

71 Dolerorthis

Description: ⅞" (22 mm) long, 1⅛" (28 mm) wide. Semicircular, with convex to concave ventral valve, and convex dorsal valve. Hinge line wide. Folds absent; fine to coarse radial ribs. Interarea on ventral valve high, notches open, dental plates

receding, cardinal process ridgelike.
Small ridges demarcating dorsal notch
continue inside valve as brachiophores,
becoming bladelike and divergent.
Ventral muscle scars bounded by ridge

Age: Middle Ordovician through Upper
Silurian.

Distribution: Widespread in North America.

Comments: *Dolerorthis* may be confused with
Hesperonomia, which is found in Lower
Ordovician strata in the southern
Appalachian Mountains, the Ozark
Mountains of Arkansas, and in west
Texas, the Great Basin, and the
Canadian Rocky Mountains. The dorsal
valve of *Hesperonomia* is flat to concave,
and has finer radial ribs with new ribs
arising in the furrows, a weak fold and
groove, and no ridge around the ventral
muscle scars. The marginal drawing
shows the exterior of a ventral valve of
Dolerorthis.

72 Hesperorthis

Description: ⅜″ (10 mm) long, ½″ (13 mm) wide.
Round, with long, straight hinge line.
Ventral valve convex, dorsal valve flat
to slightly concave. Radial ribs strong,
with fine ridges between them;
concentric ornamentation absent.
Interarea on ventral valve high, notch
closed at apex; notch on dorsal valve
open. Cardinal process a small ridge.
Brachiophores bladelike.

Age: Lower Ordovician through Upper
Silurian.

Distribution: Ottawa Valley, Appalachian
Mountains, Ohio, Indiana, Illinois,
Iowa, Missouri, Nebraska, Kansas,
Nevada.

Comments: A very successful genus, *Hesperorthis*
was abundant through its age range. It
species had large populations, and
evolved slowly. This slow evolution
renders it a poor guide fossil, since
good guide fossils evolved rapidly and

had very short age ranges that enable paleontologists to gain precise information on the age of the rocks in which they are found.

66 Glyptorthis

Description: ⅝" (16 mm) long, ¾" (19 mm) wide. Oval. Both valves convex, but dorsal valve deeper. Distinctive ornamentation of strong radial ribs crossed by small, concentric, sheetlike ridges to produce small frills. Hinge line wide, straight. Ventral interarea with open notch; ventral closing-muscle scar on raised platform. Ridgelike cardinal process.

Age: Middle Ordovician through Lower Silurian.

Distribution: Widespread in North America.

Comments: *Glyptorthis* means "engraved orthis" and refers to the characteristic ornamentation on this orthid's valves. Plate 66 shows 2 articulated specimens of *Glyptorthis,* with the dorsal side forward, on the left, and the ventral valve on the right.

67 Dinorthis

Description: 1¼" (32 mm) long, 1⅜" (35 mm) wide. Almost square, to rounded. Ventral valve flat or concave, dorsal valve convex. Radiating ribs fairly coarse, rarely with new ribs arising in hollows between. Shallow groove on anterior of ventral valve in some species. Hinge line wide. Open notches present on both valves. Muscle scar area on interior of ventral valve squarish, with 2 anterior lobes; opening-muscle scars enclose central, oval closing-muscle scars. Dental plates short; dorsal valve with oval cardinal process, upper part ridged.

Age: Middle Ordovician.

Distribution: Widespread in North America.
Comments: *Plaesiomys,* from the Middle and Upper Ordovician, and *Retrorsirostra,* from the Upper Ordovician, both widespread in North America, are closely related genera. *Plaesiomys* has finer radial ribs than *Dinorthis.* The ventral interarea of *Retrorsirostra* is bent anteriorly at a high angle, while the ventral interareas of *Plaesiomys* and *Dinorthis* are directed slightly toward the posterior.
Several other common genera occur in the same family as *Dinorthis.* *Multicostella,* from the Middle Ordovician, is found in the Lake Champlain area of New York and Vermont, the Appalachian Mountains, Oklahoma, and Nevada, and has finer ribs. *Campylorthis,* also from the Middle Ordovician, is found in the Appalachian Mountains and the Mississippi Valley, and is more finely ribbed, with covers over both the ventral and dorsal notches in the interareas. *Valcourea* is found throughout North America in the Lower and Middle Ordovician. The notches in both its valves are partially closed by shell growth and there is a fold in the ventral valve anterior.

102 Finkelnburgia

Description: ¼" (6 mm) long, ⅜" (10 mm) wide. Small; oval. Valves almost equal in convexity; covered with fine radial ribs. Ventral valve has small teeth and dental plates uniting with a thickening of valve floor to form raised platform bearing ventral muscle scars. Dorsal valve has shallow middle anterior groove, ridgelike cardinal process, and strong brachiophores. Open interarea notches present in both valves.
Age: Upper Cambrian through Lower Ordovician.
Distribution: Widespread in North America.

Comments: A common genus with more than 23 described species, *Finkelnburgia* often occurs in Lower Ordovician limestones with stromatolites, the gastropod *Maclurites,* and trilobites. *Maclurites* may have grazed on the stromatolites, the trilobites probably ate bits of organic matter found on the sea floor, and *Finkelnburgia* filtered tiny organisms from the water.

The marginal drawings show the dorsal and ventral valves of *Finkelnburgia.* *Diparelasma,* found throughout North America in the Lower Ordovician, is in the same family as, and similar to, *Finkelnburgia.* It differs by having rounder valves, a shorter hinge line, and a ridge on the inside of the dorsal valve.

Plectorthis

Description: ⅝″ (16 mm) long, 1″ (25 mm) wide. Oval. Valves almost equally convex; shallow groove in middle of dorsal valve; coarse radiating ribs; hinge line wide. Open notches in both interareas; ventral interarea forms obtuse angle with plane of commissure. Ventral valve has strong dental plates for teeth; cordlike muscle area with opening-muscle scars enclosing linear closing-muscle scars. Dorsal valve interior with ridged cardinal process; brachiophore bases converge.

Age: Middle and Upper Ordovician.
Distribution: Widespread in North America.
Comments: In the same family with *Plectorthis* are *Hebertella* and *Platystrophia,* as well as *Desmorthis,* which comes from the Lower Ordovician and is found in the western United States. *Desmorthis* is distinguished by hollow ribs and a small plate closing off the apical tip of the ventral notch.

76 Hebertella

Description: 1¼" (32 mm) long, 1¾" (44 mm) wide. Oval. Ventral valve faintly convex to concave, dorsal valve deeply convex; groove in ventral valve, wide fold in dorsal valve. Radiating ribs fine to coarse. Open interarea notches in both valves. Ventral muscle field oval. Cardinal process has fine ridges.

Age: Middle and Upper Ordovician.

Distribution: Widespread in eastern North America.

Comments: Closely related to *Hebertella* is the genus *Doleroides,* from the Middle Ordovician, which can be distinguished by its scattered, fine, hollow radiating ribs and the single ridge dividing its ventral closing-muscle scar. Plate 76 shows an articulated specimen of *Hebertella,* with the ventral valve forward.

51 Platystrophia

Description: ¾" (19 mm) long, 1¼" (32 mm) wide. Sturdy. Both valves deeply convex, almost circular in profile. Ribs very heavy and of varying thickness, with fine bumps; deep fold in dorsal valve and groove in ventral valve, with 2 or 3 ribs in groove. Open notches in both valves; muscle area in ventral valve long and rounded anteriorly. Hinge line with extremities usually extended into sharp points. Cardinal process a low ridge; brachiophore bases converge.

Age: Middle Ordovician through Upper Silurian.

Distribution: Widespread in North America.

Comments: Externally, *Platystrophia* closely resembles *Spirifer* and other spiriferids. This superficial resemblance between genera that are, in fact, not closely related is known as homeomorphy. The brachiopods include many homeomorphic genera, making classification difficult. Plate 51 shows 3 specimens of

Platystrophia: on the left, a ventral valve
with the central groove; on the right, a
dorsal valve with the corresponding
fold; and at the top, a rear view with
the dorsal valve up, showing the hinge.

96 Enteletes

Description: ¾″ (19 mm) long, ⅞″ (22 mm) wide.
Rounded, convex, almost spherical.
Dorsal valve slightly more convex than
ventral. Heavy, raised radial ribs
developed toward anterior margins and
covered with fine radial lines. Open
notches in both interareas; thin, almost
parallel dental plates. High median
septum on interior of ventral valve;
dorsal valve with long, curved
brachiophores and tiny cardinal process.
Shell pierced by numerous tiny pits.

Age: Pennsylvanian through Permian.

Distribution: Mississippi Valley, Oklahoma, Texas,
southwestern United States.

Comments: This is one of the most distinctive of all
Upper Paleozoic brachiopods, with its
highly convex valves, its nearly parallel
dental plates and median septum on
the ventral valve, and its long, curved
brachiophores.
Illustrated in Plate 96 are 3 articulated
specimens, that is, specimens with
both valves preserved together. The
left-hand specimen is shown from the
side, with the dorsal valve to the left.
The top specimen shows a ventral
valve. The lower specimen is oriented
with the anterior forward and the dorsal
valve up, to show the raised folded ribs
that form a zigzag margin.
Articulation is more common in
brachiopods than in bivalves because of
the different muscle arrangements in
the 2 groups. Brachiopods must
contract muscles to open their valves;
when their muscles are relaxed, as after
death, the valves stay closed. Bivalves,
on the other hand, open when the

closing muscles relax. Their ligaments are compressed when the valves are closed, but force the valves open when the muscles die. Bivalves, then, pop open after death and the valves are more easily scattered.

77 Schizophoria

Description: ¾" (19 mm) long, 1" (25 mm) wide. Valves almost circular to elliptical; low dorsal fold and ventral groove usually present, but uncommonly both valves have groove. Dorsal valve strongly convex, ventral valve less convex, usually concave at anterior. Ventral valve has strong teeth; open notches in both interareas. Fine radial ribs present. Ventral muscle scar area has 2 lobes separated by median ridge. Cardinal process has large muscle-attachment area in young, less prominent area in adults. Inside of shells covered with tiny pits.

Age: Silurian through Permian.

Distribution: Widespread in North America.

Comments: *Schizophoria* shells are often found bored by the sponge *Cliona*. *Schizophoria* is very abundant locally, and is one of the most common Devonian genera; it becomes rarer in succeeding periods. Illustrated in Plate 77 are 2 articulated specimens; the specimen at the left shows the dorsal valve, while in the one at the right the ventral valve and hinge are visible. Both valves have a central groove.

73 Dalmanella

Description: ½" (13 mm) long and wide. Small; almost circular in outline. Ventral valve slightly more convex than dorsal. Shallow groove in anterior middle of dorsal valve; radial ribs fine to medium size. Interior of dorsal valve has median ridge. Open notches in both interareas; ventral muscle scars heart-shaped. Cardinal process small and lobed. Interiors covered with fine pits.

Age: Middle Ordovician through Lower Silurian.

Distribution: Illinois and Missouri.

Comments: Many species formerly classified in this genus have been reclassified. Plate 73 shows *D. edgewoodensis,* one of the few *Dalmanella* species in North America. A dorsal valve is on the left, a ventral valve on the right.

68 Onniella

Description: ⅜" (10 mm) long, ½" (13 mm) wide. Oval to almost rectangular in outline, wider than long; both valves slightly convex. Dorsal valve has shallow central groove. Ornamentation of fine radial ribs. Open notches on both interareas. Ventral muscle scars small, heart-shaped; closing-muscle scars on dorsal valve bounded by low ridges. Cardinal process with 2 lobes; brachiophore bases strong, divergent. Interior of valves covered with tiny pits.

Age: Middle and Upper Ordovician.

Distribution: Widespread in North America.

Comments: Like *Onniella,* about ¼ of all articulate brachiopods have tiny pits in their shells. This condition is termed punctate, and is important in classification. The function of the pits is unknown. In life, mantle tissue extended into the pits, which terminate just below the external surface of the shell. A view held by many paleontologists, although considered debatable by some, is that punctate shells evolved independently many times—perhaps as many as 7. In the same family as *Onniella,* and very similar to it, is *Resserella,* found in the Silurian throughout North America.

Onniella was once considered a subgenus of *Resserella*. *Resserella* can be distinguished by its more circular shieldlike shape, its deeply convex ventral valve, and the ridges in its teeth and sockets.

2 articulated specimens of *Onniella* are illustrated in Plate 68. The ventral valve is visible on the left; the dorsal valve is shown on the right.

95 Dicoelosia

Description: ¼" (6 mm) long, ⅜" (10 mm) wide. Small; maximum width near anterior margin. 2-lobed, with deep groove in middle of each valve; short auricles at hinge line, fine radial ornamentation. Ventral valve deeply convex, dorsal valve less convex to slightly concave. Open notches on interareas. Cardinal process has shaft and 2-lobed muscle attachment; brachiophores long, bladelike. Interior of valves covered with tiny pits.

Age: Upper Ordovician through Lower Devonian.

Distribution: Eastern Canada, New York, Appalachian Mountains, Ohio, Indiana, Tennessee, Illinois, Iowa, Missouri, Nebraska, Kansas, Oklahoma.

Comments: *Dicoelosia* was abundant in many offshore communities. The distinctive design of its shell probably helped separate water coming into it at the side lobes from water leaving at the groove between the lobes. Plate 95 shows 3 articulated specimens: the lower specimens have the ventral valve forward; the upper specimen, the dorsal valve. Formerly called *Bilobites*, and sometimes incorrectly spelled *Dicaelosia*.

70 Rhipidomella

Description: 1¼" (32 mm) long, 1⅜" (35 mm) wide. Circular to oval; ventral valve less convex than dorsal valve. Ornamented with fine, hollow radial ribs and concentric growth lines. Notch on ventral interarea always open; dorsal notch may have cover in some species. In ventral valve, opening-muscle scars oval, broad, with strongly scalloped margins; central ridge present. In dorsal valve, cardinal process has short shaft and lobate muscle attachment. Brachiophores short. Interior covered with tiny pits.

Age: Lower Silurian through Upper Permian.

Distribution: Widespread in North America.

Comments: *Rhipidomella* is very abundant in the Middle Devonian, where it is found in most types of sediment. It is particularly common in the limy sediments of western New York State. A widespread, closely related genus is *Platyorthis,* from the Upper Silurian and Lower Devonian, found in New York, the Appalachian Mountains, and Tennessee. It can be distinguished from *Rhipidomella* by its cardinal process, which has 2 or 3 lobes instead of only 1.

69 Tropidoleptus

Description: 1" (25 mm) long, 1¼" (32 mm) wide. Oval. Ventral valve convex, dorsal valve concave. Ornamented with broad, rounded radial ribs. Narrow, shallow ventral fold and dorsal groove present; hinge line wide. Ventral interarea large, with open notch; dorsal notch covered near umbo. Massive, ridged teeth on stout dental plates; high cardinal process. Interior covered with tiny pits.

Age: Middle and Upper Devonian.

Distribution: New York, Appalachian Mountains,

Ohio, Indiana, New Mexico.

Comments: *Tropidoleptus* is one of the guide fossils for the Devonian. Guide fossils are those that are both widespread in and restricted to a narrow part of the geological column. They enable paleontologists to identify the age of rocks in many different areas.
4 specimens are illustrated. The top specimen is shown with the dorsal valve visible; the convex ventral valves of the other 3 specimens are shown. The smaller specimen on the left appears to have been attached in life to the larger specimen.

ORDER STROPHOMENIDA
(Lower Ordovician through Lower Jurassic)

This is the largest of all brachiopod orders, with more than 400 genera, many highly specialized. This order includes the largest and heaviest of all brachiopod shells. Most commonly the dorsal valve is concave or flat, and the ventral, convex; rarely, both are convex. Strophomenids are typically wider than long, with the greatest width at the straight hinge line. The triangular notch in the ventral valve is almost always covered, but there may be a tiny hole at the umbo for the stalk to emerge from, at least in juvenile shells. The cardinal process often has 2 lobes. The interiors of the valves are marked by tiny bumps.
Early in the history of this group, attachment by the stalk in adults was lost, and strophomenids either lay free, balanced by spines that sank into a soft substrate, or cemented the ventral valve at the umbo to some firm substrate.

49 Sowerbyella

Description: ⅜″ (10 mm) long, ⅝″ (16 mm) wide. Semicircular in outline, with very wide hinge line. Flat in profile, with ventral valve slightly convex, dorsal valve slightly concave. Ornamentation of fine radial ribs, with smaller ribs grouped between larger ones, sometimes with small bumps or spines on ribs. Ventral interarea notch has small cover; cardinal process fused to brachiophores to form an inverted V. Closing-muscle scars inside ventral valve form 2 long lobes that diverge from the middle of the shell toward the anterior margin; dorsal valve has 2 slightly diverging ridges down center. Interiors of both valves covered by tiny bumps.

Age: Ordovician through Lower Silurian.

Distribution: Widespread in North America.

Comments: *Plectodonta* is a similar and closely related genus. It is found from the Upper Silurian through the Middle Devonian in New York, Ohio, Kentucky, Tennessee, and Alabama, and is distinguished from *Sowerbyella* by the 4 ridges down the center of the dorsal valve interior. *Leptellina,* found in New York, the Appalachian Mountains, and Nevada in the Lower and Middle Ordovician, also resembles *Sowerbyella* externally, but has only 1 median ridge inside the dorsal valve and has raised areas to each side where the viscera were held.

2 articulated specimens are illustrated in Plate 49; the left specimen shows the dorsal valve, and the right one, the ventral valve.

55 Strophomena

Description: ⅝″ (16 mm) long, 1″ (25 mm) wide. Semicircular to almost square, widest at hinge line, with fine radiating ribs. Young have convex ventral valve and concave dorsal valve; in adults, ventral valve becomes concave and dorsal valve convex. Small stalk opening at apex of ventral valve; large covering over notch of ventral interarea; teeth not supported by dental plates. Ventral muscle scar circular, bounded by strong lateral ridges that do not unite with median ridge. Interiors covered with fine, dense bumps.

Age: Middle and Upper Ordovician.

Distribution: Widespread in North America.

Comments: Its very small stalk opening indicates that the stalk would have been inadequate to hold *Strophomena* after a very early stage in its life. The animal probably broke free and lay on its ventral valve at the time that the valve shape was changing. Experiments with working models show that if *Strophomena* was turned over by currents or predators, it could have righted itself by vigorously flapping its valves.

65 Rafinesquina

Description: ⅞″ (22 mm) long, 1⅛″ (28 mm) wide. Semicircular. Valves flat, thin. Ventral valve convex, dorsal valve flat to concave. Radial ribs of unequal thickness; usually 1 or more conspicuously thick ribs down centers of valves. Notch on ventral interarea covered; teeth supported by divergent dental plates. Ventral interior with small, circular muscle scar; bounding ridges absent. Small muscle scars on dorsal valve divided by single short ridge. Cardinal process with 2 lobes. Small bumps arranged in radiating rows on interiors.

Age: Middle and Upper Ordovician.

Distribution: Widespread in North America.

Comments: *Rafinesquina* was named to honor Professor Constantine Rafinesque, who was born near Constantinople and settled in the United States in 1815.

Long before Darwin's *Origin of Species* was published, Rafinesque argued persuasively that organisms could evolve, and that the earth was very old. A common genus that is externally very similar to *Rafinesquina* is *Oepikina,* also from the Middle and Upper Ordovician, found in the Appalachian Mountains, the Ottawa Valley, the Mississippi Valley, Tennessee, and the Arbuckle Mountains of Oklahoma. *Oepikina* can be distinguished by its anterior margin, which is gently bent upward toward the dorsal valve. Internal differences include 5 ridges, instead of only 1, down the middle of the dorsal valve.

61 Leptaena

Description: ¾″ (19 mm) long, ⅞″ (22 mm) wide. Semicircular, sometimes with hinge line extended outward into auricles. Ventral valve gently convex, dorsal valve gently concave; both valves sharply bent dorsally at margins. Fine, even radial ribs present, as well as concentric ornamentation of thick, rounded ridges or wrinkles. Oval muscle scar present on ventral valve surrounded by strong ridges; ridge bisects muscle scars on dorsal valve. Interior covered with small bumps.

Age: Middle Ordovician through Devonian.

Distribution: Widespread in North America.

Comments: Most articulate brachiopods lived above the bottom, but *Leptaena* appears to have been adapted for living in the sediment. It probably lay on its ventral valve, while the sediment settled over it. The upward-projecting anterior margin would have allowed water to circulate within the valves. As more and more sediment accumulated, the valve edges grew up and away from the sediment. The heavy concentric ribbing probably helped to stabilize the shell.

Plate 61 shows both a dorsal and a ventral valve.

Dactylogonia, from the Middle Ordovician of eastern North America, is a closely related genus that differs from *Leptaena* in not having regular wrinkles or a sharply upturned anterior.

63 Strophodonta

Description: 1″ (25 mm) long, 1¼″ (32 mm) wide. Semicircular, widest at hinge line, with small auricles. Ventral valve convex, dorsal valve concave; both valves covered with fine radiating ribs. Small ridges that function as teeth present along full hinge line. Dental plates for supporting teeth absent; ventral interarea notch partly or completely covered. Ventral muscle scar area elongate-oval; opening-muscle scars wrapped around closing-muscle scars on sides and anterior; scars deep. Interior densely covered with small bumps.

Age: Upper Ordovician through Upper Devonian.

Distribution: Widespread in North America.

Comments: Adult *Strophodonta* did not attach; their wide, thin shells were adapted for lying on a soft substrate without sinking. *Strophodonta* could probably flap their valves like a scallop to remove sediment. Plate 63 shows 2 articulated specimens, with the dorsal valve visible on the left, and the ventral valve shown on the right. The name of this genus is often incorrectly spelled *Stropheodonta.* A related genus that is common in the Lower Silurian through the Lower Devonian of eastern Canada, New York, the Appalachian Mountains, Ohio, Tennessee, Missouri, Oklahoma, and Nevada is *Strophonella.* Externally it is very similar to *Strophodonta,* except that the dorsal valve becomes convex at the anterior margin and the ventral valve becomes concave. The small ridges on its hinge are restricted to an area near the hinge's center. Furthermore, the ventral muscle scar is circular to square, not oval. *Strophonelloides,* present in the Upper Devonian of New York, the Appalachian Mountains, Iowa, Missouri, New Mexico, Arizona, and Nevada, has the same reversal in the shape of the valves as does *Strophonella,* but the interiors of the valves are like those of *Strophodonta.*

64 Douvillina

Description: ⅝" (16 mm) long, ¾" (19 mm) wide. Semicircular, widest at hinge. Ventral valve slightly to strongly convex, dorsal valve slightly to deeply concave. Ornamentation of fine radial ribs of unequal thickness. Fine vertical ridges that function as teeth present along hinge line; other teeth and sockets absent; notch on ventral interarea partly or completely covered. Rounded muscle scar completely bounded by well-developed ridges on ventral valve; 2 elevated plates arising from floor of dorsal valve just interior to closing-muscle scar. Interiors covered with small bumps.

Age: Upper Silurian through Upper Devonian.

Distribution: New York, Appalachian Mountains, Ontario, Ohio, Michigan, Indiana, Iowa, Arizona.

Comments: In the course of evolution, *Douvillina* and many related genera with wide, flattish valves, such as *Strophodonta,* lost their teeth and sockets and developed, instead, small ridges—the denticles—along the insides of both hinges. These denticles helped to prevent rotation, but could not have been nearly as effective as teeth and sockets in holding the valves together. The denticles must have had some added advantage for animals like *Douvillina,* but in just what way is still a mystery.

Douvillinaria and *Leptostrophia* are 2 closely related genera that resemble *Douvillina. Douvillinaria,* from the Upper Devonian of New York and Iowa, can be distinguished by the way the dorsal valve becomes convex and the ventral valve concave toward the anterior margin. *Leptostrophia,* from the Silurian through the Devonian in Quebec, New York, Ontario, Ohio, the Appalachian Mountains, the Mississippi Valley, and Tennessee, often has wide concentric ridges on the sides near the hinge line, very flat valves, a triangular ventral muscle scar, and ridges bounding the sides of the dorsal muscle scar.

The dorsal valve of *Douvillina* is visible in Plate 64.

92 Meekella

Description: ⅝″ (16 mm) long, ¾″ (19 mm) wide. Ventral valve deep, conical; dorsal valve much smaller, slightly convex, caplike; umbo on ventral valve may be twisted. Ornamentation of fine radial ribs superimposed over wide, strong ribs. 2 thin, high dental plates on ventral valve. Ventral interarea very high, with covered notch. Muscle scar small, oval. High, 2-lobed cardinal process on dorsal valve. Interior covered with small, dense bumps.

Age: Mississippian through Permian.

Distribution: Kentucky, Illinois, Iowa, Missouri, Kansas, Oklahoma, Texas, southwestern United States, Nevada, Oregon.

Comments: In life, *Meekella* lost its stalk at an early stage and cemented itself at the ventral umbo to a hard substrate. Most specimens will show the attachment scar as an irregularly shaped area without ornamentation but with pits, suggesting that it was attached by a byssus.

The illustration of *Meekella* in Plate 92 shows the interior and exterior of the ventral valve at the top, and of the dorsal valve at the bottom.

57 Schuchertella

Description: ½" (13 mm) long, ⅝" (16 mm) wide. Semicircular, widest at hinge, sometimes irregular in shape. Ventral valve gently convex, dorsal valve flat to slightly convex; fine radial ribs cover both valves. Teeth lack dental plates; well-developed ventral interarea has covered notch. Ventral muscle scar area oval, with fan-shaped opening-muscle scars and elongate-oval closing-muscle scars. Dorsal valve has low cardinal process fused with short, diverging ridges.

Age: Lower Devonian through Permian.

Distribution: Widespread in North America.

Comments: *Schuchertella* is a large genus, containing many species that are probably not all closely related. The irregularity in shape sometimes found in this genus occurred because, in life, some of the animals cemented themselves to the substrate. Similar in shape and ornamentation, *Schellwienella* is found throughout North America from the Devonian through the Mississippian, and can be distinguished by the presence of short dental plates and a ridge surrounding muscle scars in the ventral valve interior. *Schellwienella* was also cemented to the substrate, at least in its young stages.

Plate 57 shows, at the left, the exterior of a dorsal valve at top and of a ventral valve below. On the right are the interiors of 2 dorsal valves.

56 Derbyia

Description: 1″ (25 mm) long, 1¼″ (32 mm) wide.
Triangular to almost circular. Both
valves convex, ventral valve irregular in
shape. Ornamentation of fine radial ribs
present. Ventral interarea with covered
notch; teeth without dental plates.
Ventral interior with high, strong
median septum fusing with notch
cover; muscle scar area circular, with
fan-shaped opening-muscle scars.
Dorsal valve has high, 2-lobed cardinal
process attached to high supporting
ridges that form an inverted V. Interior
covered with small bumps.

Age: Mississippian through Permian.

Distribution: Southern Appalachian Mountains,
Ohio, Indiana, Kentucky, Iowa,
Missouri, Nebraska, Kansas,
Oklahoma, Texas, Nevada.

Comments: *Derbyia* appears to have passed through
2 stages in life: first, attachment by
byssuslike threads after the larva
settled; then, free-living when the shell
was large and heavy enough to be
stable on the substrate.
Plate 56 shows the interior and exterior
of the ventral valve above, and of the
dorsal valve below.

54 Devonochonetes

Description: ⅝″ (16 mm) long, ⅞″ (22 mm) wide.
Semicircular. Ventral valve slightly
convex, dorsal valve flat to slightly
concave. Ornamentation of many fine
radial ribs; few tiny spines on surface.
Ventral and dorsal interareas usually
have covers over notches; oblique,
hollow spines present on posterior
margin of ventral valve. Ventral
interior has large, fanlike opening-
muscle scar, and long ridge down
center. Dorsal interior has large 2- or
3-lobed cardinal process supported by
median ridge. Small bumps on valve

Age: interiors, becoming large near margins.
Middle Devonian.

Distribution: Widespread in North America.

Comments: The spines that projected from the ventral posterior margin of *Devonochonetes* are rarely preserved on the shell, but, since they were hollow, fine holes can usually be seen along the posterior margin where they were attached. These spines may have served to help stabilize the shell as it lay loose on the soft bottom. Plate 54 shows several valves of *Devonochonetes*, mostly ventral, with the bases of the spines visible on the posterior margin. *Neochonetes*, in the same family, is restricted to the Pennsylvanian and Permian throughout North America. It can be distinguished from *Devonochonetes* by the inclination of the spines, which are almost parallel to the hinge line, and by the absence of a cover on the ventral interarea notch. The fine spines over the surface are more numerous than in *Devonochonetes*.

58 Mesolobus

Description: ⅜″ (10 mm) long, ⅝″ (16 mm) wide. Small; oval, wider than long, with wide, straight hinge. Ventral valve convex, dorsal valve concave; most species have ventral groove bearing median fold, and dorsal fold bearing median groove. Smooth or ornamented with very fine ribs. Tiny spines scattered over surface, usually represented only by small bumps that are their bases. Row of larger spines along ventral posterior margin that recline 45° away from umbo. Dorsal valve has 2 lobes on cardinal process; 2 sets of interior ridges begin at cardinal process, 1 set parallels hinge and 1 set descends obliquely; median ridge extends almost to anterior margin. Inner surface of both valves covered with small bumps.

Age: Pennsylvanian.

Distribution: Widespread in North America.

Comments: Plate 58 shows 2 articulated specimens of *Mesolobus.* The one on the left has the dorsal valve forward; the one on the right shows the ventral valve covered with the tiny bases of fine spines; the bases of the larger posterior spines are also visible. These spines were probably used to anchor the valves in the mud, just as in *Devonochonetes.* Another chonetid with hinge spines and tiny spines covering the body is *Chonetinella,* found in the Lower Permian in the midwestern, central, and southwestern United States. *Chonetinella* has the same general shape as *Mesolobus,* but the dorsal fold and ventral groove are much more strongly developed, and the ventral valve is more strongly convex.

111 Productella

Description: ⅞" (22 mm) long, 1¼" (32 mm) wide. Semicircular to square. Ventral valve deeply convex, dorsal valve concave; ventral beak curved over hinge line. Ventral valve has scattered bumps bearing spines, and rows of spines near hinge; dorsal valve has few spines or none. Ventral valve has small hinge teeth; notch in interarea absent. Cardinal process has 2 high lobes.

Age: Middle and Upper Devonian.

Distribution: New York, Appalachian Mountains, Ontario, Great Basin, Mackenzie Basin, Canadian Rocky Mountains, Rocky Mountains.

Comments: Very similar externally to *Productella* is *Leioproductus,* a genus that has lost its teeth, sockets, and interareas, and is found in the Upper Devonian in western North America. The ventral valve of *Leioproductus,* which is very convex, has fewer and larger spines than that of *Productella,* and has a median ridge from near the umbo to the anterior margin, which bears broadly spaced spines.

The spines around the margins of *Productella, Leioproductus,* and other genera probably served mainly as sensory devices. Hollow and filled by extensions of the mantle tissue, these spines reached well beyond the commissure and could give early warning if the animal was disturbed, so that the valves could snap shut quickly. The spines also served as stabilizing devices to keep the shell from being dislodged.

The illustration of *Productella* shows a ventral valve. The bumps that held the spines are visible, although the spines themselves have broken off.

93 Diaphragmus

Description: ¾" (19 mm) long, ⅝" (16 mm) wide. Rounded to elongate-oval. Ventral valve deeply convex, with inflated umbo curved over hinge area; dorsal valve concave. Ornamentation of medium-size radial ribs covers surface; concentric undulations restricted to umbones. 2 rows of spines near hinge on ventral valve, more spines on flanks; dorsal valve has spines on flanks and near anterior margin. Semicircular incised area on dorsal valve anterior to umbo, where anterior margin begins to grow toward dorsal side; incision corresponds to flangelike ridge on interior. Interareas, teeth, and sockets absent. Cardinal process with 2 lobes.

Age: Upper Mississippian.

Distribution: Appalachian Mountains, Tennessee, Illinois, Arkansas, Oklahoma, Wyoming, Great Basin.

Comments: *Diaphragmus* is named for the flangelike ridge, called a diaphragm, that encircles the visceral area internally on the dorsal valve. The diaphragm must have functioned to restrict access to the visceral cavity. Anterior to the diaphragm, both valves form an extension in the shape of a half-cylinder, which was probably used to keep the anterior margin above the mud surface so that the animal could feed. *Diaphragmus* and other brachiopods with such extensions probably lay buried in the sediment with the disclike posterior part of the dorsal valve horizontal and the extension projecting vertically upward into the water. The spines would stabilize the shell.

Plate 93 shows 3 ventral valves of *Diaphragmus*. Smaller *Diaphragmus* were attached near the anterior margins of the 2 lower specimens. The larger shells grew around them, leaving an external mold of the small shells in the ventral valve of the larger shells. The marginal drawing shows a side view of *Diaphragmus*.

62 Echinoconchus

Description: 1½″ (38 mm) long, 2″ (51 mm) wide. Round to rectangular. Ventral valve long, curving, with large umbo and strongly incurved umbo tip; dorsal valve concave. Radial ornament absent; strong concentric bands on both valves contain spines of several sizes. Spines recline, forming dense coat over shell. Interareas, teeth, and sockets absent. Cardinal process 3-lobed, with long shaft.

Age: Mississippian.

Distribution: Southern Appalachian Mountains, Indiana, Kentucky, Tennessee, Mississippi Valley, Arkansas, Oklahoma, Texas, Nevada.

Comments: A similar, very spiny brachiopod is *Waagenoconcha,* found in the Upper Pennsylvanian through the Upper Permian of Indiana, Texas, the southwestern United States, and Wyoming. *Waagenoconcha,* like *Echinoconchus,* has spines covering both valves, but the spines are largest near the umbones and become smaller toward the margins.

The spines of *Echinoconchus* and *Waagenoconcha* may have functioned in 3 ways: first, to help stabilize the shells as they lay free on the substrate; second, to provide sensory information through internal mantle tissue; and third, to serve as straining devices near the commissure, so that large particles would not be drawn into the valves. Plate 62 shows the ventral valve of *Echinoconchus* on the left and the dorsal valve on the right. Although the spines have broken off, their bases are visible as small bumps on some of the ribs.

60 Juresania

Description: ⅝″ (16 mm) long, ⅞″ (22 mm) wide. Round in outline; ventral valve convex, dorsal valve concave. Both valve surfaces covered with small spines; weak, wide, concentric ridges present on posterior areas of both valves; early stages with small spines, later stages with concentric bands carrying both small and large reclining spines. Ventral valve has attachment scar near umbo. Cardinal process high, 2-lobed; median ridge present in dorsal valve; branching grooves present in closing-muscle scars. Interiors also covered with spines.

Age: Pennsylvanian through Lower Permian.

Distribution: Nebraska, Kansas, Oklahoma, Texas, Nevada.

Comments: A similar genus is *Reticulatia,* which is longer and wider than *Juresania,* and shows a pronounced, beadlike pattern over the posterior area of both valves. Shell layers overlap on the ventral valve near the margin, and the spines are restricted to the flanks and the area near the hinge. The dorsal valve has no spines, although related genera may have spines on the dorsal valve. 2 articulated specimens of *Juresania* are illustrated in Plate 60; the dorsal valve is visible on the left, and the ventral valve on the right.

48 Reticulatia

Description: 1⅝" (41 mm) long, 2⅜" (60 mm) wide. Round, with high ventral umbo and extended hinge line. Ventral valve deeply convex, dorsal valve concave. Both valves covered by cordlike radial ridges, densely beaded by concentric ribs on ventral posterior and over most of dorsal valve. Ventral valve has overlapping sheets of shell on anterior. Spines present near hinge and on flanks of ventral valve; absent on dorsal valve. Teeth sockets absent. High cardinal process with 3 lobes, median lobe broad; median ridge down center of dorsal valve. Closing-muscle scar with branching grooves. Inside of valves covered with small, rodlike bumps.

Age: Upper Pennsylvanian through Lower Permian.

Distribution: Widespread in North America.

Comments: In life, *Reticulatia* was not attached by stalk or by cementation, but lay in the soft sediment with its deep, heavy ventral umbo buried, and its anterior margin high off the bottom.

Plate 48 shows 2 articulated specimens of *Reticulatia*; on the left, the dorsal valve is forward, and on the right, the ventral valve is forward. *Reticulatia* was formerly called *Dictyoclostus,* a genus now restricted to Europe.

59 Linoproductus

Description: ⅞" (22 mm) long, 1¼" (32 mm) wide. Round to very elongate; ventral umbo curving over wide hinge. Both valves covered with fine radial ribs that may be variously distorted. Spines in 1 or 2 rows near hinge and scattered over ventral valve. Valve surface wrinkled on flanks and auricles; concentric ridges present on dorsal valve. Dorsal valve has 3-lobed cardinal process that is finely ribbed; closing-muscle scar has branching grooves.

Age: Pennsylvanian through Permian.

Distribution: Widespread in North America.

Comments: When very young, *Linoproductus* was probably attached by small, ringlike spines near the hinge. Attachment may have been to the spines of mature *Linoproductus*. At a later stage *Linoproductus* probably became too heavy for the attachment spines and broke off to become free-living. The heavy ventral posterior would have sunk into the sediment with the anterior margin projecting upward. Plate 59 shows 2 articulated specimens of *Linoproductus;* the ventral valve is visible on the left, and the dorsal valve on the right. The bumps that held long spines are visible on the ventral valve.

ORDER PENTAMERIDA
(Middle Cambrian through Upper Devonian)

This small order has fewer than 100 genera. The typical pentamerid has a short, curved hinge line, convex valves, an open notch in the ventral interarea, and a raised concave platform in the ventral valve interior for the attachment of the opening and closing muscles. The interiors of the valves never have pits or tiny bumps.

119 Pentamerus

Description: 4" (102 mm) long, 3¾" (95 mm) wide. Large, smooth, elongate, often tongue-shaped in outline, moderately convex; inverted-teardrop-shaped in profile. Ventral valve with raised platform for muscle attachment in posterior half of shell; platform supported by high, strong median septum. Brachiophores supported by high, thin plates.

Age: Silurian.

Distribution: Anticosti Island (Quebec), New York, Ontario, Ohio, Indiana, Kentucky, Wisconsin, Illinois, Iowa.

Comments: Since *Pentamerus* has large, strongly convex, swollen valves, the purpose of the muscle-attachment plates must have been to shorten the distance the muscles had to travel to connect the 2 valves.

Stricklandia, common in the Lower Silurian of Ontario and the southern Appalachian Mountains, looks very similar to *Pentamerus,* but has a few weak to strong radial ribs, and a smaller ventral muscle-attachment plate, with a shorter median ridge support. *Virgiana,* from the Lower Silurian, is found most commonly on Anticosti Island, Quebec, in the mouth of the St. Lawrence River. It is also similar, and has a smooth or radially ribbed surface with a ventral fold, a short ventral interior median ridge, and small plates supporting the brachiophores.

Plate 119 shows a cluster of *Pentamerus,* which is commonly found preserved this way, and therefore believed to have been gregarious.

82 Conchidium

Description: 2" (51 mm) long, 1⅝" (41 mm) wide. Large, tongue-shaped, with very long, curved ventral umbo. Valves strongly

convex, with thick radial ridges.
Muscle-attachment plate in ventral
valve supported by median septum that
extends forward more than half length
of valve.

Age: Upper Ordovician through Lower
Devonian.

Distribution: New Brunswick, Quebec, Appalachian
Mountains, Ontario, Ohio, Tennessee,
Mississippi Valley, Nevada, Alaska.

Comments: Some species of *Conchidium* are among
the largest of all known brachiopods.
C. alaskense measures more than 4"
(102 mm) long.

Plate 82 shows a ventral view of 2
specimens. The marginal drawing
shows an articulated specimen from the
side.

106 Gypidula

Description: 1" (25 mm) long and wide. Circular to
elongate-oval, with ventral umbo
arched well over hinge line. Ventral
valve more convex than dorsal. Strong
ventral fold and dorsal groove present.
Ornamented with medium-thick, faint
to pronounced radial ribs. Raised
muscle-attachment plate in ventral
valve; bladelike brachiophore in dorsal
valve.

Age: Middle and Upper Devonian.

Distribution: New Brunswick, Quebec, New York,
Appalachian Mountains, Illinois, Iowa,
Missouri, Manitoba, Mackenzie Valley
of the Northwest Territories, Rocky
Mountains, Arizona, Nevada, Alaska.

Comments: A genus closely related to *Gypidula*,
Antirhynchonella (formerly called
Barrandella), from the Silurian of
Anticosti Island (Quebec), and New
York, Ohio, Indiana, Kentucky,
Tennessee, and Arkansas, has the dorsal
fold and ventral groove, and is smooth.
Plate 106 shows an articulated
specimen of *Gypidula* with the larger,
ventral valve to the left.

ORDER RHYNCHONELLIDA
(Middle Ordovician to Recent)

These are small, distinctive shells, often with highly inflated valves. The ornamentation is of coarse radial ribs that usually form a zigzag margin. Rhynchonellids commonly have a strong fold on the dorsal valve and a matching groove on the ventral. The hinge line is curved, and the interareas are reduced or absent. The ventral umbo is commonly pointed and prominent, and the notch in the ventral valve is partially closed, leaving a circular opening for the stalk. Dental plates generally support the teeth in the ventral valve. 2 ridges extend down from the umbonal area inside the dorsal valve and function to support a spiral lophophore, with the apex of the spire toward the anterior margin. Most rhynchonellids have neither pits nor bumps on the inside of the shells; all have a stalk.

91   Rhynchotreta

Description: ½" (13 mm) long, ⅜" (10 mm) wide. Small, triangular, longer than wide; ventral umbo erect, forming sharp posterior point. Ornamented with strong radial ribs. Ventral valve may have anterior groove; dorsal valve may have anterior fold. Strong teeth have dental plates. Stalk opening at apex of ventral valve; notch partially covered. Dorsal valve lacks cardinal process, has long, curved brachiophores and median septum.

Age: Silurian.

Distribution: New Brunswick, Quebec, New York, Appalachian Mountains, Ohio, Indiana, Kentucky, Tennessee, Wisconsin, Illinois, Missouri, Oklahoma.

Comments: *Rhynchotreta* is a member of the

rhychonellids, a group with many similar-looking genera. Classification to genus in most cases is dependent on the interiors of the shells. Since most rhynchonellids have strong interlocking teeth and sockets, the valves are usually preserved still locked together. Therefore, determining the interior morphology is often a problem, and paleontologists commonly make serial sections by cutting across the shell, at regular intervals, in a plane perpendicular to the commissure. Peels or photographs are made of each section and the 3-dimensional morphology reconstructed. This is seldom a job for the amateur fossil collector, who may have to settle for tentative identifications for many genera in this group.

Plate 91 shows 2 articulated specimens of *Rhynchotreta:* the one on the left with the ventral valve forward, and the one on the right with the dorsal valve forward.

85, 89 Rhynchotrema

Description: ⅜″ (10 mm) long, ½″ (13 mm) wide (pl. 85); 1¼″ (32 mm) long, 1½″ (38 mm) wide (pl. 89). Small, triangular in outline; ventral umbo not pointed. Valves convex, margins almost circular. Ornamented with strong, angular radial ribs beginning at apex and crossed by closely spaced concentric lines. Small stalk opening at ventral apex; notch partially closed by plates; muscle scars triangular. Dorsal valve has slender to thick cardinal process, short, thick median septum.

Age: Middle and Upper Ordovician.

Distribution: Widespread in North America.

Comments: The angular ribs of *Rhynchotrema* interlock at the margins, resulting in a zigzag-shaped commissure, advantageous to the animal because it caused the valves, when closed, to be held firmly together. Perhaps even more important, the area of the gape was increased, and the valves needed to open only a short distance. The resultant small height of the opening could act as a sieve to keep out large particles that might damage the lophophore.

Plate 85 shows the anterior side of an articulated specimen of *Rhynchotrema* with the dorsal valve on top. Plate 89 shows the ventral valve.

80 Cupularostrum

Description: ½" (13 mm) long and wide. Triangular in outline, with shallow ventral valve and strongly convex dorsal valve. Tip of ventral umbo prominent, extending beyond hinge. Ventral groove and dorsal fold well developed anteriorly, forming tonguelike extension of ventral valve at anterior margin. Ornamented with large, simple ribs beginning at tip of umbones. Short dental plates present in ventral valve; stout median septum extending to ½ shell length in dorsal valve. Cardinal process absent.

Age: Middle Devonian.

Distribution: Widespread in North America.

Comments: Many species included in *Cupularostrum* used to be called *Camarotoechia*. Plate 80 shows a bedding plane on which many valves are preserved. The convex fossils are internal molds, the concave ones, external molds. In the upper right of the photograph is a large internal mold of a ventral valve. Just to its left is an internal mold of a dorsal valve. The 2 valves with a deep groove down their centers represent a different genus. The marginal drawing shows an anterior view of *Cupularostrum* with the valves closed.

84 Uncinulus

Description: ⅝" (16 mm) long and wide. Small, round to almost triangular in outline; highly convex, profile round. Ventral valve usually less convex than dorsal. Fold and groove usually present. Tongue on ventral valve fits into opening on dorsal valve. Ornamented with numerous rounded ribs, with straight furrows that project internally beyond margin of valves as slender spines. Stalk opening tiny or absent; ventral muscle scars divided by delicate ridge, with oval opening-muscle scar enclosing small, round closing-muscle scar. Dorsal median septum high; cardinal process low and broad; muscle-attachment area has numerous tiny vertical ridges.

Age: Devonian.

Distribution: Quebec, eastern United States, Mississippi Valley, Oklahoma.

Comments: *Uncinulus,* like *Rhynchotrema* and *Rhynchotreta,* has a zigzag commissure, but with an improvement. *Uncinulus* eliminated the greater gape at the crests of the slit by secreting spines along the margin. The spines at each furrow tuck beneath the crest of the opposite valve. In life, when the valves were open, the spines acted as a sieve, with very little restriction of water flow.

The related *Plethorhyncha,* found throughout North America in the Lower Devonian, is similar to *Uncinulus,* but lacks the pronounced tongue and slot of the latter.

Plate 84 shows 2 articulated specimens of *Uncinulus,* the dorsal valve on top in both. The specimen on the left shows the posterior with the hinge; the one on the right shows the anterior margin with the ventral tongue and dorsal slot.

88 Eatonia

Description: ⅞" (22 mm) long, 1" (25 mm) wide. Round, with bluntly pointed ventral umbo. In side view, lateral dorsal valve margins bend around and partly cover ventral valve margins. Ornamented with broad radial ribs, sometimes with fine radial and concentric lines superimposed. Large plates cover ventral interarea notch. Dorsal median septum short. Cardinal process has stout stem and 2 strong, posterior-projecting lobes, concave on ends.

Age: Lower Devonian.

Distribution: Quebec, New York, Appalachian Mountains, Tennessee, Oklahoma, Texas.

Comments: *Eatonia* was named for Amos Eaton, a man so important in the early history of American geology that the decade from 1820 to 1830 is still called the Eatonian Era. Eaton studied law and was at one time legal advisor to the van Rensselaers (see *Rensselaeria*). His real interest, however, was natural history, and especially botany, and his home soon became a museum. He lectured not only on botany, but also chemistry, medical law, natural philosophy, mineralogy, and geology. Eaton was instrumental in winning legislative approval for the Erie Canal and was the force behind the founding of the Rensselaer Polytechnic School.
Plate 88 shows a ventral valve of *Eatonia* on the left and a dorsal valve on the right.

ORDER SPIRIFERIDA
(Middle Ordovician through Jurassic)

This is a large and diverse group of brachiopods united by the presence of a spiral lophophore that was supported by 2 calcareous spires, either directed laterally, or with ends directed or toward the dorsal valve. The shells are almost always both convex, and the hinge line may be straight or curved, wide or narrow. Most spiriferids have radial ornamentation, but some are smooth. The ventral valve generally has a large interarea (although it may be much reduced or absent), but the dorsal valve usually has no visible interarea. The triangular notch on the ventral interarea is either closed or open. The insides of the valves are smooth or, rarely, have tiny pits.

81 Zygospira

Description: ¼" (6 mm) long and wide. Small; round, with strong radial ribs forming zigzag commissure. Ventral valve slightly deeper than dorsal valve; ventral fold and dorsal groove present. Notch of ventral interarea partly closed by plate; stalk opening occurs partly in notch cover, partly in umbo; dental plates absent. Brachiophore consists of 2 spiral cones directed toward dorsal center.

Age: Middle and Upper Ordovician.

Distribution: Quebec, New York, Ontario, Appalachian Mountains, Ohio, Mississippi Valley, Tennessee, Oklahoma.

Comments: *Zygospira* is an early, primitive member of the large, important Paleozoic order Spiriferida. The spirifers had spiral lophophores, and the supporting bands for these are sometimes preserved in the fossils. Spirifers are almost always convex, with relatively large body cavities.

2 other primitive spirifer genera that are common in North America could be confused with *Zygospira*. *Protozyga*, from the Middle Ordovician of New York, the Appalachian Mountains, and the Mississippi Valley, is more triangular, and either is smooth or has

a few partial radial ribs. *Atrypina,* from the Lower Silurian through the Lower Devonian of Quebec, New York, the Appalachian Mountains, Indiana, Tennessee, Missouri, and Oklahoma, is also more triangular than *Zygospira,* with a few wide, low, rounded radial ribs, and sheetlike concentric growth lines crossing the ribs.

Plate 81 shows an accumulation of many articulated *Zygospira* specimens. They may have been living clustered together when they died, or may have been dislodged by a storm, swept together, and quickly buried.

74 Desquamatia

Description: 1⅜" (35 mm) long and wide. Rounded, but usually slightly longer than wide; hinge wider than anterior margin. Ventral valve less convex than dorsal, to almost flat. Ventral umbo curves slightly over hinge. Valves may have ventral groove and dorsal fold. Fine radial ribs increase in number by forking and by arising in interspaces. Concentric growth lines not prominent. Dental plates well developed in ventral valve. Dorsal valve has widely diverging brachiophore and dorsally directed spiral lophophore cones.

Age: Devonian.

Distribution: Widespread in North America.

Comments: *Desquamatia* is one of the most common genera in the Devonian. It was formerly considered part of the genus *Atrypa.* Plate 74 shows an articulated specimen, dorsal valve forward. The marginal drawing shows a side view.

161 Spinatrypa

Description: 1" (25 mm) long and wide. Valves round in outline; ventral valve flat to slightly convex, dorsal valve strongly convex. Ornamentation of coarse, rounded ribs, with new ribs arising toward anterior margins by forking or by forming between old ribs. Concentric ornamentation of coarse growth lines with blunt spines. Ventral umbo curves slightly over hinge line; ventral groove and dorsal fold may be present. Interareas absent, dental plates short or absent. Ventral opening-muscle scars large, fan-shaped. Brachiophore is 2 spiral cones directed toward center of dorsal valve.

Age: Devonian.

Distribution: Widespread in North America.

Comments: *Spinatrypa* species were formerly included in *Atrypa,* which was one of the most common Paleozoic brachiopod genera. Now *Atrypa* has been split into about 20 separate genera, and *Atrypa* in the new, restricted sense is not common in North America. Other species formerly in *Atrypa* have been placed in the genus *Desquamatia,* also widespread in North America in the Devonian. *Desquamatia* has no spines and has finer ribs than those of *Spinatrypa.*

Plate 161 shows 2 articulated specimens of *Spinatrypa.* On the left the ventral valve is visible; on the right the dorsal valve is shown. The spines were broken off these specimens, probably when the shells were removed from the matrix, and may show on the external molds that formed around these fossils.

90 Hustedia

Description: ⅜″ (10 mm) long, ¾″ (19 mm) wide. Small, triangular; posterior pointed. Ventral valve slightly more convex than dorsal. Fold absent. Ornamented with even, thick, rounded radial ribs. Prominent opening for stalk at ventral apex; interarea notch covered by plates.

Teeth present, but dental plates absent. Cardinal process distinctive, with 4 prongs projecting into ventral valve, single small spine at base. Lophophore supports form laterally directed spirals. Interior covered with tiny pits.

Age: Mississippian through Permian.

Distribution: Widespread in North America.

Comments: 2 common North American genera are related to *Hustedia,* and have very similar interiors, but are easily distinguished externally. *Parazyga,* from the Middle Devonian of New York, Pennsylvania, Ontario, Michigan, and Indiana, is usually wider than long (although it may be elongate-oval), and has a deep fold and groove, and ribs that bear fine spines. *Trematospira,* from the Lower Devonian of New Brunswick, Quebec, New York, the Appalachian Mountains, Ontario, Indiana, Kentucky, Tennessee, Illinois, and Missouri, is always wider than long and has a diamondlike shape. It has fine, regular ribs, and dental plates, but no spines.

Plate 90 shows a dorsal view on the left, and a ventral view on the right, of 2 articulated specimens of *Hustedia.*

109 Meristella

Description: ¾″ (19 mm) long, ⅝″ (16 mm) wide. Triangular to round. Large ventral umbo curving over hinge line may conceal opening for stalk. Valves convex; ventral valve usually deeper. Ventral groove and dorsal fold usually present; valves smooth, with irregularly spaced growth lines. Dental plates absent. Ventral muscle scars triangular, deeply impressed, striated. Dorsal valve has concave, triangular or square plate on interior of apex; median septum extends from plate to anterior margin.

Age: Lower and Middle Devonian.

Distribution: New Brunswick, Quebec, Maine, New York, Appalachian Mountains, Ontario, Tennessee, Illinois, Missouri, Oklahoma, Nevada.

Comments: *Meristina* somewhat resembles *Meristella* externally. It occurs in the Silurian in Quebec, New York, Indiana, Tennessee, Wisconsin, Illinois, Missouri, and Oklahoma. *Meristina* has dental plates extending forward as ridges that surround the muscle area. *Merista,* another similar and closely related genus, is found in the Upper Silurian through the Middle Devonian of Quebec, New York, the Appalachian Mountains, western Tennessee, and Oklahoma, and rarely in the Mississippian of Texas. *Merista* can be distinguished by the unusual muscle-attachment plate inside the ventral valve. The plate consists of an anchored rooflike sheet of shell that forms an inverted V-shape, with the apex attached just below the dental plates and the sides attached laterally. Plate 109 shows internal molds of 2 articulated specimens of *Meristella*. On the right, the dorsal median septum can be seen near the umbo as a groove.

---

110 Athyris

Description: 1⅛" (28 mm) long, 1¼" (32 mm) wide. Round to oval, with narrow hinge line. Valves convex, ventral valve grooved anteriorly, dorsal valve folded. Concentric growth lines narrowly to widely spaced. Stalk opening circular, dental plates short, stout. Brachiophore is 2 spirals directed toward sides of valves.

Age: Lower Devonian through Mississippian.

Distribution: Widespread in North America.

Comments: *Athyris* was one of the earliest brachiopods capable of reabsorbing its shell around the stalk opening to allow the stalk to expand in size. In most

other brachiopods, the growth of the
stalk diameter was limited once the
stalk opening was formed. As the
animal grew larger and heavier, the
stalk could not increase in diameter. In
many brachiopods, this restriction must
have limited the maximum size of the
valves. In other groups, the valves
commonly broke off to become free-
living in adulthood.

Plate 110 shows a dorsal view on the
left, and a ventral view on the right, of
2 articulated specimens of *Athyris*.

133 Cleiothyridina

Description: ⅝″ (16 mm) long, ¾″ (19 mm) wide.
Circular to oval; umbones small. Valves
convex, usually with ventral groove and
dorsal fold. Surfaces covered with
broad, sheetlike concentric expansions
of shell projecting toward anterior as
broad, flat spines. Ventral umbo has
circular opening for stalk. Dental plates
present; cardinal process small and
triangular.

Age: Mississippian through Permian.

Distribution: Appalachian Mountains, Indiana,
Kentucky, Illinois, Missouri, Arkansas,
Kansas, Oklahoma, Texas.

Comments: *Cleiothyridina* had one of the thickest
coverings of spines of any brachiopod.
The spines were not hollow, and thus
could not have been sensory as they
were in some other genera. The stalk
opening is large, and probably held a
functioning stalk at all stages. It is
therefore unlikely that the spines served
to stabilize the shell on the substrate,
so they must have helped to strain out
large particles from the water before it
entered the mantle cavity.
Cleiothyridina is almost always preserved
with most of the spines broken off, but
traces of spines can usually be seen,
especially around the margins.
In Plate 133, the bases of some of the

spines of *Cleiothyridina* can be seen, especially on the dorsal valve on the right. The ventral valve is shown on the left.

134 Composita

Description: 1" (25 mm) long and wide. Smooth, round to triangular, with pronounced ventral groove and dorsal fold. Valves convex. Stalk opening at tip of ventral umbo prominent; dental plates short; square plate inside dorsal umbo with small hole near apex. Median ridge absent.

Age: Upper Devonian through Permian.

Distribution: Widespread in North America.

Comments: With its smooth, relatively small shells, *Composita* would have been well adapted to live in a turbulent environment. In fact, it is often found in rock with much broken shell debris and large compound corals turned upside down.

Plate 134 shows 2 articulated specimens of *Composita*. On the left is a side view, with the ventral valve to the left. On the right only the ventral valve is visible.

98 Eospirifer

Description: 1¼" (32 mm) long, 1⅜" (35 mm) wide. Round to almost oval. Ventral umbo curved toward hinge; ventral valve strongly convex, dorsal valve equally or less convex. Well-developed groove in ventral valve, fold in dorsal valve. Dental plates long; notch in ventral interarea covered by plates. Fine radiating ribs present.

Age: Silurian.

Distribution: Widespread in North America.

Comments: *Macropleura,* from the Silurian and Devonian of Maine, New York, the

Appalachian Mountains, and Tennessee, is so similar to *Eospirifer* that it was included in that genus until 1963. *Macropleura* is larger, with each of its flanks marked by 3 rounded folds. Plate 98 shows 2 articulated specimens of *Eospirifer*. The ventral valve is visible on the left, the dorsal valve on the right. The marginal drawing shows a side view of an articulated specimen.

97 Ambocoelia

Description: ⅜″ (10 mm) long and wide. Small; semicircular. Ventral valve strongly convex, but with high ventral umbo that curves toward hinge. Dorsal valve weakly convex. High, open interarea notch on ventral valve. Hinge line narrow. Valves smooth, with very fine concentric growth lines. Dental plates absent; cardinal process large.

Age: Lower Devonian through Mississippian.

Distribution: New York, Appalachian Mountains, Michigan, Ohio, Indiana, Kentucky, Rocky Mountains.

Comments: When one of the most common *Ambocoelia* species was first described in 1842, it was called *Orthis umbonata.* This name placed it in a different order from the one in which it is now classified, and grouped it with orthids like *Billingsella* and *Eoorthis.* Such an example illustrates the great difficulty paleontologists have always had in classifying brachiopods, especially when internal structure is not known. Plate 97 shows 2 ventral valves of *Ambocoelia*, an exterior view on the left and an interior on the right.

94 Cyrtina

Description: ¼″ (6 mm) long, ⅜″ (10 mm) wide. Ventral valve strongly conical; dorsal valve small, caplike, slightly convex, with median fold. Ornamented with broad radial ribs. Very large, high, triangular interarea present on ventral valve, often somewhat deformed; notch covered by convex hood, but open at hinge. Dental plates unite with median ridge to form raised muscle-attachment area. Interior covered with tiny pits.

Age: Silurian through Mississippian.

Distribution: Widespread in North America.

Comments: The large interarea on *Cyrtina* probably served as a flat, stable base for the shell to rest on.

Cyrtina might be confused with *Spondylospira*, a closely related genus from the Triassic of Idaho, Nevada, the Yukon, California, and Alaska; but the interarea notch of *Spondylospira* is open, and little teeth run all along the hinge line.

2 articulated specimens of *Cyrtina* are shown in Plate 94. On the left is a posterior view with the broad interarea of the ventral valve visible. The ventral valve is shown on the right.

50 Mucrospirifer

Description: ½″ (13 mm) long, 1⅝″ (41 mm) wide. Much wider than long, with extremities of hinge line extended out into sharp points. Valves convex. Ventral groove may have single ridge down center; dorsal fold smooth, pronounced. Ornamented with fine concentric growth lines and radial ribs on flanks. Interior pits absent.

Age: Middle Devonian.

Distribution: New York, Appalachian Mountains, Ontario, Michigan.

Comments: The name *Mucrospirifer* means "sharply pointed spire-bearer." The hinge line is extremely elongated in some species, and probably served 2 functions. The free-lying shell would have had a large surface area to help prevent it from sinking into the mud, and the "wings" would help separate in-flowing and out-flowing currents. Water would enter at the "wings," pass through the spiral lophophore, and exit at the anterior fold.

Another spirifer genus with an extended hinge line is *Cyrtospirifer,* from the Upper Devonian through the Lower Mississippian of New York, the Appalachian Mountains, Iowa, the Mackenzie Valley, the Canadian Rockies, Montana, Idaho, Arizona, and Nevada. In *Cyrtospirifer,* both the fold and groove are ribbed, and there are no interior pits.

Plate 50 shows 2 articulated specimens of *Mucrospirifer;* the dorsal valve is visible on the left, and the ventral valve on the right.

53 Spinocyrtia

Description:	1⅝" (41 mm) long, 2⅜" (60 mm) wide. Wider than long, with hinge extended into small "wings." Both valves convex; ribs broad, low, radial, unforking. Ventral groove and dorsal fold smooth, but sometimes with slight groove in middle of fold. Entire surface covered with minute, teardrop-shaped granules. Ventral valve has short, stout dental plates and cover on interarea notch. Median ridge absent. Cardinal process is roughened area under umbo. Brachiophore has many laterally directed coils. Interior without pits.
Age:	Middle and Upper Devonian.
Distribution:	New York, Pennsylvania, Appalachian Mountains, Ontario, Michigan, Indiana, Iowa.
Comments:	Many *Spinocyrtia* species were formerly placed in the genus *Spirifer,* but *Spirifer* has now been split into many separate genera and, in this restricted sense, is no longer a common genus in North America. 2 common North American

genera that are similar to *Spinocyrtia* are
Mediospirifer and *Paraspirifer*.
Mediospirifer is found in the Lower
Devonian of New York, Pennsylvania,
Maryland, Virginia, Ontario,
Michigan, Ohio, Indiana, Kentucky,
and Missouri. It has no granules on its
ribs, and has a large interarea with a
wide notch and a comblike cardinal
process. *Paraspirifer* is found in the
Lower and Middle Devonian of New
York, Pennsylvania, Maryland,
Ontario, Ohio, Indiana, Kentucky, and
Illinois. *Paraspirifer* is widest at mid-
length, and has a highly convex dorsal
valve with a less convex ventral valve,
radial ribs that fork, and its surface
covered by fine concentric lines bearing
minute spines.
Plate 53 shows an articulated specimen
of *Spinocyrtia* with the dorsal valve
forward. The small granules show
clearly, especially on the fold.

75 Costispirifer

Description: 1⅞" (48 mm) long, 2" (51 mm) wide.
Large; wider than long, with maximum
width at hinge line; valves convex,
with ventral groove and dorsal fold
present. Medium-coarse, flat-topped
radial ribs covered with very fine radial
ribs. Dental plates short, thick.
Age: Lower Devonian.
Distribution: Quebec, New York, the Appalachian
Mountains, Ontario, Tennessee,
Illinois, Missouri, the Great Basin.
Comments: *Costispirifer* has thick deposits of shell
inside the cavities under the umbones
that probably served as weights to hold
the shell firmly on the substrate.
Plate 75 shows the dorsal side of an
articulated specimen of *Costispirifer*.

52 Neospirifer

Description: 1⅜" (35 mm) long, 2⅛" (54 mm)
wide. Large; with hinge width equal or
almost equal to maximum shell width;
ends of hinge line angular. Both valves
convex; ventral groove and dorsal fold
present. Small, ridgelike teeth along
inside of hinge. Ornamented with
strong radial ribs that differ in width
and are covered with fine growth lines.
Ribs present on groove and fold.
Ventral interarea distinct, dental plates
short.

Age: Pennsylvanian through Permian.

Distribution: Widespread in North America.

Comments: *Neospirifer,* along with the 18 other
genera in the family Spiriferidae, all
became extinct just at the end of the
Permian Period. There has been endless
speculation about the causes of this
extinction, one of the greatest in the
fossil record, but it is still a mystery.
Perhaps as many as 90% of all species
became extinct at that time.
2 articulated specimens of *Neospirifer*
are illustrated in Plate 52; the ventral
valve is visible on the left, and the
dorsal valve on the right.

ORDER TEREBRATULIDA
(Lower Devonian to Recent)

Terebratulids are the most abundant
and diverse of living brachiopods. They
attach with the stalk to rocks and other
shells. The valves are always convex,
pointed beyond the curved hinge, and
most often teardrop-shaped. There is
usually a circular or oval opening at the
ventral umbo for the stalk, and the
notch in the ventral interarea is
generally closed. The surface of the
valves is most often smooth, but may
bear faint to pronounced radial ribs;
a fold and complementary groove
extending usually from midvalve to the

anterior margin are common. The support for the lophophore is a simple short or long loop extending anteriorly from the umbonal region of the dorsal valve. The interior of all terebratulids is covered with tiny pits.

83 Rensselaeria

Description: 3⅜" (86 mm) long, 1⅝" (41 mm) wide. Large; almost circular to elongate-oval, with convex valves, ventral valve somewhat deeper. Dental plates reduced or absent; muscle-support platform absent. Covered with narrow radiating ribs. Brachiophore is broad loop that attaches at midshell to long ridge fastened to valve floor.

Age: Lower Devonian.

Distribution: Quebec, Maine, New York, Appalachian Mountains, Ontario, Tennessee, Illinois, Missouri.

Comments: One of the first and most famous of all American paleontologists, James Hall, named this genus *Rensselaeria* in 1859 to honor Stephen van Rensselaer, an early patron of paleontology and the founder of the Rensselaer Polytechnic Institute, the first American institution for technical and scientific training. *Amphigenia* is very similar in appearance to *Rensselaeria* and is found in the Lower and Middle Devonian of Maine, New York, the Appalachian Mountains, Ontario, Michigan, Kentucky, Tennessee, and Illinois. *Amphigenia* has finer radial ribs than *Rensselaeria*, so that the shell appears almost smooth, and the dental plates unite to form a raised muscle-attachment platform.

 Plate 83 shows the ventral valve of an articulated specimen of *Rensselaeria*, with the hinge on the left. The marginal drawing shows a side view of *Rensselaeria*.

103 Beecheria

Description: ⅝″ (16 mm) long, ⅜″ (10 mm) wide.
Smooth, elongate-oval, with ventral
umbo curving over hinge line. Both
valves convex. Ventral groove and
dorsal fold may be present near anterior
margin. Ventral valve has prominent
stalk opening at umbo; dental plates
present. Brachiophore is short loop
attached posteriorly to 2 triangular
plates whose apices are at center of
hinge line. Dental ridges separate from
triangular plates.

Age: Lower Mississippian through Upper
Permian.

Distribution: Widespread in North America.

Comments: Plate 103 shows 2 articulated
specimens of *Beecheria*. The one on the
left shows the dorsal valve and the stalk
opening, which is plugged with the
rock that fills the space between
the valves. The specimen on the right
shows the ventral valve.

105 Oleneothyris

Description: 2″ (51 mm) long, 1¼″ (32 mm) wide.
Large; elongate-oval, with sides almost
parallel, ventral umbo curved over
hinge line. Valves convex, ventral valve
deeper. 2 folds and median groove on
dorsal valve reflected on ventral valve
by 2 grooves and median fold. Valves
smooth except for growth lines. Stalk
opening large, notch on ventral valve
covered. Dental plates and dorsal
median ridge absent. Brachiophore loop
is ⅓ valve length, with V-shaped
transverse band pointed ventrally.

Age: Paleocene.

Distribution: New Jersey, Delaware, Maryland,
South Carolina.

Comments: *Oleneothyris* is one of the few
brachiopods common in eastern North
America after the Paleozoic. It belongs
to the same group as most living

articulate brachiopods—the terebratulids.

Plate 105 shows a side view of a single articulated specimen of *Oleneothyris*, with the dorsal valve on the left.

99 Terebratulina

Description: ½" (13 mm) long, ⅜" (10 mm) wide. Teardrop-shaped to almost pentagonal. Both valves gently convex. Ventral umbo extends beyond dorsal umbo, with large, rounded stalk opening. Ornamented with very fine ribs; young individuals may have fine bumps on ribs. Inside dorsal valve, widely separated socket plates bearing short, descending ring.

Age: Upper Jurassic to Recent.

Distribution: Widespread in North America.

Comments: Plate 99 shows a ventral valve of *Terebratulina*. *Terebratulina* is one of the few common Mesozoic and Cenozoic brachiopods. It is still found along the Atlantic coast from Labrador to New Jersey, and is called the Northern Lamp Shell. It attaches to rocks from the low-tide line to a depth of 12,500' (3810 m). The common West Coast form is *Terebratalia*, which is found from the Oligocene to the Recent. Living *Terebratalia* are also found attached to rocks, from the low-tide line to a depth of 6,000' (1829 m). *Terebratalia* is quite different from *Terebratulina* in shape. It is more fan-shaped, with a groove in the dorsal valve and a fold in the ventral valve. The surface is either smooth or covered with radial ribs.

108 Cranaena

Description: ⅝" (16 mm) long, ⅜" (10 mm) wide. Smooth, elongate-oval. Ventral umbo curves over hinge line. Valves convex.

Anterior margin not bent. Ventral valve with prominent stalk opening at umbo; dental plates present. Small teardrop-shaped hole in plate covers inside of dorsal valve under umbo. Brachiophore forms short loop.

Age: Lower Devonian through Upper Mississippian.

Distribution: Widespread in North America.

Comments: *Cranaena* is abundant in the Devonian, less so in the Mississippian. It is often marked by 6 red radial bands. Color-marked shells are rarely found in Paleozoic strata.

Plate 108 shows an articulated specimen of *Cranaena* with the dorsal valve in front. The stalk opening can be seen on the ventral umbo, and has been plugged by the rock that fills the valves.

107 Waconella

Description: ⅞″ (22 mm) long, ¾″ (19 mm) wide. Round, but with suggestion of 5 sides Valves convex, smooth except for covering of tiny bumps. Stalk opening large, cardinal process small. Complex brachiophore, with loop descending from attachment near hinge, then doubling back toward posterior; wide transverse band across loop.

Age: Cretaceous.

Distribution: Texas.

Comments: *Waconella* was named for Waco, Texas, where the species *Waconella wacoensis* is found. *Waconella* is called *Kingena* in older books.

2 articulated specimens of *Waconella* are shown in Plate 107, with the dorsal valve visible on the left, and the ventral valve on the right.

Phylum Echinodermata
(Lower Cambrian to Recent)

Echinoderms are one of the most important invertebrate phyla. Living sea cucumbers, sea urchins, starfishes, and brittle stars are members of most ocean-floor communities. All echinoderms are marine; almost all live on the bottom, attached or free, but never colonially. The echinoderm body is not segmented and a head is not differentiated. Echinoderms are basically bilaterally symmetrical, with 5-part radial symmetry superimposed. Some burrowing echinoderms have reverted to bilateral symmetry. All have a calcite skeleton, secreted under the skin and consisting of porous plate that are either fused together, interlocked, or free. In the sea cucumbers, the skeleton is reduced to microscopic plates or rods.

Echinoderms have a unique *water vascular system* consisting of water-filled sacs and canals that project at the surface as *tube feet*, and are used for sensing, feeding, locomotion, and respiration.

There are 4 classes of living echinoderms; only the Holothuroidea, or sea cucumbers, are omitted here because of their poor fossil record. There are about 15 classes of extinct echinoderms; most are known from only a few specimens. The 3 most important extinct classes—the Cystoidea, the Blastoidea, and the Edrioasteroidea—are discussed below.

CLASS CYSTOIDEA
(Lower Ordovician through Upper Devonian)

Cystoids were diverse, usually stemmed echinoderms. The globular or pear-shaped *theca* is covered by plates and pierced by slits or pores. The slits are continuous or disjunct across the sutures between plates; the pores pierce the plates. At the top of the theca are inconspicuous *food grooves* leading to a central mouth; the food grooves and mouth are sometimes covered over by plates. Slender *arms* for feeding were attached along the food grooves, but they are almost never preserved. The *anus* is located at the top of the theca, near the mouth.

---

332 Caryocrinites

Description: 1⅜" (35 mm) high, 1¼" (32 mm) wide. Almost spherical to egg-shaped; base slightly tapered; top gently convex. Theca composed of small number of large, regular plates. Base has 4 unequal plates that surround attachment for stem: 2 of these are 6-sided, 2 are 5-sided. Next circlet of 6 plates contains 2 that are 5-sided, 2 that are 6-sided, and 2 that are 7-sided. Next circlet has 8 polygonal plates. Top of theca has many small plates containing ring of 12–13 circular scars where arms, seldom preserved, were once attached. All plates have distinctive pattern of fairly large pores radiating from centers of plates to corners, except for lowest plates, where pores radiate from stem attachment. Smaller pores form concentric patterns on plates. Small pyramid of 5 triangular plates covers anus just inside arm bases. Stem, if preserved, composed of cylindrical plates.

Age: Middle Ordovician through Middle Silurian.

Distribution: New York, Ontario, Indiana, Kentucky, Wisconsin, Iowa, Missouri.

Comments: The pores on the theca of *Caryocrinites* are the openings of canals that cross below the sutures connecting to the pores on opposite plates. The canals may have been lined with tissue that held fluid and could balloon out at the pores for gas exchange, or they may have had some other function. *Caryocrinites* has arms with small branches, and probably fed in the same way that crinoids do.

No other cystoid can be easily confused with *Caryocrinites* because of its large plates and distinctive pore pattern. But another fairly common cystoid, *Echinosphaerites*, found in the Ordovician of Pennsylvania, Virginia, and Tennessee, has the same overall shape. *Echinosphaerites* has many more plates than *Caryocrinites*—from 2 to several hundred—but these plates are only visible on worn specimens. If the plates are visible, they show grooves radiating from the centers. Sometimes the only structure that can be seen on an *Echinosphaerites* theca is a small pyramid on the upper 1/3 of the sphere, which was probably the anus.

335 Holocystites

Description: 2¾" (70 mm) high, 1⅝" (41 mm) wide. Elongate, egg-shaped, with tapering base and rounded top. Composed of large, mostly 6-sided plates, or of alternating rows of large and small plates. Number of small plates increases with age of individual. Plates may have pores around peripheries. Mouth at top; anus also at top, covered with 5–6 triangular plates.

Age: Upper Ordovician through Middle Silurian.

Distribution: Ohio, Indiana, Tennessee, Wisconsin, Illinois.

Comments: *Holocystites* may have been stemless, or may have had a short stem. Little is known of its life habits. Some specimens show grooves leading out from the mouth, presumably to arms where food was collected, as in *Caryocrinites*. The pores probably functioned in respiration, again as in *Caryocrinites*.

Trematocystis, from the Middle Silurian of east-central North America, is very similar to *Holocystites*, but the mouth is larger and almost square, with short grooves leading out from the corners. There are fewer plates in the theca, and these are more equal in size. The shape of the theca is more rounded.

CLASS BLASTOIDEA
(Silurian through Permian)

Blastoids were small, stemmed echinoderms, abundant and diverse in the Middle and Late Paleozoic. Usually only the globular theca, covered by regularly arranged plates, is preserved. 5 petal-shaped areas extend from the mouth, at the center of the upper side of the theca, partway down the sides of the theca. In life, threadlike arms projecting upward from the sides of the petals were used for feeding. The anus is located between the petals, near the mouth. Additional, round or slitlike openings near the mouth were for the discharge of water.

334 Troosticrinus

Description: ⅞″ (22 mm) high; ⅜″ (10 mm) wide. Conical, with apex of cone at base and top slightly rounded. 3 circlets of plates; lowest circlet is of 3 plates; middle circlet of 5 plates that are forked near top to hold rays of starlike grooves radiating from center of top; upper circlet of 5 tiny plates almost completely covered by middle circlet. 1 large upper plate contains large, circular anus. Mouth at center of top is star-shaped. 4 small holes between rays near mouth.

Age: Silurian.

Distribution: Widespread in North America.

Comments: The marginal drawings show a reconstruction of *Troosticrinus* as it probably looked when alive, and a view of the top surface of the theca. Like *Pentremites, Troosticrinus* had small, delicate arms attached along the sides of the rays. These collected food particles—perhaps principally the small floating plants and animals called plankton—and moved the food to the rays and along the central groove to the mouth.

Blastoids are often found preserved with rugose corals, bryozoans, brachiopods, and crinoid debris. They probably lived in fairly well-agitated water of normal marine salinity. Since they were stationary, they needed some water movement to bring food to them. *Metablastus,* found in the Mississippian of Indiana, Kentucky, Illinois, Iowa, and Missouri, looks almost identical with *Troosticrinus* in side view, except that the rays descend farther along the sides. When viewed from the top, *Metablastus* shows clear differences. The anus is very large, and the holes, or spiracles, near the mouth are much larger also, and are shaped like wings. *Heteroschisma,* found throughout North America in the Devonian, has a shape similar to that of *Troosticrinus,* although it is not quite so long and slender. The rays, however, do not join together at the top, but instead terminate individually. The upper circlet of 5 plates is well exposed, and the rays are bordered by triangular fields of fine slits that functioned like the spiracles on *Troosticrinus* and *Pentremites*. *Heteroschisma* is a more primitive blastoid than the others.

333 Pentremites

Description: 1¼″ (32 mm) high, 1″ (25 mm) wide. Almost conical, widest near base and narrowing to gently rounded top. 5-sided in cross section. Base has circlet of 3 plates around stem attachment, 2 large plates and 1 half the size of others. Above basal plates are 5 U-shaped plates, forking around lower ends of 5 deep, wide rays that form flower shape over top and sides of body. 3rd circlet of plates is 4-sided, ending in sharp angle where 5 rays come together. Mouth at top, surrounded by 5 circular holes, 1 larger than the others. Rays are finely, transversely grooved; in perfectly preserved specimens, grooves covered over by small transverse plates; fine arms extend from both sides of grooves.

Age: Mississippian through Permian.

Distribution: Widespread in North America.

Comments: *Pentremites* is particularly common in the Upper Mississippian of the midwestern United States. Many beautifully preserved specimens are found in Indiana, Kentucky, and Illinois. When alive, the animal was held upright by a stem, its delicate arms used for feeding. Under each ray was an elaborate system for respiration, consisting of a complexly folded calcite sheet, like a curtain. Water entered at pores near the bases of the arms, bathed the tissues lining the folds, and exited through holes near the mouth. The marginal drawings show a side, a top, and a bottom view of a theca. 2 similar blastoids are *Schizoblastus* and

Cryptoblastus. Schizoblastus, from the Mississippian of Illinois, Iowa, and Missouri, has an upper circlet of plates that covers more than ½ the length of the body. The body is more spherical than that of *Pentremites,* and the rays narrower and longer, ending just short of the stem attachment. *Cryptoblastus,* from the Mississippian of Iowa and Missouri, has the same ellipsoid shape as *Schizoblastus,* but the upper plates are small, like those of *Pentremites. Cryptoblastus* has 10 small openings near the mouth instead of 5, 2 of which merge with the anus. The upper plates reach all the way to the mouth, and are not pinched off by the expanded rays.

CLASS CRINOIDEA
(Lower Ordovician to Recent)

Crinoids, called sea lilies if stalked and feather stars if unstalked, were diverse and abundant in the Paleozoic Era, when many rocks were formed chiefly of their skeletons.

Most fossil crinoids were stemmed. The stem was attached in various ways to the sea floor or to other animals or plants. The stem bore a crown consisting of a theca covered with plates, and 5 or more branched or unbranched arms. The lower part of the theca, the *dorsal cup,* is covered with plates arranged in 2 or 3 rings. The lowest ring forms the base where the stem attached. Another ring of basal plates, usually 5, is sometimes present. The next ring consists of 5 radials, shown in black on the marginal drawings. The radials support the *ray plates,* which lie between the radials and arms; or the arms may rest directly on the radials. In most genera, there is an anal plate in the radial ring. This plate is marked with an X in the drawings of plate arrangements. It may support other anal plates above, and these sometimes form an anal tube on the upper surface of the theca. Many crinoids have small plates between the ray plates. The upper surface of the theca, covered with plates that are fused or flexible, contains food grooves. These food grooves, continuous with food grooves on the arms and arm branches, lead to the mouth.

323, 343 Crinoid stems

Description: ¼″ (6 mm) long, ⅛″ (3 mm) wide (pl. 323); ½″ (13 mm) long, ⅛″ (3 mm) wide (pl. 343). Small discs, usually round, but may be elliptical or 5-sided; isolated or stacked together to form stems. Hole in center of discs may be small or large, round or shaped like 5-petalled flower. Flat part of disc covered with radiating grooves and ridges that may extend from hole to periphery, or may cover only part of surface. Edges of discs may have keel or circular scars where rootlets were attached.

Age: Lower Ordovician to Recent.

Distribution: Widespread in North America.

Comments: Most crinoid fossils are formed only from the plates of crinoid stems. These may still be articulated (Plate 343) or they may be scattered (Plate 323). Whole limestone formations may be composed chiefly of such fossilized plates. Almost all Paleozoic genera of crinoids have been described from parts of the crown, and only a few isolated plates or stems have been identified. In the Mesozoic and Cenozoic, however, most stalked genera have been described only on the basis of stems. The plates of the stem interlock by means of the radiating ridges, and contain in the central hole an extension of the body cavity. The cirri, or rootlets, that are sometimes attached at the scars on the plates are also permeated by the body cavity. These have a sensory function, as well as providing anchorage by wrapping around objects or growing down into the sediment.

SUBCLASS CAMERATA
(Middle Ordovician through Permian)

The camerates are characterized by the rigid union of all plates on the dorsal cup. The arms are commonly covered with 2 rows of plates, but there may be only 1 row. The arms of most genera bear small branches. The more primitive genera have anal plates in the dorsal cup, but more advanced genera lack them.

319 Dorycrinus

Description: 1″ (25 mm) high, ¾″ (19 mm) wide. Rounded; all plates rigidly joined. 3 plates form lowest circlet; 6 plates in next circlet, including 5 radials and 1 anal plate, that supports several anal plates above. Radials and lowest anal plate with bumps. Ray plates above radials support 16–20 pairs of arms in 5 groups; arms have 2 rows of plates and many small branches. Upper surface of theca rigid, covers mouth and food groove. Center of upper surface of theca bears single spine that may be long or short, smooth or barbed. 5 peripheral spines may also be present. Anal tube absent.

Age: Lower Mississippian.

Distribution: Indiana, Kentucky, Tennessee, Illinois, Iowa, Missouri, New Mexico.

Comments: *Dorycrinus* is illustrated in Plate 319 by a small theca without arms or stem. The 3 basal plates circle the stem attachment like small petals. In the species illustrated, there is a long spine on the upper surface which appears pointing down to the left in the photograph. The drawing in the margin shows a species of *Dorycrinus* with 6 spines on the upper surface of the theca. The spines project through the arms, and were probably used for defense, making *Dorycrinus* unpleasant eating as well as an uninviting resting place for other animals.

318, 325 Batocrinus

Description: 1⅛" (28 mm) high, 1" (25 mm) wide (pl. 318); 4⅜" (111 mm) high, 1¾" (44 mm) wide (pl. 325). Theca globular; dorsal cup low, broad, dish-shaped. All plates rigidly joined. 3 small, equal plates around stem attachment; 5 radial plates above, forming bases for rays. Many other small plates, mostly equal in size, in dorsal cup. 18–22 large, horizontally directed arm facets around periphery of theca bear free arms, each with 2 rows of plates and small branches. Upper surface of theca rigid, high, conical, with prominent spines or bumps on some plates. High anal tube extends up from middle of upper surface of theca. Mouth and food grooves covered over by plates.

Age: Upper Mississippian.

Distribution: Virginia, Indiana, Kentucky, Tennessee, Illinois, Missouri.

Comments: Plate 318 shows the theca of *Batocrinus* with the arms broken off. The spines on the upper surface of the theca are prominent, and the base of the anal tube can be seen, although most of it is broken off. The arm facets can be seen around the periphery. Plate 325 shows an entire crown, preserved with part of the stem. The arms are intact and hide the upper surface of the theca, the spines, and the anal tube. The marginal drawing shows the arrangement of plates in the dorsal cup of *Batocrinus*. In the same family as *Batocrinus* is a similar genus, *Abatocrinus,* found in the Lower Mississippian of Indiana, Kentucky, Tennessee, Illinois, Iowa, and Missouri. *Abatocrinus* has a higher dorsal cup than *Batocrinus,* with straight or gently convex, or concave sides. The theca may have small bumps on the upper surface, but no spines.

---

329 Eutrochocrinus

Description: 4″ (102 mm) high, 1¾″ (44 mm) wide. Dorsal cup narrow at base, abruptly expanding to arm facets; upper surface of theca contracting to long, narrow anal tube. 3 large plates form base of dorsal cup; all plates rigidly united. 5 radial plates form ring above basal plates, along with 1 anal plate that supports 9–13 more anal plates above. 18–40 short, free arms, curving in when closed, single or paired, each bearing small branches. Upper surface of theca rigid, covers mouth and food grooves.

Age: Lower Mississippian.

Distribution: Illinois, Iowa, Missouri.

Comments: The very high anal tube is a distinctive characteristic of this genus and can be clearly seen to the left in the specimen illustrated. The primitive crinoid had an anus that opened on the upper surface of the theca and was not elevated. This arrangement was sufficient as long as crinoids lived in areas where currents were strong and could wash wastes off the upper surface of a theca. But in quiet areas, wastes probably would have fouled the upper surface of the theca. So anal tubes may have evolved to carry the wastes high up before they were discharged. Several crinoid genera have dorsal cups shaped like that of *Eutrochocrinus,* and are closely related to it. *Eretmocrinus*—found in the Lower Mississippian of Indiana, Kentucky, Tennessee, Illinois, Iowa, Missouri, and New Mexico—has basal plates that form a projecting rim at the base of the dorsal cup, with a short, sometimes curved anal tube, and 12–26 free arms. Most characteristic, the outer halves of the arms are widened, look paddlelike, and have no small branches. *Uperocrinus* is found in the Lower Mississippian of Indiana, Kentucky, Tennessee, Illinois, Iowa, and Missouri. It also commonly has projecting basal plates and 10–22 short arms that curve in and bear small branches. The plates between rays are depressed and in contact with the upper surface of the theca.

331 Glyptocrinus

Description: 2⅜″ (60 mm) high, 3⅛″ (79 mm) wide. Dorsal cup high, conical, expanding regularly to arm base. All plates in dorsal cup rigidly joined together. 5 basal plates form small pentagon at stem attachment. Circlet of 5 radials present, all in contact. Each radial supports series of plates culminating in 20 free arms that are simple or branch once, and have only 1 row of plates, and fine, long, narrow branches. Dorsal cup plates have radiating, starlike ridges that join across sutures to divide surface into small triangles. Anal tube absent.

Age: Middle Ordovician through Silurian.

Distribution: Widespread in North America.

Comments: Plate 331 shows a specimen of *Glyptocrinus* with the dorsal cup almost free of the matrix and with many of the arms visible. The triangular pattern of the ridges is clearly visible and obscures the actual outlines of the plates. The arms can be seen with fine, small branches extending from the single row of plates. These branchlets increased the potential food-gathering area of the crinoid enormously. Along the length of each branchlet was an extension of the food groove, with tube feet for respiration and catching food. Food particles were caught by the tube feet and dumped into the food grooves. Cilia then moved the particles down the food grooves to the upper surface of the theca, and across under the plates to the mouth, which was also covered. The marginal drawing shows the arrangement of the plates in the dorsal cup of *Glyptocrinus*.

Another crinoid with a similar pattern on the plates is *Periglyptocrinus*, found in the Middle Ordovician in North America. Its basal plates are larger, and its arms bear 2 rows of plates.

330, 336 Eucalyptocrinites

Description: 5½" (140 mm) high, 4" (102 mm) wide (pl. 330); 1⅜" (35 mm) high, 1" (25 mm) wide (pl. 336). Theca egg-shaped; plates rigidly fused. Dorsal cup has pronounced concave base bearing 4 basal plates and part of radials, which are in contact all around. Sides of cup expand gently. Upper surface of theca bears 10 riblike plates, forming high cylinder rising from edge of cup, with compartments on sides that hold arms.

Age: Middle Silurian through Middle Devonian.

Distribution: Widespread in North America.

Comments: Plate 336 shows only the dorsal cup of *Eucalyptocrinites*, with the stem, arms, and riblike plates missing. Plate 330 shows the root system that held *Eucalyptocrinites* to the substrate. The stem that supported the crown is broken near its base. Such a root system represents only one of many ways that crinoids moored themselves to the bottom. Other methods included stems that ended in a kind of grappling hook that would dig into the bottom as it was dragged across, or a disc or irregular lump on the bottom of the stem to hold the crinoid to the bottom. Some crinoids attached by wrapping their stems around the stems of other crinoids or similar objects. The riblike plates that held the arms clearly functioned to protect the arms when they were closed. One marginal drawing is of a reconstruction of the crown of *Eucalyptocrinites*. The other marginal drawing shows the arrangement of the plates in the dorsal cup.

327 Platycrinites

Description: 4″ (102 mm) high, 1⅜″ (35 mm) wide Dorsal cup bowl-shaped; base has ring of 3 large basal plates, unequal in size and sometimes partly or completely fused. Above basals are 5 radial plates, each very large, with horseshoe-shaped arm facets that hold arms. Each arm has 1 row of plates initially, then 2 rows; arms in each ray branch 3–10 times. Small branches are fine and dense. Upper surface of theca flat or pyramidal, composed of numerous plates. Anus opens either directly through upper surface of theca or through short or long tube. Stem plate elliptical near dorsal cup, and twisted into spiral. Many species have bumps on basal and radial plates.

Age: Lower Devonian through Upper Permian (possibly also in the Upper Silurian).

Distribution: Widespread in North America.

Comments: Plate 327 shows a well-preserved specimen of *Platycrinites* with its characteristic flattened and twisted stem. Note the few large plates in the dorsal cup, covered with bumps, and the arms that branch repeatedly near the cup. The arrangement of plates in the dorsal cup is shown in the marginal drawing. *Platycrinites* and similar genera have elliptical stem plates that are easily recognized. *Eucladocrinus* is a similar Lower Mississippian crinoid from Illinois, Iowa, and Montana. *Eucladocrinus* has a stem and dorsal cup like those of *Platycrinites*. But each ray has only 2 massive arms that bear many short, stout armlets with small branches.

SUBCLASS INADUNATA
(Lower Ordovician through Upper Permian)

The dorsal cup has rigidly fused plates, with 1 or 2 rings of plates below the radials. Plates above the radials are not incorporated into the cup. The mouth and food grooves on the upper surface are covered by plates.

326 Cyathocrinites

Description: 3⅛″ (79 mm) high, 1¼″ (32 mm) wide. Dorsal cup bowl-shaped to globular, with only 3 circlets of plates. Base a 5-pointed star composed of 5 diamond-shaped plates. Next circlet of 5 large plates, all touching, 1 of these larger than the others. Final circlet of 5 radials plus one 5-sided anal plate above the larger basal plate. Radial plates notched at upper side for arm facets. Arms branch regularly several times and have only 1 row of plates; small branches absent. Upper surface of theca stout; anal tube narrow, cylindrical, set to 1 side of center. Stem large, round.

Age: Middle Silurian through Lower Mississippian (possibly also in the Upper Permian).

Distribution: Widespread in North America.
Comments: Illustrated is a complete crown of
Cyathocrinites with some of the stem
preserved. The few large plates that
make up the dorsal cup can be seen
clearly, as well as the arms with their
many branches. The diagram in the
margin shows all the plates in the
dorsal cup.

Barycrinus is a Mississippian genus,
found in the central United States; it
belongs to a closely related family that
may have evolved from cyathocrinitids
during the Silurian. *Barycrinus* has a
much smaller anal plate than that of
Cyathocrinites, with 4 sides plus another
very small plate that is part of the anal
series just to the left of the larger one.
The radials support 2 ray plates that
branch immediately to form 10 massive
arms bearing armlets.

321 Delocrinus

Description: 2¾" (70 mm) high, 1⅝" (41 mm)
wide. Dorsal cup shaped like shallow
bowl; bottom of cup has deep and
narrow cavity; lowest ring of 5 tiny
plates may be covered by stem
attachment; next circlet has 5 large
plates that project above base. 5 radial
plates are next, with 2 plates separated
by small anal plates. Ray plates above
radials may bear large spines. 10 heavy
unbranched arms present, covered with
2 rows of plates. Small branches also
present. Anal sac on upper surface of
theca slender, cylindrical. Stem round.
Age: Middle Pennsylvanian through
Permian.
Distribution: Widespread in North America.
Comments: The specimen of *Delocrinus* in Plate 32
is almost complete, with only the tips
of the arms missing. Large spines, the
middle one broken, can be seen
projecting from just below the arms.
The spines may have functioned to

deter predators. Studies of living crinoids indicate that most of their enemies are parasites and small snails that bore through the crinoid body. No animals are known to bite pieces out of crinoids, the type of predation against which spines might be useful, but some paleontologists have suggested that extinct sharks like the hybodonts, which had batteries of flat, crushing teeth, may have fed on Paleozoic crinoids. A similar crinoid, *Phanocrinus* from the Upper Mississippian of Indiana, Kentucky, Alabama, Illinois, Arkansas, Oklahoma, and possibly Nevada, can be distinguished from *Delocrinus* because the bowl is somewhat restricted at its rim, its arms have only a single row of plates, and there are 3 anal plates in the dorsal cup instead of 1. Another similar genus is *Endelocrinus,* from the Lower Pennsylvanian through the Lower Permian of Pennsylvania, Ohio, Illinois, Missouri, Nebraska, Kansas, Oklahoma, Texas, Arizona, and Nevada. *Endelocrinus* has the same arrangement of plates in the dorsal cup as *Delocrinus,* but the concavity in the base is shallower, there are grooves along the sutures between the plates, and pits at the angles between plates.

SUBCLASS FLEXIBILIA
(Middle Ordovician through Upper Permian)

The Flexibilia have 3 plates in the lowermost ring, one smaller than the others. Above these, there is always another ring of plates before the radials. The ray plates are incorporated into the dorsal cup, but not rigidly; the whole cup is flexible. The separation of the dorsal cup and the free arms is not clearly marked. The upper surface of the theca is also flexible, with mouth and food grooves covered by small plates. The arms are covered with 1 row of plates, and do not carry fine branches.

322 Taxocrinus

Description: 4⅝" (117 mm) high, 1¾" (44 mm) wide. Crown elongate; lower branches of arm incorporated into dorsal cup, but not rigidly. Lower circlet of 3 plates completely or mostly hidden by stem. Next circlet is of 5 basals, 1 elongated with series of small plates above, forming anal tube. 5 plates in next circlet are radials, all in contact except where elongated basal plate protrudes. 3 ray plates above each radial plate support arms, which branch regularly, have 1 row of plates and no small branches, and curl inward when closed. Many small plates present between ray plates. Upper surface of theca flexible, with exposed food grooves and mouth. Stem round, without armlike extensions.

Age: Middle Devonian through Mississippian.

Distribution: Ohio, Indiana, Kentucky, Illinois, Iowa, Missouri.

Comments: *Taxocrinus* is a typical flexible crinoid. The short arms without small branches functioned in food gathering. The arms may have been held out to form a collecting bowl for catching food that fell from above. The stem also may have flexed so that the arms could be held vertically to catch food particles carried parallel to the sea floor by currents. A complete crown of *Taxocrinus* with part of its stem still attached is shown in Plate 322. The arrangement of plates in the dorsal cup is shown in the marginal drawing. *Eutaxocrinus* is a similar and closely related flexible crinoid from the Upper Silurian through the Lower

Mississippian of New York, Ontario,
Illinois, Iowa, and Montana.
Eutaxocrinus has only 2 ray plates above
the radials.

324 Onychocrinus

Description: 2⅜″ (60 mm) high, 1⅝″ (41 mm)
wide. Crown egg-shaped or round,
with arms either curved in when closed
or spread outward like starfish arms
during feeding. 1st circlet of plates in
base of dorsal cup nearly or completely
hidden under stem. Next circlet is of 5
basals, 1 of which is elongated, with
top edge hollowed for holding first anal
plate that supports anal tube. Next
circlet is of 5 radial plates, each
supporting 3–6 ray plates. Arms
branch irregularly, with small side
branches that branch again.

Age: Mississippian.

Distribution: Canada, Indiana, Kentucky, Alabama,
Illinois, Iowa, Oklahoma.

Comments: Plate 324 shows a complete crown of
Onychocrinus with a bit of the stem still
attached. The highly branched arms
look as though they had been woven
together. The marginal drawing shows
the arrangement of the plates in the
dorsal cup.

SUBCLASS ARTICULATA
(Triassic to Recent)

This is the only subclass with living
crinoids, and includes, in addition,
nearly all Mesozoic and Cenozoic
crinoid fossils. The theca is relatively
small, containing 3 rings of 5 plates
each. There are no ray plates in the
dorsal cup. The arms are long and
flexible. The upper surface of the theca
is also flexible, with exposed food
grooves and mouth.

328 Uintacrinus

Description: 4¼" (108 mm) high, 2" (51 mm) wide. Large, globular, stemless, with extremely long arms. Single 5-sided plate may be present on base of dorsal cup; circlet of 5 plates present above base; next circlet of 5 radial plates supports 2 ray plates. Rays branch into long, straight series of plates that support 10 arms. Many plates present between ray plates and between branches. Arms unusually long, dividing only once, with many small branches; may reach length of 79" (2 m). Upper surface of theca is flexible skin studded with small plates; mouth and food grooves exposed. Large, conical anal tube located centrally.

Age: Upper Cretaceous.

Distribution: Kansas, Montana, Wyoming, Utah, British Columbia.

Comments: Plate 328 shows a single specimen of *Uintacrinus* from the Niobrara Chalk Formation of Kansas. Slabs of this chalk have been found containing as many as 250 complete specimens of *Uintacrinus*, with the arms interlocked. These assemblages have been interpreted as once-floating mats of crinoids up to 35' (11 m) wide. The extremely long arms would have swept through the water, gathering up small floating plants and animals with the tube feet. The globular body may have held gas bubbles to keep the animal afloat.

CLASS STELLEROIDEA
(Lower Ordovician to Recent)

Stelleroids are the common starfishes and brittle stars of modern seas. They have probably been abundant since the Ordovician, but at death the plates typically separate and disintegrate, so whole fossils are rare. Only in cases where an animal was suddenly buried is the entire skeleton preserved.

All stelleroids have a central disc, or body, from which the arms extend. The mouth is in the center of the lower surface; the anus, if present, is on the upper surface.

SUBCLASS ASTEROIDEA
(Lower Ordovician to Recent)

The arms of asteroids are broadly connected to the central body, and the outline overall is generally like a 5-pointed star. The number of arms and their length vary, however. Some asteroids have very short arms, giving the body a pentagonal outline. On the lower surface, there is an open groove from the mouth out to the end of each arm.

371 Hudsonaster

Description: 1" (25 mm) long and wide. Star-shaped, with 5 short, broad, triangular rays. Top and bottom surfaces covered with many regularly arranged, domed plates. Upper surface of each ray covered by 5 radiating rows of plates; center covered by 3 circles of plates. Bottom surface has open center extending into each ray as deep grooves bordered by small, rectangular plates. Larger, more rounded plates form margin.

Age: Middle Silurian.
Distribution: Widespread in North America.
Comments: Starfishes usually disintegrate quickly after death, since their plates are held together only by soft tissue. *Hudsonaster* is one of the few genera fairly often found intact. Living asteroid starfishes are either predators or suspension feeders. As a predator, a starfish uses its

tube feet to grasp a gastropod or bivalve while it inserts its stomach inside the prey's shell and begins digesting the soft parts, sometimes also swallowing the entire shell. As a suspension feeder, a starfish secretes over its top surface a mucus covering that picks up small bits of food. The mucus is moved by cilia to the mouth on the animal's underside. Some starfishes supplement predation with suspension feeding, which may be the more primitive method. Plate 371 shows the bottom of a *Hudsonaster* specimen. The top is shown in the marginal drawing. A starfish similar to *Hudsonaster* is *Palaeaster*, found in the Middle Silurian in North America. The rays of *Palaeaster* are relatively longer and narrower than those of *Hudsonaster*, and the centers of the rays have a row of many tiny plates on their top surfaces, instead of a single series as in *Hudsonaster*. The central area in *Palaeaster* is also covered with many tiny plates.

SUBCLASS OPHIUROIDEA
(Lower Ordovician to Recent)

Ophiuroids, or brittle stars, differ from asteroids in that the arms are sharply delineated from the small, disclike body. The arms are long and thin, move in a snakelike fashion, and, in some specimens, are branched. Ophiuroids lack an open groove on the lower surface of each arm.

372 Ophiura

Description: ¾″ (19 mm) long and wide. Small, circular disc bearing 5 narrow, snakelike arms. Upper surface of disc covered with small, scalelike plates except for 2 large, oval plates located just inside each arm joint. Disc notched at juncture with arm. Arms covered on both sides by many small plates; spines inconspicuous or absent. Lower surface of disc has star-shaped mouth and 5 large, almost triangular plates near mouth and between arm joints.

Age: Upper Cretaceous to Recent (possibly also in the Jurassic).

Distribution: Widespread in North America.

Comments: *Ophiura* is one of the most common brittle stars still living, and is a fairly common fossil as well. Brittle stars are rarely preserved complete. After death, the soft tissue usually decays quickly and the plates fall apart and scatter. The plates are made of calcite and can fossilize, but since they are porous, they often disintegrate before this process can occur. Plate 372 shows a top view of a complete *Ophiura*. Living brittle stars feed in a variety of ways, and the fossil ones probably did too. *Ophiura* may have eaten both living and dead material, plant or animal. The plates around the mouth functioned as jaws and could tear off bits of food. Another relatively common fossil brittle star is *Onychaster*, found in North America in the Mississippian. *Onychaster* has a relatively smaller disc without covering plates, and its arms are long, thick, and rounded. They are often preserved curled under the disc in clawlike fashion. The arms are studded either with many tiny plates or with larger double rows of plates on the upper surface. They may branch, and they have no spines.
Ophiura was formerly known as *Ophioglypha*.

CLASS EDRIOASTEROIDEA
(Lower Cambrian through
Mississippian)

These are small, generally disc-shaped
animals that usually cemented
themselves to a solid object. The upper
surface is covered with small plates,
and there is a 5-rayed star pattern. The
rays hold the pores from which the tube
feet protruded, and at the center of the
star is the mouth, generally covered by
plates. The rays are commonly curved,
and may be branched.

Carneyella

Description: ¾" (19 mm) long and wide. Disc-
shaped, attached by broad base to
foreign object. Overlapping, curved
plates, becoming progressively larger
toward center of disc, cover broad,
marginal zone of upper surface in
several concentric rows. Marginal zone
encloses 5-rayed star; rays curved at tips
and commonly pressed against inner
border of marginal zone. 4 rays curve
counter clockwise, and 1, clockwise.
Each ray covered by 2 rows of small
polygonal plates. Center of star covered
with 1 large plate and 2 smaller plates.
Area between rays covered with
overlapping, scalelike plates. Small
pyramid of 7 or more triangular plates
covers anus, located to left of
clockwise-curving ray.

Age: Middle and Upper Ordovician.
Distribution: Widespread in North America.
Comments: Certain edrioasteroids could live
unattached on the sea floor, and some
may have been able to move about, but
Carneyella was always permanently
attached to some hard object, most
often a brachiopod. The small plates
over the rays could open outward,
exposing the underlying floor plates,
pierced by small holes that must have

held tube feet. These tube feet may have secreted mucus nets to catch small particles of food falling onto the rays. The marginal drawing shows a reconstruction of the upper surface of *Carneyella*.

There are several fairly common edrioasteroids that are similar to *Carneyella*. *Hemicystites* is found in North America in the Ordovician through the Middle Devonian. It has a more irregular marginal zone than does *Carneyella*, and the tips of the rays are less curved, but they make indentations in the marginal zone. The rays may branch. The plates between the rays are also more irregularly arranged, and more polygonal than scalelike.

Agelacrinites, from the Lower Devonian through the Lower Mississippian of New York, Ontario, Iowa, and Missouri, has long slender rays, all of which curve counterclockwise except for the 2 rays to the left of the anus, which curve clockwise. The plates between the rays are polygonal, and the mouth is also covered by numerous small plates.

Cooperidiscus is found in the Upper Devonian of New York and Pennsylvania. It has a star that is almost smooth and is covered with very tiny plates. Its rays are extremely slender and strongly curved clockwise at the margins of the disc, which has no border.

CLASS ECHINOIDEA
(Ordovician to Recent)

The echinoids are the sea urchins, sand dollars, and heart urchins—common near-shore animals today. Their fossil record is poor in the Paleozoic, but better in the Mesozoic and Tertiary, when their shells are abundant and well preserved in many formations.

Echinoids are globular or disc-shaped. Primitive echinoids have 5-part symmetry; more advanced echinoids, like the heart urchins, have secondary bilateral symmetry because the anus has shifted from its position on top to the side or bottom of the shell. The mouth is always on the bottom, and generally, but not always, central. The shell consists of small, thin, interlocking calcareous plates that are polygonal in shape. These plates are arranged in 10 narrow bands that radiate from the top of the shell around to the mouth on the bottom. 5 of the bands contain plates with holes or slits from which the tube feet project. These may form a 5-petalled flower pattern on top of the shell. Alternating bands contain plates without such holes. The plates bear movable spines or, in most fossils, bumps where the spines were attached. At the top of the shell, where the 10 bands come together, are special small plates with holes from which the eggs and sperm emerge, and a sievelike plate for water intake. The mouth at the bottom of the shell is generally armed with a set of jaws consisting of 5 teeth —called "Aristotle's lantern"—which can be protruded out of the mouth and is used to scrape at algae, seaweed, or detritus. In spite of these impressive teeth, echinoids are not predators, since they move very slowly on their tube feet and spines and lack the ability of starfishes to wrap around their prey. Echinoids are divided informally into 2 groups—regular and irregular. Regular echinoids are radially symmetrical, and irregular echinoids are secondarily bilaterally symmetrical.

The measurements given with the illustrations and in the text are length and width, taken looking down on the top of the specimen. The length is the distance from the anterior to the posterior (determined by the position of the anus). The width is taken at right angles to the length. If the anus is central, as in regular echinoids, length and width are equal.

ORDER CIDAROIDEA
(Upper Silurian to Recent)

Cidaroids are regular echinoids with 2 columns of plates in each of the 10 bands. Each plate in those bands that lack pores has a single large bump that bore a substantial spine.

373, 374 Stereocidaris

Description: 3⅜" (86 mm) long and wide. Almost spherical, top and bottom slightly flattened; robust; radially symmetrical. Covered with 5 double rows of large, polygonal plates, each with large spine base: a smooth, circular area sunk into plate, and raised, knoblike bumps encircled by groove in middle, with pit in center. Ring of smaller bumps surrounds large spine base. Remaining area of plate covered with irregular bumps. Between rows of large plates are 5 sinuous double rows of tiny plates bearing 2 pores each. Small hole at top, surrounded by small plates, held anus. Larger hole in center of base held mouth.

Age: Cretaceous.

Distribution: Southern United States.

Comments: *Stereocidaris* still lives in the Indo-Pacific, and the living animal has a creamy white skeleton. The largest bumps bear blunt spines up to 4" (102 mm) long, which are used for defense and walking. The rings of smaller bumps around the large spine bases bear small spines that cover the bases, protecting the muscles that move the large spines. The other bumps bear secondary spines, usually less than ¼" (6 mm) long. Some very small bumps bear either very tiny spines or pedicellariae, a kind of spine that has been modified into a claw and serves mainly to snap at larvae attempting to settle on the echinoid. Spines and pedicellariae are rarely preserved attached to the skeleton, but are commonly found separately. A side view of a *Stereocidaris* specimen is shown in Plate 373, and a top view of the same specimen in Plate 374.

ORDER HOLECTYPOIDA
(Lower Jurassic to Recent)

Holectypoids are variable in shape and as a group are midway between regular and irregular echinoids. Many are hemispherical or globular, and would be radially symmetrical except that the anus is never in the center of the upper surface—it may be just off-center, marginal, or on the lower surface. Some holectypoids are egg-shaped, having become slightly elongated front to rear.

362, 363   Holectypus

Description: 2" (51 mm) long and wide (pls. 362, 363). Conical, with circular outline, flat bottom, and domed top. Skeleton consists of 20 rows of plates: 5 double rows of plates, each plate with 2 tiny pores near outer edge, and 5 double rows of plates in between that have no pores. Areas with pores, narrower than areas without pores, form points just before reaching center of skeleton top, which is composed of 10 tiny plates; 4 of these have small holes for emission of sperm or eggs. All plates very narrow, extended horizontally. Plates without pores covered with transverse row of small bumps; plates with pores have single bump each. Large opening in center of base has scalloped edge of 10 small lobes and contains mouth; large, oval anus on base or posterior margin.

Age: Lower Jurassic through Upper Cretaceous.

Distribution: Widespread in North America.

Comments: Only the off-center placement of the anus destroys the apparent radial symmetry of *Holectypus*. The anus shifted from the primitive position on the top of the body, as in *Stereocidaris*, when the ancestors of *Holectypus* moved into quieter waters. As long as the anus was on top, strong currents were needed to prevent the wastes from fouling the body and clogging the gills. Plate 362 shows the base of a *Holectypus* specimen, with the anus in the upper right. Plate 363 shows the top of another specimen. The marginal drawing shows a side view of *Holectypus* from the rear.

ORDER CLYPEASTEROIDA
(Upper Cretaceous to Recent)

Clypeasteroids include the sand dollars, which are flat and rounded in outline. Other clypeasteroids are biscuit-shaped. In all, the pores are arranged in a 5-petalled flower shape on the upper surface of the shell. The anus varies in position from just above the margin to near the small, round mouth on the bottom. The spines are small, short, and numerous.

364 Periarchus

Description: 2¾" (70 mm) long and wide. Circular in outline, base flat, top slightly to markedly domed. Top covered by small, symmetrical flower pattern with 5 petals outlined by border of tiny, transverse slits; pores on inside of slits. Petals extend halfway to margin. 5 pores located between petals in center of flower. Base has food grooves that fuse near small, round mouth in center. Anus located midway between posterior margin and mouth. Interior complexly partitioned.

Age: Upper Eocene.

Distribution: Southeastern United States.

Comments: Like most sand dollars, *Periarchus* probably lived in fairly shallow water where the bottom was sandy. It may have buried itself slightly, but probably kept the domed top and petals exposed, so the elongated tube feet in the slits around the petals could function in gas exchange. The body was covered with tiny spines that could move *Periarchus* slowly along the bottom. Mucus was secreted over the surface of the animal to collect food particles that fell on it, and the strands of mucus were moved to the food grooves and then to the mouth by cilia. Plate 364 shows the top view of a *Periarchus* specimen. The marginal drawing is a side view.

Other similar sand dollars are *Protoscutella, Mortonella,* and *Encope*. *Protoscutella*, found on the Gulf Coast and in the southeastern United States from the Middle and Upper Eocene, has only 4 pores at the center of the flower, food grooves that do not branch, and an anus located on the posterior margin, where there is a slight notch. *Mortonella*, of the same age and with the same distribution as *Periarchus*, has wider, longer petals, a heavier skeleton with thicker margins, and a flatter top. *Encope*, found from the Pliocene to the Recent in coastal areas of North America, has a large, notched margin, and a large keyhole-shaped opening passing through the posterior part of the body.

368 Dendraster

Description: 1¾″ (44 mm) long and wide. Disc-shaped, slightly wider than long in outline; flat on base and slightly domed on top. 5 petals form asymmetrical flower pattern on top, with center shifted toward posterior margin. 2 posterior petals shortest, anterior petal longest. Petals crossed by tiny transverse slits; outer ends of petals rounded, closed. 4 tiny pores in center of flower. Base has small, round, centrally located mouth; small anus near posterior margin. Food grooves begin at margin, except near anterior margin, where they begin closer to mouth, and fuse toward mouth.

Age: Pliocene to Recent.

Distribution: Puget Sound to Gulf of California.

Comments: *Dendraster* is a common sand dollar living on the West Coast today. It inhabits sandy bottoms below the surf zone, and often occurs in enormous numbers within a small area. *Dendraster* sometimes stands vertically, with ⅔ of the disc buried in the sand. This habit may account for the absence of food grooves near the anterior margin. *Dendraster* is a suspension feeder, like *Periarchus,* and its petals bear elongated tube feet for gas exchange. In life the animal's surface is covered with tiny spines used for locomotion. Plate 368 shows the top of a *Dendraster* specimen, with the posterior side down. The marginal drawing is a side view, with the posterior to the right.

3 other sand dollars occur commonly as fossils on the West Coast. *Astrodaspis,* found from the Middle Miocene through the Pliocene in California, has a more rounded skeleton than *Dendraster,* with broader, equally sized and spaced petals that are not closed at the ends. *Merriamaster,* from the Middle and Upper Pliocene and found from central to southern California, has petals placed off-center, like *Dendraster,*

but the petals are longer and wider. *Echinarachnius,* found from the Miocene to the Recent all along the West Coast, has centered petals ⅗ the length of the radius. The ends of the petals open widely toward the margins, and the slits form only a narrow border. The food grooves are very regular, with 15 grooves that begin at the margin and fuse to form 5 grooves at the mouth.

ORDER CASSIDULOIDA
(Jurassic to Recent)

Cassiduloids are biscuit-shaped and generally elongated along the anterior-posterior margin. The anus is commonly located at the posterior margin or just above it. There are 4 separate, small plates, each with a pore, on the upper surface of the shell where 5 petals come together. Around the mouth on the lower surface are 5 small petals, and 5 small bumps between the petals.

369 Hardouinia

Description: 1⅞" (48 mm) long, 1" (25 mm) wide. Almost pentagonal in outline, highly inflated, with flat base. Area on top shaped into 5 broad petals bordered by short transverse slits and closed at tips. Petals meet at highest part of skeleton, which is shifted slightly toward anterior. Area continues beyond petals, with 1 pore on each plate. Anus located on upper surface near margin; shallow groove extends from anus to margin. Mouth on base shifted toward anterior. Flowerlike pattern of humped plates around mouth, small petals outlined by pores. Surface covered with bumps, best developed on lower surface but absent on area containing anus.

Age: Upper Cretaceous.
Distribution: Atlantic, Pacific, and Gulf coastal plains.
Comments: *Hardouinia* was probably a shallow burrower like most of its living relatives. The concentration of bladelike tube feet in the slits of the petals was an adaptation for burrowing. The animal could be covered with mud or sand except for the petals and still respire. The nonburrowing ancestors of *Hardouinia,* which were probably similar to *Holectypus,* had smaller tube feet over more of the body surface. Plate 369 shows a top and a bottom view of *Hardouinia.* A side view, with the posterior to the right, is shown in the marginal drawing.

366 Rhyncholampas

Description: 1⅝" (41 mm) long, 1⅜" (35 mm) wide. Oval in outline, flat on base and domed on top. 5 long, narrow petals extend almost to margins of domed surface, forming flower pattern. Center of flower, which is shifted slightly to anterior of skeleton, has 4 prominent pores in single plate. Petals bordered by transverse slits and small pores on inside of borders. Ends of petals open. Anus located on posterior margin. Mouth shifted anteriorly, wider than long, with 5 small, wide petals around it. Areas anterior and posterior to mouth smooth; other areas on base covered with prominent bumps. Bumps smaller on top of body.
Age: Paleocene to Recent.
Distribution: East and West coasts of North America.
Comments: The life habits of *Rhyncholampas* are similar to those of *Hardouinia,* to which it is closely related. *Rhyncholampas* can be distinguished from *Hardouinia* by the longer, narrower petals, the more oval body shape, and the oval mouth

that is farther forward on the base.
Plate 366 shows a top and a bottom
view of *Rhyncholampas,* with the
posterior side down. The marginal
drawing shows a side view, the
posterior to the left.

ORDER SPATANGOIDA
(Lower Cretaceous to Recent)

These are the heart urchins, which are
specialized for burrowing. The anus is
near the posterior margin, the pores are
arranged in a flower pattern, and the
area where the petals come together is
not directly opposite the mouth, which
is always on the lower surface and lacks
jaws. There are generally 5 small petals
around the mouth, but no bumps
in-between.

367 Hemiaster

Description: 1½″ (38 mm) long, 1⅜″ (35 mm)
wide. Heart-shaped, longer than wide,
highly domed, with flat base; posterior
margin truncated. Upper surface has 4
small petals of unequal size; posterior
petals small and close together; anterior
petals longer and diverging forward.
5th area not petal-like, but expanded
to fill shallow groove extending from
center of upper surface of skeleton to
mouth on lower surface. 2 rows of tiny
pores border each petal and 5th area.
Narrow band of small bumps circles
petals and crosses 5th area; other areas,
except for petal areas and their
extensions, covered with larger bumps.
Anus located on posterior side above
margin. Mouth kidney-shaped, wider
than long, located on base near anterior
margin.

Age: Lower Cretaceous to Recent.
Distribution: Widespread in North America.

Comments: Judging from living species of
Hemiaster, the band that passes around
the petals probably had specialized
spines that were ciliated at the bases
and flattened at the ends. Glands
around the spines secreted large
amounts of mucus that was plastered by
the spines to the inside of the burrow.
The cilia set up currents for ventilating
the burrow. Plate 367 shows a top view
of a *Hemiaster* specimen, with the
anterior to the upper left. The marginal
drawing shows a side view, the
posterior to the left.

Heteraster, found in the Cretaceous in
North America, is another heart-shaped
urchin. The groove from the top of the
skeleton around to the mouth is
narrower and shallower than in
Hemiaster, and the pores outlining the
petals are elongated into slits.
Linthia, found from the Upper
Cretaceous through the Pliocene in
North America, is also similar, but the
band on top curves inward to follow the
outline of the petals, and an additional
band extends from the anterior-lateral
petals down and along the margin to a
point below the anus. The posterior of
the skeleton is not truncated.

365 Eupatagus

Description: 2⅛″ (54 mm) long, 1¾″ (44 mm) wide
(left); 1⅝″ (41 mm) long, 1¼″ (32
mm) wide (right). Almost oval in
outline, but slightly truncated at
anterior; rounded at posterior. Posterior
narrower than anterior. Rounded on
top, highest at posterior end; flat on
base. 4 long, narrow petals on top,
meeting slightly to anterior. Each petal
outlined by 2 rows of pores set in
shallow groove; ends of petals closed.
Row of plates on anteriormost area
covered with 2 rows of inconspicuous
pores; anterior of skeleton slightly

indented where row of plates crosses margin. Narrow band of small bumps curves around petals and cuts across anterior row of plates. Another band circles anus, located above posterior margin. Bumps above band fairly large; bumps below band much smaller. Mouth scoop-shaped, elongated sideways, located on lower surface toward anterior. Triangular area posterior to mouth slightly raised and smooth, and pierced by several holes near mouth.

Age: Eocene to Recent.

Distribution: Southeastern United States.

Comments: A living *Eupatagus* burrows deeply and maintains mucus-lined tubes at the surface of the sand. Extremely elongated tube feet from the anterior row of plates reach to the surface, where they pick up small bits of food and bring them back down to the mouth. When alive, *Eupatagus* is covered with long, slender, sharp spines, and resembles a small porcupine. The bands on the top of the skeleton bear specialized spines and cilia, like those of *Hemiaster*. Plate 365 shows a top and a bottom view of *Eupatagus*. The marginal drawing shows a side view, with the posterior to the right.

Phylum Hemichordata
(Middle Cambrian to Recent)

The only familiar living hemichordates are the tubeless and burrowing acorn worms and the pterobranchs, small colonial worms living in horny organic tubes usually in deep water. Hemichordates are closely related to the phylum Chordata, which includes the vertebrates. They have a larval stage, a basic body plan, and gill slits similar to those of chordates. Only 1 group of hemichordates, the class Graptolithina, has an important fossil record.

CLASS GRAPTOLITHINA
(Middle Cambrian through
Mississippian)

Graptolites are colonial animals known
only from their horny organic tubes,
which are arranged in 1 or 2 rows along
branches of the colony. The tubes have
growth lines and are banded. The
branches, 1 or more per colony, are
straight, curved, or spiralling. In the
most common preservation, they are
flattened on black shale and reduced to
a thin film. The branches appear
serrated on 1 or both edges, depending
on whether there were 1 or 2 rows of
tubes.

460 Graptolite Fragments

Description: 1¾" (44 mm) long, ⅟₁₆" (2 mm) wide.
Long, narrow streaks, straight or
branching, usually black or white,
most often on dark gray or black shale.
Outline may appear serrated on one or
both sides of streak.

Age: Cambrian to Mississippian.

Distribution: Widespread in North America.

Comments: Unless a graptolite's tubes are well
preserved or most of a colony has been
fossilized, graptolite fragments usually
cannot be identified to genus.
Graptolites are found only in Paleozoic
rock. While fragments of graptolites
may resemble blades of grass, grass
does not appear as a fossil until the
Miocene, and even then is very rare.
The branching forms of graptolites may
resemble seaweed, but uncalcified
seaweed is more delicate than
graptolites and is almost never
fossilized.

459 Didymograptus

Description: 2⅜″ (60 mm) long, ⅟₁₆″ (2 mm)
wide. 2 branches forming straight or
V-shaped colony. Branches may hang
down from short threadlike support,
with single row of tubes on insides of
branches; tubes straight, oblique;
apertures opening downward. Angle
between branches from 10° to 180°, or
angle may exceed 180°, with tubes on
outside of V inclined upward.

Age: Ordovician.

Distribution: Widespread in North America.

Comments: The drawing in the margin shows a
typical colony of *Didymograptus* with its
branches extended like the tongs of a
tuning fork. Plate 459 shows several
branches of *Didymograptus* arranged on a
bedding plane of black shale, the form
of preservation in which most
graptolites are found. Such shale
commonly has a high carbon and pyrite
content, an indication that the water
just above the sea floor was deficient in
oxygen when the mud that later
hardened to shale was accumulating.
This lack of oxygen makes it very
unlikely that *Didymograptus* lived on
the sea floor. The threadlike support
could have been attached to floating
seaweed, and *Didymograptus* would have
settled to the sea floor when the
seaweed died and sank.

461 Tetragraptus

Description: 2″ (51 mm) long, ⅟₅₀″ (0.5 mm) wide.
4 joined branches. Branches may all
hang down together from a threadlike
support, or form a cross or an X with 2
branches up and 2 down, or an H-
shape. All branches have single row of
straight, obliquely oriented,
overlapping tubes.

Age: Lower Ordovician.

Distribution: Widespread in North America.

Comments: Some specimens of *Tetragraptus* have a
             weblike structure where the 4 branches
             intersect, which may have acted as a
             flotation device. Any specimen that
             hangs down from a threadlike support
             may have been attached to a floating
             object such as seaweed. Some very
             young specimens have been found with
             a small disc, probably for attachment,
             still at the end of the threadlike
             support. Plate 461 shows a typical
             X-shaped specimen of *Tetragraptus.* The
             drawings in the margin show another
             colony form of *Tetragraptus* at the top,
             and an enlargement of the tubes at the
             bottom. *Dichograptus,* found in the
             Lower Ordovician of Quebec and New
             York, is similar, but has 8 branches.

Dicranograptus

Description: 1¾" (44 mm) long, ¹⁄₁₆" (2 mm) wide.
               Y-shaped; arms of Y have single outer
               row of tubes; stem of Y has double
               row. Tubes oblique, overlapping, bent
               upward near opening so outer margins
               become parallel to axis of branch.
               Openings directed straight up within
               notches in branch formed by bends in
               tubes.
        Age: Ordovician.
Distribution: Widespread in North America.
   Comments: The marginal drawing shows a nearly
               complete colony of *Dicranograptus,*
               which has never been reported to have a
               threadlike support. Just how it lived is
               a mystery, but it must have drifted,
               since it was dispersed worldwide.
               Whether it provided its own means of
               flotation or somehow attached to
               another organism is not known.

462 Dicellograptus

Description: ¾" (19 mm) long, ⅟₃₂" (1 mm) wide. 2 branches joined in V or U shape. 1 row of tubes on outside of each branch. Tubes S-shaped, curving inward near openings so margins of openings are inwardly inclined.

Age: Ordovician.

Distribution: Widespread in North America.

Comments: Several graptolite genera are shown in Plate 462. *Dicellograptus* appears as the small, V-shaped form on the left. The marginal drawings show a complete colony of *Dicellograptus* and an enlargement of the tubes.
Leptograptus, found in the Upper Ordovician, and possibly also in the Lower Ordovician, throughout North America, has a colony form like *Dicellograptus* and also has S-shaped tubes. However, the tubes do not curve inward toward the axis near the apertures.

457 Orthograptus

Description: ⅝" (16 mm) long, ⅟₁₆" (2 mm) wide. Single branch suspended by threadlike support, with 2 rows of alternating, oblique, straight tubes, inclined upward and overlapping. Each opening bears single spine. First tube in branch grows downward.

Age: Middle Ordovician through Lower Silurian.

Distribution: Widespread in North America.

Comments: Plate 457 shows 5 fairly complete colonies of *Orthograptus,* 1 still attached to its threadlike support, showing white against dark gray shale—the most common kind of preservation for graptolites. The graptolites settled on the sea floor in mud rich with organic matter. When this mud compressed, the graptolites were flattened and most of their skeletal materials leached away,

leaving behind only a carbonaceous
film. But a slight cavity remained
where the skeletons had been, which
later filled with a white clay mineral.
The drawings in the margin show both
a complete colony of *Orthograptus* and
some tubes that have been enlarged and
cut open.

2 other genera have the same colony
form as *Orthograptus. Lasiograptus,* found
in the Ordovician of Quebec, Maine,
Vermont, New York, Ohio, and
Arkansas, has S-shaped tubes with the
openings inwardly inclined, and 2
spines on the outer margin of each
opening. *Glossograptus,* found in the
Ordovician of Quebec, Maine, New
York, Ohio, Kentucky, Alabama,
Oklahoma, British Columbia, and
Nevada, has tubes similar to those of
Orthograptus, but the margin of each
outer opening bears 2 long spines, and
other spines, sometimes quite long,
extend from other parts of the colony.

458   Climacograptus

Description: 2⅛″ (54 mm) long, ¹⁄₁₆″ (2 mm) wide.
Colony a single bladelike branch with a
double row of tubes. Branch nearly
circular in cross section. Tubes have
angular outline, growing up and then
abruptly angling out and then up
again, so outer margins are parallel
with axis of branch. Openings located
within notches in colony outline.

Age: Lower Ordovician through Lower
Silurian.

Distribution: Widespread in North America.

Comments: Plate 458 shows many *Climacograptus*
colonies on a slab of dark shale. Some
colonies have been found in which each
individual is attached to a threadlike
support with an oval structure at its
end that may have functioned as a float
or attachment disc. The first marginal
drawing shows such a colony. The

other drawing shows part of a colony enlarged and cut open to reveal the structure of the tubes.

463 Monograptus

Description: 1″ (25 mm) long, ⅟₃₂″ (1 mm) wide. Single branch with tubes on 1 side only; branch may be coiled, spiralling, gently curved, or straight. Tubes straight, oblique, overlapping, sometimes with openings drawn out into spines.

Age: Lower Silurian through Lower Devonian.

Distribution: Widespread in North America.

Comments: Plate 463 shows many straight specimens of *Monograptus* preserved as carbon films on dark shale. *Monograptus* is the most common Silurian graptolite, and often occurs in enormous numbers on such dark shale bedding planes. It is thought to have evolved from a form like *Orthograptus* by losing 1 row of tubes. The drawings in the margin show a colony of *Monograptus* with a curving branch, and an enlargement of the tubes.

Cyrtograptus

Description: Coil 3⅛″ (79 mm) long and wide; branch ⅟₃₂″ (1 mm) wide. Main branch coiled in spiral, with side branches arising at regular intervals, curving in same direction as main branch. Single row of tubes on outside of spiral and secondary branches. Tubes straight, oblique, overlapping.

Age: Middle Silurian.

Distribution: Widespread in North America.

Comments: *Cyrtograptus* is distributed worldwide except in South America and is thus an ideal guide fossil for the Middle Silurian. The colony form of

Cyrtograptus is distinctive and not readily confused with other genera. The only similar North American genus is *Nemagraptus,* which occurs in the Ordovician of Newfoundland, Quebec, Maine, New York, New Jersey, Tennessee, Alabama, Arkansas, and Oklahoma. *Nemagraptus* has an S-shaped main branch, with secondary branches attached at the convex parts of the S.

Trace Fossils

Trace fossils are traces in rocks of animal activity. They are among the most interesting fossils because they reveal the life habits of ancient and often extinct animals, and give clues to ancient environments. Certain trace fossils are always terrestrial, others indicate freshwater origins, and still others are always marine. The depth of the water when a particular sediment was accumulating can often be estimated from the kinds of trace fossils preserved in the rock.

Many rocks will contain trace fossils but no actual remains of organisms because body fossils are often destroyed by solution or recrystallization of the rock. For example, Triassic sandstones in the Connecticut Valley contain abundant dinosaur tracks, but almost no dinosaur bones.

Trace fossils are most apt to be found in sandstones and siltstones, where the rocks commonly split along bedding planes. They are also abundant in shales. In limestone, they are found less often.

Trace fossils are the predominant record of animals that lacked hard parts. These animals have been abundant for the last 600 million years, but without trace fossils there would be virtually no evidence of their early existence.

It is seldom possible to identify the animal that left a given trace, except where the animal's remains have been preserved with the trace, or where the trace is especially distinctive. Since many different kinds of animals could have left identical traces, trace fossils are not usually classified according to the kinds of animals that made them, but according to the kinds of activities that they record. The 3 main groups of trace fossils are described here.

DWELLING STRUCTURES

These are typically simple vertical burrows. Helical, U-shaped, and elaborately branching burrows also occur.

470 Skolithos

Description: Burrows 2¾" (70 mm) long, ⅛" (3 mm) wide. Burrows straight, cylindrical, perpendicular to bedding plane and parallel to each other; commonly crowded together.

Age: Cambrian through Upper Pennsylvanian.

Distribution: Widespread in North America.

Comments: *Skolithos* frequently occurs in sandy sediments, as illustrated in Plate 470. Burrows as long as 39" (1 m) and as wide as ⅝" (16 mm) have been found. These were the dwelling places of soft-bodied, wormlike animals that lived near shore, perhaps between the tide lines of an open beach. Today the sandy surf zone of many beaches is packed with similar small burrows of polychaete worms. The deep vertical burrows are needed to protect the animals from being washed away and to safeguard them when the tide is out.

CRAWLING TRACES

These were made by animals moving from one place to another, and are typically straight or only slightly sinuous.

466 Cruziana

Description: 11¼" (28 cm) long, 1⅜" (3.5 cm) wide. Elongate, ribbonlike furrows covered by 2 low ridges with groove between; ridges marked with herringbone pattern, V-shapes of pattern meeting in groove between ridges. Sides of furrow may be marked with narrow grooves.

Age: Upper Precambrian through Devonian.

Distribution: Widespread in North America.

Comments: *Cruziana* fossils are the crawling traces of trilobites and other, similar arthropods. The V-shapes are scratch marks made by the legs of the arthropods. The grooves on the sides of the tracks may have been caused by the dragging of the animals' head shield spines. Plate 466 shows several tracks crossing each other at various levels. The V-pattern points in the direction opposite the one in which the animal was traveling.

Kouphichnium

Description: Tracks of horseshoe crab; rows 1⅜" (35 mm) apart, ¼" (6 mm) long, ⅛" (3 mm) wide. There are 2 kinds of tracks in each fossil: chevronlike, in paired rows, each track composed of 4 oval or round holes or V-shaped impressions (made by 1st 4 pairs of feet); and, less frequent, larger toe-shaped, or variable tracks (made by birdfootlike 5th pair of feet).

Age: Devonian through Jurassic.

Distribution: Widespread in North America.

Comments: For years, tracks such as these were ascribed to amphibians, birds, pterodactyls, dinosaurs walking on 2 legs, or jumping animals. About 40 years ago, a paleontologist watching a horseshoe crab walk across wet sand realized that its tracks were identical with these problematic trace fossils that may extend 33' (10 m) or more. The horseshoe crab that made the tracks shown in the margin was moving from top to bottom.

FEEDING STRUCTURES

These are the most common trace fossils. Many organisms ate sediment, and this feeding activity could leave a trace. Such traces may be simple, elaborately branching, or spiral burrows; lobed structures; elaborately folded surface traces; or honeycomb patterns, all made as different organisms ate their way through sediment.

465 Oldhamia

Description: Trace 1″ (25 mm) long, ⅝″ (16 mm) wide. Bunches of very fine rills radiating from slight depression in semicircular or circular pattern.

Age: Lower and Middle Cambrian.

Distribution: Widespread in North America.

Comments: In the past, *Oldhamia* was variously interpreted as the remains of algae, hydrozoans, or bryozoans. Not until 1942 was it described as a trace fossil. It is now believed to be the feeding trace of a worm. A very similar trace is made by living polychaete worms on sand flats. Such worms live in vertical burrows and extend long tentacles out onto the sand to pick up small bits of organic matter that have been brought in by the tide. As the tentacles extend and contract, they create a trace in the sand identical with *Oldhamia*.

464 Chondrites

Description: Trace 1⅜″ (35 mm) long, 1⅝″ (41 mm) wide. Single tunnel 1/16″ (2 mm) wide. System of small, cylindrical tunnels that do not cross or interpenetrate. Tunnels descend through bedding planes, then lie parallel to bedding planes and branch in regular or irregular patterns.

Age: Ordovician through Tertiary (possibly also in the Cambrian).

Distribution: Widespread in North America.

Comments: This is the feeding structure of a sediment-eating animal, but precisely what kind of animal is still unknown. It has been suggested that an animal with a long, extensible proboscis, like a sipunculid worm, could have made the tunnels while extending its proboscis to feed on sediment. The digested sediment would have been voided at the surface. The width of the tunnels varies from ⅟₅₀″ (0.5 mm) to ¼″ (6 mm), but is constant in any single tunnel system.

467, 468 Planolites

Description: Tunnels ¾″ (19 mm) long, ¼″ (6 mm) wide (pl. 467); 2¾″ (70 mm) long, ¼″ (6 mm) wide (pl. 468). Straight to gently curved, cylindrical or nearly cylindrical burrows filled with material different from that composing the matrix. Burrows more or less horizontal, or oblique to bedding planes, penetrating sediment in irregular courses; burrows may cross, but do not branch.

Age: Precambrian to Recent.

Distribution: Widespread in North America.

Comments: These burrows, which may reach a width of ⅝″ (16 mm), were made by wormlike animals that ate the sediment as they burrowed. As the sediment passed through their gut, they digested organic matter from it. The sediment was extruded from the rear of the animals as feces, and back-filled the burrows as the animals travelled along. *Planolites* is one of the most common trace fossils and occurs in sedimentary rocks of all kinds. It is also one of the oldest of all fossils. In fact, the first trace of a multicellular animal is a

Planolites burrow in Precambrian rock. The 2 *Planolites* fossils illustrated appear different, but both were formed in the same way, perhaps even by very similar animals. In Plate 468 the burrows show as a darker pink against a light pink background. The pink color comes from iron in the rock.

Zoophycos

Description: 3½" (89 mm) long, 2⅜" (60 mm) wide. Complex 3-dimensional structure with 2 main parts: central vertical tunnel that descends perpendicular to bedding planes; and very flat, wide tunnel that forms spiral from top of vertical tunnel down through bedding planes, becoming wider as it descends and always connected to vertical tunnel. Insides of flat tunnel marked with irregular or regular ridges that radiate out from vertical tunnel and sweep around periphery of flat tunnel. Outer margin of flat tunnel may change distance from vertical tunnel several times per revolution, producing lobed shape.

Age: Ordovician through Tertiary.

Distribution: Widespread in North America.

Comments: This is a difficult fossil to collect because it lies along many bedding planes. *Zoophycos* must have been made by an animal that lay in the vertical tunnel and extended part of its body, perhaps a proboscis, in sweeping circles out into the sediment around the vertical tunnel, eating the sediment as it descended. The marginal drawing shows a complete *Zoophycos* fossil. *Spirophyton* is a smaller example of the same kind of feeding structure. It occurs throughout North America in the Devonian through the Pennsylvanian. The whorls are more rounded and not lobed, and they bend upward at the periphery.

Fossils of Plants, Insects, and Vertebrates

While most of the fossils in this guide are described in detail, with a separate account for each genus, the plants, insects, and vertebrates are very difficult for the average fossil-collector to identify precisely, and are discussed here in more general terms. The purpose of this section is to provide the reader with information on the major groups of fossil plants, insects, and vertebrates, their evolutionary history, and the conditions under which they are preserved as fossils.

Plants

Plants manufacture the food that supports animal life on earth. By taking carbon dioxide from the air or water, and combining it with water using the energy from sunlight, plants produce sugars that are then burned to supply the energy for synthesizing all the compounds in the plant body. Animals eat plants, and use the sugars and other compounds for energy and growth. The fossil record bears out the theory that plants appeared on earth before animals. The first plant fossils are the simple cells of blue-green algae, 3.4 billion years old. The first stromatolite mounds are 3 billion years old and these mounds are common in rocks 2 billion years old. Multicellular plants appeared about 1 billion years ago, and became common about 600 million years ago.

Adaptation to Life on Land: All known Precambrian plants lived in the ocean. Not until the Silurian Period is there any evidence of plant life on land. It is possible that, before this, the level of oxygen in the atmosphere was still too low to screen out ultraviolet light, which is damaging to the cells of plants and animals. Because of the advantages for plants of living on land —abundant room, light, and freedom from predation— it is likely that they moved onto land as soon as conditions were favorable. But life on land required a few structural modifications. As long as plants stayed in the water, there was no need for any transport system other than diffusion. Water bathed the tissues and supplied them with carbon dioxide, oxygen, and minerals. No specialized structures such as roots, stems, or leaves were required, either. A few gas-filled bladders could keep a plant floating near the surface, and all parts of the plant could photosynthesize.

Structural Support and Feeding in Plants: When plants moved ashore, the upper part of the plant was no longer bathed in water, so it had to get water and minerals from the lower part, which in the earliest land plants grew in water. This conveying of nutrients required a vascular system. The upper parts of the plant could manufacture food, but the lower parts could not, so sugars had to be transported downward. The first plants to develop a simple vascular system were the psilophytes, which appeared in the fossil record in the Upper Silurian, but had no roots or leaves. Photosynthesis took place in the stems. These plants were only a few inches high.

The first plants with true leaves and roots appeared in the fossil record soon thereafter. These were the lycopods, sphenophytes, and ferns, some as high as 100' (30 m).

Reproduction in Plants: Still, plants were not fully terrestrial. Spores borne in capsules on stems or under leaves had to fall into water in order to germinate, so the spore-bearing plants were still tied to open water. The first fully terrestrial plants were the seed ferns. They had mastered the trick of retaining their spores and having them germinate within the tissues of the plant, where they could be protected from drying out. The cordaites were another group that evolved seeds in the Upper Paleozoic. In the Mesozoic, the conifers, ginkgos, and cycads flourished. But these seed-bearing plants, while more advanced in their reproductive methods, were now dependent on the wind to blow the pollen to the eggs, which would then form seeds. This system required the production of enormous amounts of pollen. By the Cretaceous, the anthophytes, or angiosperms, had overcome this disadvantage by developing flowers. Flowers attracted insects, and insects spread the pollen.

Fossilization of Plants: Most marine plants have no decay-resistant parts and have left virtually no fossil record. The blue-green algae formed stromatolites in the Precambrian and Early Paleozoic, and some calcified multicellular algae have left fossils. But most marine plant fossils are microscopic. These are the diatoms—single-celled, golden-brown algae that have a siliceous shell—and the coccolithophorids of uncertain affinity, with a calcareous shell. Some rock formations are composed mostly of the shells of diatoms or coccolithophorids.

The fossil record of land plants resembles that of insects in that it is fairly good during the Pennsylvanian and Permian, sparse through most of the Mesozoic, but good again in the Tertiary. Plants and insects are often preserved in the same rocks, but plants are more resistant to decay than insects, and are usually larger. Plants are frequently preserved as compressions. The plant, generally a leaf, falls into water, sinks, is buried by sediment, and is compressed. Most of the plant material leaches away and decays, leaving a film of carbon. When a rock is split open, the plant shows as a dark impression on one side, and usually as a colorless impression on the other side. Compressions record much of the important information for identifying plants—the outline of the leaf and the venation. The cuticle may be preserved, recording the fine details of the leaf surface, such as the shape and location of the tiny openings that let air in and water out of a leaf. Plant material is sometimes preserved as a petrifaction. This happens when water seeps into the spaces between and within cells, carrying minerals, usually silica but sometimes calcite or pyrite, that precipitate in these places, encasing the original material and preserving it.

488 Division Cyanophyta
(Precambrian to Recent)

Cyanophytes are commonly called blue-green algae. They are not members of the plant kingdom, but are included with bacteria in the kingdom Protista. The cells of protists are more primitive than those of plants and animals because they lack a nucleus and certain other specialized structures. However, blue-green algae are like plants in that they manufacture their own food from sunlight, carbon dioxide, and water, rather than having to obtain it externally, as animals do.

Early Stromatolites: Blue-green algae are common today in a wide range of environments; during the last half of the Precambrian and the beginning of the Paleozoic, they were the dominant organisms on Earth and left an extensive fossil record. The fossils, called stromatolites, are large or small layered mounds that may be domed, conical, columnar, spherical, or fingerlike. *Collenia* (Plate 488) is an example of one of these mounds. The mounds were built up layer by layer, usually by a combination of calcium carbonate precipitated by the algae, and fine grains of sediment held by the mucilaginous sheaths surrounding the algal cells. The fossil's distinctively layered structure may be related to the tides, if a layer formed each time the mound was covered at high tide, or to the day-night cycle, if the algae photosynthesized and precipitated calcium carbonate during the day, and grew upward through the sediment and calcium carbonate layer at night. Stromatolites became very rare after the Cambrian, probably because grazing animals like gastropods, which had evolved by that time, ate the algal mats as they formed so that no mound could build up.

Stromatolites Today: Stromatolites form in a few places today, most spectacularly in Shark Bay in western Australia. Here large mounds grow in the clear, shallow water as lushly as they did in the Precambrian seas. The secret of their survival is that Shark Bay is a restricted lagoon with above-normal salinity that creates conditions intolerable for the animals that feed on stromatolites.

490   Division Chlorophyta
(Precambrian to Recent)

The cells of chlorophytes, or green algae, are advanced, and they are therefore not protists like the blue-green algae, but members of the plant kingdom. Green algae have cellulose cell walls, store food as starch, and have the same kind of chlorophyll as do mosses and higher plants. Terrestrial plants probably evolved from green algae, which are classed as algae because they lack the differentiated structures of roots, stems, and leaves that higher plants have. Most green algae are single cells that float as an important part of the plankton (the tiny organisms that live suspended in the sea), but they may also form large, calcified, multicellular structures. Some of the larger algae, like living *Halimeda* and *Penicillus,* have been of enormous importance in the formation of limestone. These plants precipitate calcium carbonate over their exteriors. When the plants die, the calcium carbonate crystals fall off and become part of the sediment. Florida Bay, between the southern tip of Florida and the Florida Keys, is today filling with lime mud largely precipitated by green algae. In time, this mud will become limestone.

Receptaculitid: Receptaculitids are the fossils of algae. They lived throughout North America from the Lower Ordovician through the Permian. Until about 20 years ago, these organisms were usually classed with sponges, but careful study of their skeleton has convinced paleontologists that they are indeed algae, probably belonging to the division Chlorophyta. Plate 490 shows only part of an organism. A whole receptaculitid specimen is globular to platter-shaped, and may measure from a few inches to 1' (30 cm) across. The surface is covered by rectangular plates arranged in intersecting sets of clockwise and counterclockwise spirals, rather like the arrangement of seeds in a large sunflower. The plates are connected, by small pillars below, to another layer of plates, which form an inner wall. The specimen in Plate 490 is an internal mold from which the skeleton itself has dissolved away, leaving only the hardened mud that filled all the spaces between the skeletal elements. The pillars are represented by hollow tubes that can be seen in the upper left. The double wall is also apparent.

Division Rhodophyta
(Precambrian to Recent)

Red algae, or rhodophytes, are the most common of the marine seaweeds. They include the coralline algae, which have heavily calcified cell walls and, thus, a good fossil record. Coralline algae are often encrusting and do much of the binding that forms the framework of modern coral reefs. Up to half of the calcium carbonate precipitated on some reefs comes from coralline algae.

Solenopora: A relative of the corallines, *Solenopora* is a nodular or bulbous algal growth. Its surface is fairly featureless, but a cross section of it will show closely packed radial or divergent rows of elongated tubes. *Solenopora* is found throughout North America in Ordovician rocks.

491, 492 Division Lycophyta
(Devonian to Recent)

The lycophytes, or lycopods, include the most primitive trees. They were dominant in the coal swamps of the Mississippian and Pennsylvanian periods, and much of the coal mined in North America today comes from the leaves, trunks, and stems of these plants, some of which grew to heights of more than 100′ (30 m). There are only 3 genera of living lycopods, and all are very small and are called club mosses. *Lycopodium* is the most familiar one in North America.

Lycopods reproduced much as ferns do. Spores were borne in cones. When the cones opened, the spores fell into water (the lycopods of the Paleozoic probably all lived in swamps), where they sprouted and grew into tiny plants that produced eggs and sperm. The sperm were shed into the water, where they swam to the eggs. The fertilized eggs then grew into new trees, which in turn reproduced asexually. This process is called alternation of generations, and is functional as a system as long as the plants live in moist places. Without water, the sexual generation cannot reproduce.

Lepidodendron: The most common lycopod fossil is probably *Lepidodendron*. Plate 491 shows part of the trunk of one of these trees. The rounded, diamond-shaped structures arranged in diagonal rows are the leaf cushions and scars.

Lepidodendron bore its leaves directly on the trunk in spiral rows. The roots, leaves, and cones of this tree are also known as independent fossils, and since they were originally described from isolated parts, each structure has a different name. Thus, the roots are called *Stigmaria,* the leaves *Lepidophyllum,* and the cones *Lepidostrobus.* The separate names are retained for convenience.

Sigillaria: Another common lycopod from the Paleozoic coal swamps is *Sigillaria*. The trunk is covered with leaf cushions and scars, but these are arranged vertically instead of spirally, and each is club-shaped.

Drepanophycus: Plate 492 shows a small branch, with leaf scars, of *Drepanophycus*, one of the earliest lycopods.

489, 493 Division Sphenophyta
(Devonian to Recent)

1 genus of Sphenophyta is still living today. This is *Equisetum*, the horsetail rush. In the Paleozoic, sphenophytes often grew as trees 50' (15 m) or more in height, but with the same form as today's small, reedy horsetail rushes.

Calamites: Parts of 2 sphenophytes are illustrated: a piece of trunk called *Calamites* (Plate 489), and some whorls of leaves called *Annularia* (Plate 493). Originally these were thought to be from separate plants, but they are now known to be of the same genus. *Calamites* was a large tree. The trunk had distinct nodes, or raised, riblike rings, spaced regularly, with vertical ribs between them. Leaves or branches were attached in whorls at these nodes. Cones containing spores were borne at the nodes also, just inside the leaf whorls. *Calamites*, like the lycopods, had alternating sexual and asexual generations.

499, 500, 502 Division Pteridophyta
(Devonian to Recent)

Pteridophytes, or true ferns, are still very common today, especially in shady, damp environments. But in the Paleozoic they commonly grew into trees, with trunks 2' (0.6 m) in

diameter and 30–45′ (9–14 m) high.
Fern leaves have a stalk and a blade.
The blade is usually compound, that is
divided into many leaflets. A leaflet, in
turn, may be divided again into smaller
leaflets, and these into even smaller
leaflets to produce the characteristic
distinctive, feathery leaf. Immature
leaves are coiled into fiddleheads. The
spores are borne on the undersides of
leaves in clusters called fruit dots. As
with the lycophytes, the spores must
fall into water to germinate and grow
into the tiny plants that produce the
eggs and sperm. When an egg is
fertilized, it grows into the spore-
producing fern plant.

Pecopteris: Plates 499, 500, and 502 show
Pecopteris, a common tree fern in the
coal swamps of the Pennsylvanian. The
short leaflets are attached to the leaf
stalk along their entire bases, and the
margins are smooth and parallel.

501, 503 Division Pteridospermophyta
(Mississippian through Jurassic)

Pteridospermophytes, or seed ferns, had
foliage like the pteridophytes, or true
ferns, but they advanced beyond the
true ferns to bear seeds that were
fertilized by wind-blown pollen. No
separate sexual plant was needed. This
freed seed ferns from dependence on
standing water, whose absence even
now hinders the spread of lycophytes,
sphenophytes, and true ferns. It is
possible that seed ferns are the ancestor
of flowering plants.

Neuropteris: A very common seed fern genus,
Neuropteris (Plate 503) has leaflets that
attach to the leaf stalk at a single
point. The larger leaflets usually form
lobes at their bases.

Alethopteris: *Alethopteris* (Plate 501) is another very
common genus. The leaflets are
attached to the stalk along their

expanded bases and are oriented obliquely. The tops are broadly rounded.

498 Division Cycadophyta
(Permian to Recent)

Cycads may have evolved from Paleozoic seed ferns. They were abundant plants during the Mesozoic, which is sometimes called the Age of the Cycads, but they declined in the Tertiary and are now restricted to 9 genera and about 100 species that live in Central and South America, southern Africa and eastern Asia. Most cycads have short, squat trunks that do not branch and a crown of large, palmlike leaves. Seeds are borne in cones.

Nilssonia: The compound leaves of *Nilssonia* (Plate 498) are linear to lance-shaped. Each leaflet is attached along its entire base to the leaf stalk. The veins rise at right angles to the leaf stalk, and curve toward the leaflet margins.

505 Division Ginkgophyta
(Permian to Recent)

A developmental pattern similar to that of the cycads appears in ginkgo trees, which arose in the late Paleozoic. Like cycads, ginkgos flourished in the Mesozoic and declined in the Tertiary. Today, only a single species of *Ginkgo* remains. *Ginkgo* leaves are fan-shaped, and may be unlobed to deeply lobed. The veins appear to be parallel, but each vein is deeply divided. The stalked pollen sacs and seeds are clustered together at the bases of the leaves. The *Ginkgo* in Plate 505 is from the Upper Cretaceous of Montana.

494, 495, 496,   **Division Coniferophyta**
497, 504   (Pennsylvanian to Recent)

The coniferophytes are divided into 2 groups: the cordaites, which flourished in the Mississippian through the Permian, forming trees that grew to a height of 100' (30 m), and the conifers, which appeared in the Pennsylvanian and still dominate the forests in northern areas and at high altitudes.

Cordaites: The cordaites are represented by the most common genus, *Cordaites* (Plate 504), which has a slender trunk, unbranched except at the crown, and long, straplike leaves with parallel veins. The cordaites were probably ancestral to the conifers.

Conifers: The conifers include among living genera the pines, spruces, hemlocks, cedars, and redwoods. Many have a conical shape, and most have a distinctive needlelike leaf with a protective coating. Most conifers do not shed their leaves annually. The pollen is borne in small cones, and the eggs in larger cones. The conifers depend on wind to blow pollen to the eggs for fertilization.

Pagiophyllum: The pointed, triangular leaves of *Pagiophyllum* (Plates 494, 495) are arranged in spirals around the branches. *Pagiophyllum* trees are found from the Triassic through the Cretaceous in western North America.

Dawn   *Metasequoia,* the dawn redwoods (Plate
Redwoods: 496), occur as fossils throughout North America from the Upper Cretaceous through the Middle Miocene. These trees were thought to be extinct until 1946, when a small grove of them was found in an area of western China. Now dawn redwoods are cultivated all over the world. They were dominant elements in western North American flora in the first half of the Tertiary. The needles are attached opposite each other on the stem. The

globular cones are fairly common fossils.

Sequoia: Closely related to *Metasequoia*, *Sequoia* (Plate 497) is the common California redwood that lives in well-drained lowland areas. The needles are alternate instead of opposite on the stems, and the scales on the cones are staggered instead of vertically arranged as in *Metasequoia*. *Sequoia* ranges from the Upper Cretaceous to the Recent.

487, 506, 507 **Division Anthophyta**
(Cretaceous to Recent)

Anthophytes, or angiosperms, as they are often called, are the flowering plants of the world. Today they constitute the dominant vegetation, with 220,000 species, and range in size from tiny floating duckweed to huge oaks. Their distinctive feature is the flower. The most common kind of flower has a vessel, or carpel, enclosing the eggs, and surrounded by stamens (the stalks that hold the pollen). The carpel and stamens are, in turn, surrounded by brightly colored petals that are modified leaves. Petals evolved to attract insects, and most angiosperms are fertilized by insects carrying pollen from the stamens to the tops of the carpels. When pollen lands on the carpel, a tube grows down to the eggs for the passage of the sperm. Once fertilized, the eggs grow into seeds. A flower is homologous to the tiny plants that produced eggs and sperm in ferns and lycopods.

Platanophyllum: A member of the sycamore family, *Platanophyllum* (Plate 507) is abundant in several western North American fossil floras. The large leaves are fan-shaped, with 5–9 projecting lobes and a wedge-shaped base.

Quercus: The leaves of *Quercus* (Plate 506), the familiar oak, are variable in size and shape, but all have a strong midvein that extends to the apex, and distinct alternating secondary veins.

Celtis: A woodland tree, *Celtis,* or hackberry, was common in western North America in the Tertiary. Plate 487 shows the polished surface of a piece of hackberry wood preserved by petrifaction, when silica was deposited within the cells. Tree rings are visible, indicating a seasonal cycle of growth. Each ring formed in 1 year.

310, 311, 312, 313, 314, 315 Insects

Insects (class Insecta) are members of the phylum Arthropoda. They are perhaps the most successful group of organisms that has ever lived. There are more species of insects living today than of all other animals and plants combined. 700,000 have been described, and this is probably less than half the number in existence. There are only about 400,000 living species of other animals and plants.

The Insect Body: Insects are similar to other arthropods, but differ from them in that most insects bear wings, or once did. Insects have segmented bodies, with sense organs concentrated on the head, and former legs modified into mouthparts for chewing, biting, or sucking. They have a median thorax that bears 3 pairs of legs. Like other arthropods, insects are covered with a nonliving exoskeleton that does not grow and therefore must be shed periodically, but unlike that of trilobites and crustaceans, the insect skeleton is not mineralized.

Adaptation to Life on Land: Arthropods are preadapted for land life, with an external skeleton to retard evaporation of body moisture, stiff legs that work on land as well as in water, and a strong external skeleton to support the body out of water. In the water, arthropods breathe with finely branched gills, thin-walled appendages on various parts of the body. But gills must be kept moist and will not serve a fully terrestrial animal. Insects evolved from millipedelike animals, which, along with scorpionlike arachnids, may have been the first animals to come ashore. Centipedes, millipedes, and insects have solved the problem of breathing with a system of branching internal tubules that carry air to all parts of the body. Because they are internal, the surfaces of these tubules can be kept moist. This solution limits the size of an insect, however, since circulating air becomes a problem as the distance the air must travel increases and the air passages in the insect's body become progressively narrower.

Evolution of Insects: No fossils representing the early evolution of insects from millipedelike animals have been found. Fossil millipedes appear for certain in the Devonian. While there are also some wingless, insectlike fossils in the Devonian, the first clearly identifiable insects do not appear until the beginning of the Pennsylvanian, and by this time they are specialized. For example, these Pennsylvanian insects already had wings, which the first insects did not. Wings developed from extensions of the upper surface of the thorax, and may first have been used for gliding. These extensions presumably became larger and larger, and eventually developed struts (the veins) to strengthen them, and muscles to move them. Wings gave the insects an important advantage for survival. They could now escape from their predators and could fly in search of food, mates, and places to lay eggs. They could readily disperse to all areas as soon as there was vegetation to attract them. Since Pennsylvanian times, insects have been the most diverse and abundant of terrestrial animals.

Insect and Plant Interactions: A majority of insects feed directly on plants. Some fossil seed fern leaves from the Pennsylvanian have edges that have been scalloped, apparently by chewing insects. From their first appearance until the Cretaceous, insects mainly ate plants. In the Cretaceous, plants evolved to take advantage of insects in a way that greatly increased the diversity of both insects and plants.

Before the Cretaceous, plants tended to be dependent on wind to blow pollen to the egg capsules for fertilization. Then angiosperms began to develop flowers specifically designed to attract insects. Insects are attracted by the bright colors and perfume of a flower, and then are lured closer by the nectar and edible pollen within it. In the process of consuming the nectar, insects rub against the flower's pollen-carrying anthers, and the pollen sticks to their bodies. As they forage, they carry the pollen and cross-fertilize the flowers in which they feed. This is a much more efficient device for fertilization than the wind, especially since plants of the same species can live far from each other and still be fertilized by insects. Of course, insects are still a scourge to plants, but plants have evolved to withstand this predation, and it is unlikely that insects alone have caused the extinction of many groups of plants.

Rarity of Fossil Insects: Insects have been abundant and diverse for hundreds of millions of years, but are rare as fossils because they are terrestrial and their skeletons are not mineralized. Most insects live on land, and a dead insect must fall into water and quickly be covered with sediment or it will disintegrate; therefore, most fossil-bearing rocks contain no insect fossils. Nevertheless, paleoentomologists have collected and described more than 13,000 species of fossil insects.

Collecting Insect Fossils: Of the 2 dozen known localities in the world where fossil insects are reasonably abundant, 5 are found in North America. The earliest beds are those of the Pennsylvanian Mazon Creek in northern Illinois. The insects here are preserved in ironstone nodules that formed around them soon after they died and were buried. At the time that the insects were alive, this area was a coal swamp, and preserved along with the scarce insects and arachnids are many leaves, as well as a few freshwater fishes and crustaceans. The only other major Paleozoic locality is in an area spanning parts of Kansas and Oklahoma, where the Wellington shale contains a large variety of insect fossils. The North American Mesozoic has left few insect fossils. On the shore of James Bay, in eastern Canada, amber containing insects can be found eroded out of a Cretaceous formation. There are 2 famous Tertiary fossil insect localities. One is in an area of Colorado, Utah, and Wyoming, in the Green River shales of the Eocene (from which the specimens in Plates 310–315 were taken). These shales were deposited in ancient lakes, and plant and fish fossils as well as insects are common here. The other locality is also an ancient lake bed, the Miocene shales at Florissant, Colorado, near Pike's Peak. Here much ash is mixed with the shale. Ash from repeatedly erupting volcanoes may have killed the insects and preserved them quickly after they sank into the lake.

Vertebrates

Vertebrates are animals with backbones. They include most of the familiar animals on earth—the fishes, frogs, lizards, snakes, birds, and mammals, including, of course, humans. Vertebrates make up 1 subphylum of the phylum Chordata, the rest of whose members are invertebrates.

Rarity of Vertebrate Fossils: Vertebrate fossils are much rarer than those from the major invertebrate groups. Nevertheless, a visit to the fossil rooms of almost any museum may convince you that "fossil" and "vertebrate" are almost synonymous. Vertebrate fossils are certainly not rare because they are recent; quite the contrary. The first vertebrate fossils occur in Lower Ordovician rocks. The most advanced vertebrate class, the mammals, first appeared in the Triassic, almost 200 million years ago. Nor are vertebrates rare as fossils because of a lack of hard parts. One characteristic of all vertebrates is an internal skeleton, and, except in most sharks, this is ossified. Even sharks have some hard parts—teeth, spines, bony denticles in their skin, and, sometimes, ossified skeletons.

Formation of Vertebrate Fossils: Vertebrates would seem to be ideally constituted to leave an abundant fossil record, yet they have not done so. The reasons are several. First, many vertebrates are terrestrial. After an animal dies on land, the bones usually lie on the ground. Vertebrate skeletons have many parts held together by ligaments and tendons that are entirely organic and decay quickly. The bones then scatter, and each bone can be attacked by destructive processes. The bones are dispersed, eaten, dissolved by acids in the soil, and destroyed by roots. If an animal does fall into water after death, the bones are still apt to be scattered, eaten, leached, dissolved, and disintegrated before they can be buried in sediment. When bones are buried, they tend to be concentrated in little pockets of vast formations, and the chances of finding one of these pockets are very small.

Secondly, most vertebrates are relatively large and have small populations. Large animals, in general, are rarer than small animals. Many small invertebrates, both terrestrial and marine, occur in enormous numbers over a small area, particularly in temperate and boreal regions where food may be concentrated in the ocean.

Thirdly, vertebrate bone is porous. When alive, bone is permeated with blood vessels and is continually being dissolved and reprecipitated. After an animal's death, water can percolate readily through the bone, dissolving it. Most vertebrate fossils are teeth, which are composed of more dense cement and enamel that dissolve less easily than bone does.

Significance of Vertebrate Fossils: But in spite of the sparse record that the vertebrates have left, their evolutionary history is known better than that of any invertebrate group. Why is this true? First of all, most people are more interested in vertebrates than in invertebrates, probably because vertebrate history is our own history. The number of scientists working on vertebrate fossils is certainly out of proportion to their abundance. Secondly, the skeleton of a vertebrate is internal and complex, and bears a more intimate relationship to the soft tissues than is true of invertebrates. Invertebrate skeletons are largely external, usually a kind of enclosure for the soft tissues, and do not record details of fine structure; the elaborate skeleton of a vertebrate tells us more about the animal's anatomy.

Vertebrate Fossils and Evolution: Our present understanding of vertebrate history is truly remarkable, and makes the best case for evolution that can be found anywhere. There are many examples of a skeletal structure being traced through a group of related animals, changing a bit here and there in form and function but preserving a record of its origins.

A classic example is found in vertebrate limbs. They began as fleshy folds on the sides of jawless fishes, used for stabilizing the animal in swimming. These folds are thought to have developed spines for stiffening, and then to have broken up into a series of small paired fins, as in the acanthodians. In 1 line of fish, many paired fins became 2 pairs, 1 forward and 1 in the rear. Later, fish evolved small bones and muscles in the fins so that they could be moved and used for steering as well. The bones became better developed in some of these fins, first to hold the fish off the bottom while they fed, and then, in times of drought, to help the fish scramble across to new streams or ponds. After more strengthening of the bones and modification of the fins into flat pads with digits, the first 4-legged vertebrate, an amphibian, appeared. Later, these land legs were modified for creeping, running, hopping, digging, climbing, gliding, flying, and, finally, for the manipulation of tools. In every case, the relationships can be clearly seen and the evolutionary pathways traced. In this section we have illustrated a range of vertebrate fossils. Some, like the bird eggs, are very rare. Others, like the shark teeth and fish scales, are more common.

Identifying vertebrate fossils is more difficult than identifying invertebrates, mainly because vertebrate fossils are apt to be more fragmentary.

Class Agnatha
(Lower Ordovician to Recent)

The agnathans, or jawless fishes, are the most primitive of all vertebrates. Today they are represented by lampreys and hagfishes. When they first appeared in the Lower Ordovician, they were quite different from the living agnathans, which are all long, slender, eel-like animals that have no scales and only a cartilaginous skeleton. The early agnathans were armored with a solid layer of bony plates over the head and thorax, and the tail was covered with thick, bony scales. These strange forms are called ostracoderms. They fed on the bottom of the ocean or in freshwater lakes or streams, taking in water and mud through the mouth—a small hole in the underside of the head —and filtering out bits of food. They had no jaws, and thus could not have been predators. One of the best-known ostracoderms is *Cephalaspis,* whose remains are abundant in Devonian marine deposits.

Class Placodermi
(Upper Silurian through Lower Permian)

This group of fishes includes the first jawed vertebrates. All placoderms are peculiar in some way, and their relationships to the rest of the vertebrates are a continual matter of debate. There are 2 main groups: arthrodires and antiarchs. All placoderms have jaws, a mainly cartilaginous internal skeleton, head and trunk armor, paired fins, and projecting spines to the sides of the trunk shield.

Arthrodires: The best-known of the placoderms are the arthrodires. They had a heavy head shield hinged to a separate trunk shield. The jaw margins formed teeth, and large spines projected from the sides of the trunk shield. They were common in the Devonian, but became extinct at the end of that period. Among the arthrodires was the largest of all Paleozoic animals, *Dunkleosteus* (called *Dinichthys* in some books). *Dunkleosteus* grew to a length of 30′ (9 m) and had enormous jaws. Nothing in the Devonian ocean could have been safe from it.

Antiarchs: Another abundant group in the Middle and Upper Devonian were the antiarchs. The antiarch remained small, and had weak jaws. It had a head and trunk shield similar to those of arthrodires, and a fishlike tail with small scales. The pectoral fins at the sides of the trunk shield were long, heavy, and armored with plates. Antiarchs may have used these spines as pushers as they floated over the bottom, feeding on small invertebrates and soft vegetation. Most antiarchs appear to have lived in fresh water. *Bothriolepis* was a typical antiarch, common worldwide in the Middle and Upper Devonian.

454, 478, 480, 483 **Class Chondrichthyes** (Devonian to Recent)

Most Chondrichthyes, or sharks, have a wholly cartilaginous skeleton, which may have evolved from an ancestor with a calcified skeleton. This lack of calcification does not appear to have been disadvantageous to sharks, which have survived and flourished for 400 million years, and are still an important part of ocean life. The disadvantage falls to vertebrate paleontologists, who must piece together shark history simply from the teeth or an occasional spine, since uncalcified cartilage is seldom fossilized.

Sharks are differentiated in ways other than the nature of their skeletons. Because sharks practice internal fertilization, the males have claspers on their pelvic fins. The eggs may be hatched internally and the young are born alive. Sharks have no lungs or air bladders like those of bony fishes. They have separate gill openings, with a reduced opening in front of the first gill slit.

Shark teeth drop out when they become worn or broken, and are immediately replaced—one reason that shark teeth are not uncommon as fossils. Early in the Paleozoic, some species may have inhabited fresh water, but today they are all marine.

Orthacanthus: A member of the order Pleuracanthodii that ranged from the Upper Devonian through the Triassic, the primitive shark *Orthacanthus* lived throughout North America from the Upper Devonian through the Middle Permian. It is known mostly from the large, serrated spine that was attached at the back of the skull and projected over the dorsal fin (Plate 454). *Orthacanthus* probably had an elongated body and a tail that extended straight back and ended in a point. The teeth have 2 long blades with a small cusp between.

Ptychodus: The modern sharks, skates, and rays are members of the order Selachii. Selachians appeared first in the Pennsylvanian and are still abundant. *Ptychodus* is a selachian that lived worldwide in the Upper Cretaceous. It may have looked like a modern skate, but is known only from its teeth, jaws, and vertebrae. The teeth (Plate 480) are broad and strong, and obviously designed to crush the shells of mollusks or other shelled invertebrates. *Ptychodus* bite marks have been found on shells of the clam *Inoceramus*. The teeth are square and flattened, with a pattern on the tops like fingerprints surrounded by tiny bumps.

Carcharodon: The teeth of *Carcharodon* (Plate 478) are the largest shark teeth known. Some of these heavy, triangular teeth reach a length of 8″ (20 cm). If *Carcharodon* had the same proportion of tooth length to body length that modern sharks show, then a *Carcharodon* with 8″ teeth would have been 45′ (14 m) long. This is about twice the length of the living Great White Shark (*Carcharodon carcharias*). *Carcharodon* fossils are found worldwide from the Miocene through the Pleistocene.

Coprolites: Occasionally the feces of sharks and other animals are fossilized. These fossils are called coprolites. The coprolites of sharks (Plate 483) are distinctive because of the banding that can be seen on the left side of the specimen. The bands result from a spiral membrane in the gut of sharks.

486 Class Osteichthyes
(Lower Devonian to Recent)

Osteichthyes are the bony fishes. They are the most successful of all aquatic animals, and the most varied and numerous of all vertebrates. They have shown a rapid increase in diversity during the latter part of the Mesozoic and the Tertiary, and may still be diversifying. They have well-ossified skeletons and their skin is usually covered with thin, bony scales. Today most bony fishes are teleosts, and it is this group that is responsible for the present success of bony fishes.

Holosteans: An earlier, more primitive group of bony fishes are the holosteans, found in North America from the Upper Permian to the Recent. These fishes were typically heavily built, with peglike teeth for crushing invertebrate shells. Plate 486 shows the scales of one of these, *Semionotus* from the Upper Triassic. The scales are rhomboid-shaped and composed of several layers of bone. The thick outer layer is a kind of enamel, called ganoine, and scales like this are called ganoid. A living relative of *Semionotus* is the American gar pike *Lepisosteus*. The gar pike has similar scales, but has a more streamlined body, and teeth that stab rather than crush.

Acanthodians:
Another group of bony fishes, the acanthodians, has sometimes been placed with the placoderms, but is more similar to the Osteichthyes. Acanthodians looked more like modern fish than the placoderms, but had many fins, both paired and unpaired, and each except the fin on the tail was stiffened by a spine. Most acanthodian fossils are spines, and some of these spines are almost as long as the whole fish. Some spines and scales that probably belong to acanthodians first appeared in the Lower Silurian. Acanthodians reached their peak in the Devonian and were rare thereafter, but survived until the Lower Permian. Some of the earlier acanthodians may have been the ancestors of bony fishes. They have the same kind of jaw and braincase, a well-ossified internal skeleton, a covering of scales, and a similar tail.

Class Amphibia
(Devonian to Recent)

The most primitive of the 4-footed vertebrates, amphibians were the first vertebrates to come ashore, but they have never become fully terrestrial. Living amphibians—the frogs, toads, and newts—are small and frequently inconspicuous. In the Paleozoic, however, many amphibians were quite large, weighing as much as 500 pounds (227 kg). They evolved from a group bony fishes called the crossopterygians

763

or lobe-fin fishes. The first amphibians, among which was *Ichthyostega*, were little changed from their fishy ancestors. The main difference was in the appendages, which had become recognizable legs and feet, but contained many of the same bones that were in the paired fins of the crossopterygians. The tail and the head were still fishlike. As with frogs and salamanders, desiccation must have been a problem for *Ichthyostega*, and it had to keep its skin moist. One of the biggest liabilities that the amphibians inherited from their fish ancestors, and one that inhibits the success of amphibians even today, was the necessity to return to the water to lay eggs. Amphibians have no protective covering for their eggs. There is a double disadvantage here. In the first place, the eggs are much more likely to be eaten in the water than they would be if they could be hidden on land; and secondly, amphibians cannot colonize areas where there is no standing water.

72, 473, 474, 484, 485

Class Reptilia
(Pennsylvanian to Recent)

While living reptiles include only the snakes, lizards, turtles, crocodiles, and the tuatara of New Zealand, reptiles were the dominant land vertebrates during the late Paleozoic and Mesozoic. Reptiles evolved in the Pennsylvanian from amphibians. They developed a thicker, more protective skin that slowed dehydration, and an egg that could be laid almost anywhere because it had a protective waterproof covering. Reptiles thus gained independence from water, and could colonize areas closed to amphibians.

Turtles: Plate 485 shows a bony plate from the turtle genus *Trionyx*, which appeared in the Upper Cretaceous and is still living. The order of turtles, Chelonia, is one of the most successful of all reptile groups. Turtles have developed the most remarkable armor of any 4-legged vertebrate. The backbone and ribs are attached to bony plates that entirely enclose the body. Over these bony plates, most turtles have horny scutes, the material we call tortoise shell. The scutes are rarely preserved a fossils, but the bones and bony plates ◀ turtles are fairly common. *Trionyx* is a soft-shelled turtle, lacking scutes, and has only a leathery covering over its plates.

Dinosaurs: Dinosaurs comprise 2 orders of reptiles, the Saurischia and the Ornithischia The more primitive order the Saurischia, includes all the carnivorous dinosaurs and the huge plant-eating brontosaurs. The other order, the Ornithischia, are the "bird-hip" dinosaurs. All these are plant-eating, and many of them were relatively light and fast-moving. In th ornithischians, the pelvic bones were modified so that the pubis grew back along the ischium, as it does in living birds. Dinosaur tracks are shown in Plates 472–474, and Plate 484 shows piece of dinosaur bone.

Characteristics   Both orders of dinosaurs evolved from
of Dinosaurs: ancestors that walked upright on their hind legs, and the first dinosaurs were also bipedal. They developed a long, powerful tail to balance the forward til of their body. The weight was largely supported on the hips, and many structural changes occurred in the pelvic and leg bones as a result. The front legs became short and weak. When some dinosaurs reverted to walking on 4 legs, the front legs were still usually shorter than the rear. The dinosaurs' ancestors were meat-eaters, and the first dinosaurs had sharp, stabbing teeth. Later, most dinosaurs became plant-eaters, with teeth modified for biting and grinding.

Dinosaurs showed enormous variation in size. Some were no bigger than a rooster; others were the largest land animals ever to live, and weighed up to 80 tons (72,575 kg).

Dinosaur Fossils: Dinosaur bones occur in Mesozoic formations with some regularity, but bones had to fall into water and be buried to be preserved, and most dinosaurs spent most of their time on land. The dinosaur bone in Plate 484 is characteristically black, but instead of filling with mud after the decay of the soft tissues, the marrow cavity remained hollow, and beautiful quartz crystals grew inside. The easiest way to tell a dinosaur bone from a mammal bone is probably to know the age of the formation in which it is found. All dinosaurs will be found in Mesozoic formations, and all mammals larger than a cat, in Tertiary and Pleistocene formations. Most dinosaur bone has a structure denser than that of mammal bone, and often shows a layering pattern in cross section that is never seen in a mammal bone.

Dinosaur tracks are among the most spectacular of all fossils and can teach paleontologists much about the behavior of dinosaurs. The tracks in Plate 472 were made by young plant-eating dinosaurs; those in Plates 473 and 474 were made by flesh-eaters. Most dinosaurs walked on 3 toes, the middle one elongated like a bird's foot.

481, 482 Class Aves
(Upper Jurassic to Recent)

Birds: Some paleontologists, noting that birds are more closely related to dinosaurs than to any other group, place birds in the same class with dinosaurs, the archosaurs.

Characteristics of Birds: Birds have few unique features, but feathers, at least as far as we know, are one of them. Birds are warm-blooded, trait they share with mammals (and perhaps with dinosaurs), and most fly, like bats and reptilian pterosaurs. *Archaeopteryx* is the earliest bird fossil, preserved complete with feathers in a fine-grained limestone that accumulated, on the probably poisonous bottom of a lagoon, in southern Germany in the late Jurassic Bird fossils are very rare throughout tl Mesozoic. Because it lived in water, tl diving bird *Hesperornis* has left some fossils in the Cretaceous of Kansas and Wyoming. Bird fossils are more plentiful in the Tertiary, but still far from common. This is not surprising since few creatures lead lives less conducive to the fossilization of their remains. Almost all birds are eaten when they die, so that their bones cannot be buried. Most bird bones are light and fragile as adaptations for flight, and if a bird escapes being eaten, its bones usually fall to the ground, where they scatter, break, an quickly decompose. Plates 481 and 4 show fossil bird eggs, which are extremely rare. These must have been laid in a hole in the ground and quickly covered by sediment before they could hatch.

475, 476, 477, **Class Mammalia**
479   (Upper Triassic to Recent)

Mammals: Today almost everyone agrees that mammals are the most successful and intelligent of all land vertebrates. However, some vertebrate paleontologists are beginning to doul that the differences between the mammals and their supposedly dull-witted ecological counterparts of the Mesozoic—the dinosaurs—are really very great. Dinosaurs were neither stupid nor slothful, and they dominated the land for 150 million years while mammals remained small, nocturnal, and insect-eating. Mammals certainly did not cause the extinction of dinosaurs. Quite the contrary; it was only after the dinosaurs were gone that mammals had a chance to diversify. Mammals did not "conquer the earth." Instead, they took over the niches that had been vacated by the dinosaurs.

Characteristics of Mammals: Most of the distinguishing features of mammals are soft parts that do not fossilize: the fur, the placenta for nourishing the unborn, the mammary glands for nursing the young, the warm blood, the 4-chambered heart, and the enlarged and folded outer layer of the brain. However, there are distinctive features of the skull that characterize mammals. The nasal and food passages are separated by a bone over the roof of the mouth, and the nostrils open through a single hole in the skull. Mammals have only 2 sets of teeth, the early milk teeth and the permanent teeth, while reptiles can replace teeth indefinitely. Mammal teeth are highly differentiated. Primitive mammals have, on each side of a jaw, 3 incisors, 1 longer piercing tooth (the canine), 4 premolars, and 3 grinders or molars. Mammals have 3 ear bones instead of the single bone found in reptile ears. Plates 475–477 show 3 mammal skulls.

Mesohippus: Plate 475 illustrates the skull of an early horse, *Mesohippus,* which lived in the Middle Oligocene. The skull is broken open and shows a cast of the brain. *Mesohippus* was about the size of a modern goat, and grazed on the western plains, which were beginning to be covered with grass. With 3 toes on each foot, it was not as swift a runner as the 1-toed horse of today. Horses are odd-toed grazers, members of the order Perissodactyla, which also includes zebras, rhinoceroses, and tapirs.

Subhyracodon: Plate 476 shows the skull of another odd-toed grazer, the rhinoceros *Subhyracodon,* found in Lower and Middle Oligocene rocks. Like *Mesohippus, Subhyracodon* has a central middle digit; the first and fifth digits have been lost.

Merycoidodon: A skull of *Merycoidodon* is shown in Plate 477. *Merycoidodon* is a member of the large order Artiodactyla, or even-toed grazers. This order includes the familiar domestic cows, goats, sheep, and pigs, as well as camels, deer and hippopotami. Even-toed grazers have lost the thumb, or first digit, and sometimes the second and fifth as well. In living forms, the feet are modified into hooves. *Merycoidodon* was a piglike oreodont that lived during the Oligocene, and had short legs, a heavy body, and 4 toes.

Mastodon: Very few mammal fossils consist of whole skulls, much less whole bodies. The most common kind of mammal fossil is the isolated tooth, as illustrated in Plate 479. This is an unusually large tooth because it comes from a primitive elephant, the mastodon *Mammut.* Mastodons roamed North America from the Middle Miocene to about 7000 years ago, when they were eliminated, perhaps by hunters. Mastodons had only 2 or 3 teeth in each half of the jaw, and the upper incisors were modified into great, long, curved tusks.

Part III
Appendices

The illustrations on the following pages show representative fossils from 8 major invertebrate groups, with labels identifying the body parts referred to in the text.

Sponge (Class Demospongea)

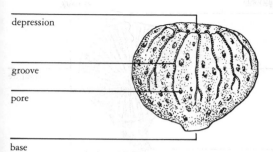

depression

groove

pore

base

Sponge (Class Hexactinellida)

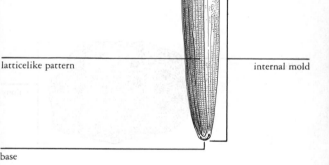

latticelike pattern

internal mold

base

Sponge (Class Calcarea)

spicule

Coral (Subclass Rugosa)

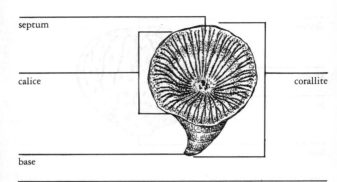

septum

calice

corallite

base

Coral (Subclass Scleractinia)

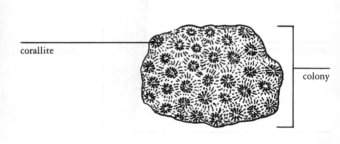

corallite

colony

Coral (Subclass Tabulata)

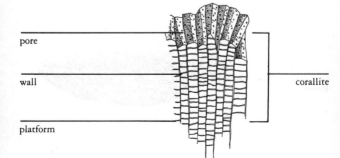

pore

wall

corallite

platform

Gastropod (Order Archaeogastropoda)

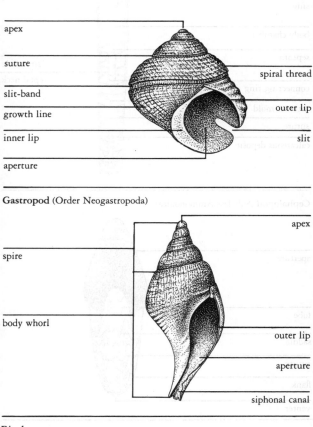

apex

suture

slit-band

growth line

inner lip

aperture

spiral thread

outer lip

slit

Gastropod (Order Neogastropoda)

apex

spire

body whorl

outer lip

aperture

siphonal canal

Bivalve

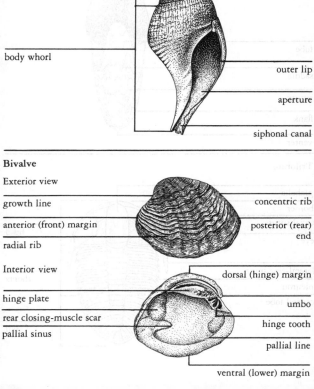

Exterior view

growth line

anterior (front) margin

radial rib

concentric rib

posterior (rear) end

Interior view

hinge plate

rear closing-muscle scar

pallial sinus

dorsal (hinge) margin

umbo

hinge tooth

pallial line

ventral (lower) margin

Cephalopod (Subclass Nautiloidea)

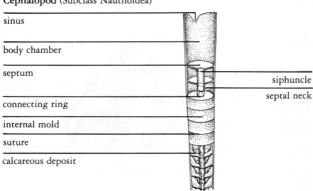

sinus

body chamber

septum

siphuncle

septal neck

connecting ring

internal mold

suture

calcareous deposit

Cephalopod (Subclass Ammonoidea)

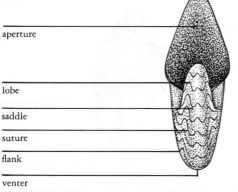

aperture

lobe

saddle

suture

flank

venter

Trilobite

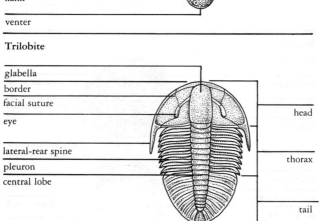

glabella

border

facial suture

eye

head

lateral-rear spine

pleuron

central lobe

thorax

tail

Bryozoan
Enlarged section

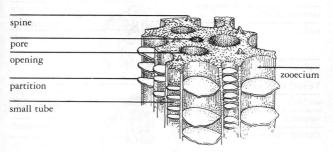

spine
pore
opening
partition
small tube
zooecium

Bryozoan

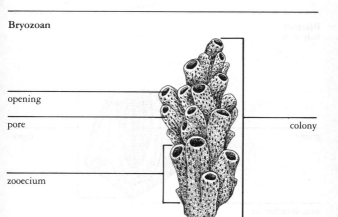

opening
pore
zooecium
colony

Brachiopod (Class Articulata)

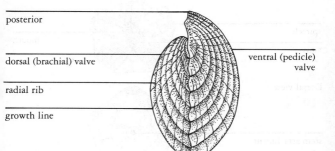

posterior
dorsal (brachial) valve
radial rib
growth line
anterior
ventral (pedicle) valve

Brachiopod (Class Articulata)
Dorsal valve interior

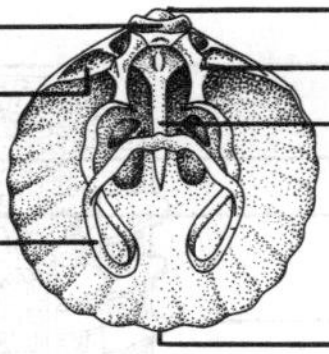

cardinal process — posterior margin

socket — brachiophore

— median ridge

lophophore support

Ventral valve interior

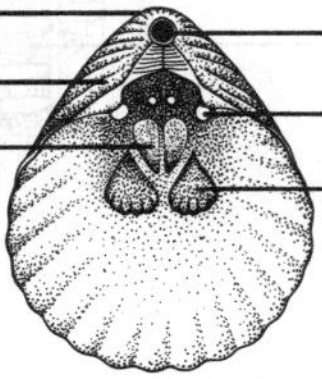

— anterior margin

umbo — opening for stalk

interarea

closing muscle scar — hinge tooth

— opening-muscle scar

Blastoid
Side view

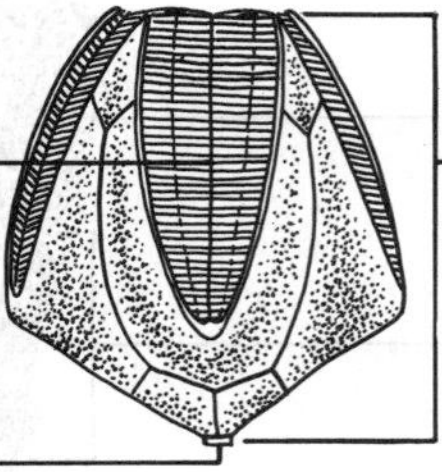

groove — theca

stem attachment

Blastoid
Ventral view

spiracle — mouth

anus

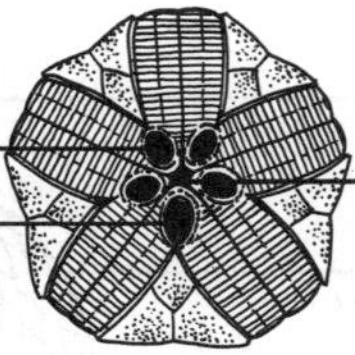

Dorsal view

stem attachment

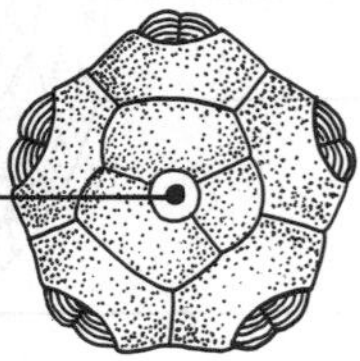

Crinoid

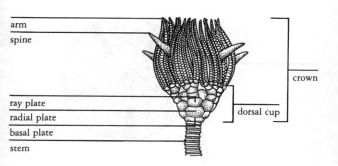

arm
spine

crown

ray plate
radial plate
basal plate
stem

dorsal cup

Crinoid
Arrangement of plates

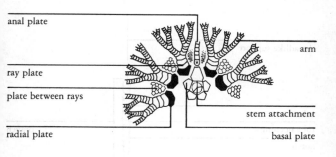

anal plate

arm

ray plate

plate between rays

stem attachment

radial plate

basal plate

Asteroid

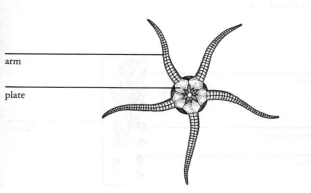

arm

plate

Echinoid

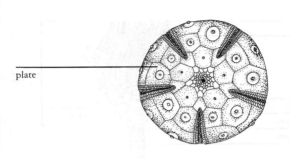

plate

Graptolite

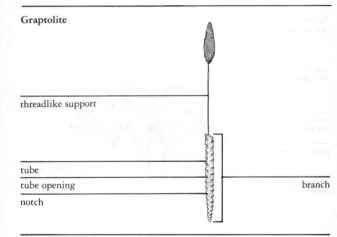

threadlike support

tube
tube opening   branch
notch

Graptolite
Enlarged section

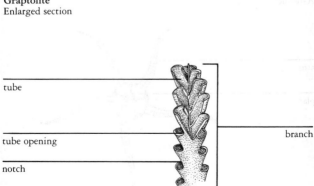

tube

tube opening   branch

notch

HOW TO COLLECT AND PRESERVE FOSSILS

Discovering traces of ancient life is a rewarding experience for amateur and professional paleontologist alike. To start collecting fossils, you need not acquire elaborate equipment and detailed maps—just begin to look. In most parts of North America, sedimentary rocks containing fossils are common, and you can learn to recognize the various types of both fossil-bearing rocks and the fossils themselves by studying the color plates in this guide. After you have begun to find fossils, you may wish to consider obtaining some equipment.

Equipment: For most fossil collecting, only a few inexpensive tools are necessary. The most important items are a field notebook, a felt-tip pen, and adhesive tape. Also useful are newspaper, tissue, or plastic bags in which to wrap your fossils. Since a fossil has little value without detailed locality data, you should record in your notebook exactly where you find each specimen. Assign a number to each locality, and write the same number on the fossil with the felt-tip pen. If a fossil is very small, write the number on a slip of paper and tape it to the fossil, or wrap the two together. With experience, and for collecting in special localities, you may decide that you need more equipment. If you are collecting some distance from a road, a backpack is helpful for carrying fossils. A geologist's hammer with a chisel claw is valuable for breaking soft sedimentary rocks and prying their layers apart. A trowel and a large, strong table knife are often useful both for soft sedimentary rock and for splitting finely bedded shales. When collecting in hard sedimentary rock, you will also want a set of chisels of different sizes. A small paintbrush can be used to brush off rock surfaces. A hand lens that magnifies 7 to 10 times is another useful item.

Where to Find Fossils: In some parts of North America, especially in mountainous regions, sedimentary rock is visible almost everywhere. But over most of the country, some searching is necessary. Sedimentary rock usually appears at artificial or natural cuts in the earth.

Roadcuts: Most important among artificial exposures are roadcuts, where a roadway has been cut through a rock outcropping. Before collecting at any roadcut, check that it is not illegal to do so. Then investigate fragments of rock at the base of the cut; if any layers in the outcropping are fossiliferous, some fallen pieces, called scree, should contain fossils. If you find fossils in the scree, try to locate the layers in the cut that they came from. You may be able to do so by matching the color or texture of the scree with that of a particular layer. Fossiliferous layers are almost always clearly distinct from the adjacent layers that contain no fossils. For example, if the outcrop is mostly sandstone, the fossiliferous layers may be thin beds of shale or limestone. Often the angle of the bedding planes will be different, with fossils coming from a few finely bedded layers.

Railroad Cuts: Railroad cuts are often productive collecting localities. Some of these are famous, and collectors visit them regularly. Their locations are given on geological highway maps. Such cuts are usually older than roadcuts, and their age is often an advantage if the rock is hard and must weather before it can be split easily.

Quarries: In most regions there are quarries for limestone, gravel, slate, shale, or sandstone. Most will allow collectors to visit, especially during hours when the quarry is not being worked, but permission must first be obtained. A quarry operator may also be able to give a visitor information about the name and age of the formation and about specific fossiliferous areas. The best places to look are usually the quarry's scrap heaps, where the rock is already split and may be weathered just enough to be broken easily into smaller pieces. If you find no fossils in a given place within half an hour, search another area.

Canals: Canals are sometimes cut through rock, and these artificial cliffs can be rich in fossils. The Erie Canal, for instance, was dug in New York State in the 1820s, just as the geology of the area was beginning to be studied. The famous paleontologist Amos Eaton used the canal as a geology laboratory, and many new species of fossils were collected along its banks.

Construction Sites: Some of the most spectacular fossil finds have been made at construction sites. When the excavation of a new library was begun in 1946 at Princeton University in central New Jersey, workmen soon discovered that some of the slabs of shale they were removing contained fish bones. Paleontologists identified many specimens of freshwater invertebrates and fishes, including a new species that was named the "library fish." It should be possible to gain permission to look through rock being removed at most construction sites.

Rivers: Among naturally exposed sedimentary outcrops, river and stream beds probably provide the best sources of fossils. If conditions are unfavorable for walking and wading, a small boat may be the most efficient means of travel. Sometimes fossils can be found in the stream bed, washed free from the surrounding rock.

Badlands: In any rugged terrain, outcrops may exist as a result of erosion. In parts of the West where the rainfall is meager and the sediment soft, the land often weathers into badlands topography— characterized by tortuous, steep-sided gullies—because there is little vegetation to hold the soft sediment. When rain does come, erosion is rapid, and new material is always being uncovered. Because most of these areas consist of terrestrial and flood plain deposits, land-living vertebrates are the fossils commonly found there.

Ocean Cliffs: Where the ocean breaks against cliffs, fossils may be washed free, accumulate along the shore, and mix with recent shells. There are fossil-bearing cliffs on both the East and West coasts.

Information on Collecting Areas: If you want specific suggestions on where to look for fossils, contact your state or provincial geological survey. Many surveys publish pamphlets listing collecting localities. Geological maps that include detailed information on productive collecting areas are available in the United States from the American Association of Petroleum Geologists, P.O. Box 979, Tulsa, Oklahoma 74101; and in Canada from the Geological Survey of Canada,

Department of Energy, Mines and Resources, 601 Booth Street, Ottawa, Ontario K1A OE8. Inquire at local natural history or science museums about fossil-collecting clubs in your area. These museums and organizations often conduct fossil-collecting trips on weekends. Some adult education schools offer courses in geology, which usually include at least one field trip. Community colleges, too, provide opportunities for studying geology, and sometimes paleontology as well.

Collecting Regulations: While a few fortunate collectors have turned up fossils in their backyards, most must search for them on private or government land. Since fossils technically belong to the landowner, a fossil hunter must obtain permission to explore private property, and must learn what federal, state, or local regulations govern fossil-hunting on publicly owned land. There are no uniform regulations for collecting on public land. Before collecting fossils on federal land, find out from the authorities at the particular national park or wildlife refuge what their restrictions are. To learn what laws govern fossil-collecting on state land, contact the appropriate state geological survey (see Appendix).

Removing Fossils From Rock: The process of removing a fossil specimen from rock varies according to the condition of the fossil and the nature of the sedimentary rock. In many areas, fossils commonly weather free from the matrix, or the matrix is so soft that it can be brushed or rubbed away easily. But more often, the fossils you find will be embedded in slabs of rock, some of which you will have to remove in order to free the specimen. If the fossiliferous rock is thin, some of it can be broken away by scoring the rock surface in 3 or 4 straight lines that frame your specimen, leaving a generous margin around the fossil itself. Set one of the scored lines at the edge of a larger rock, fossil side up, and tap the overhanging part with a hammer; the excess matrix should split off neatly and drop away. Repeat the process with each line.

If the slab is thick, breaking the rock without shattering the fossil is more difficult. Deeper scores, made with a chisel and hammer, may allow you to remove bits of the rock; or you may excavate the fossil from its rocky matrix by chipping a trench around the fossil, angling your chisel away from the fossil. When the trench is ½″ (13 mm) or more deep, depending on the size of the fossil, turn the chisel toward the fossil; give the chisel a sharp blow and the fossil should pop loose.

Protecting a Fossil: Often the enclosing shale or sandstone will be crumbly, and a fossil will be in danger of disintegrating before it can be extracted or carried away. A liquid that can be painted on a fossil and its enclosing matrix to protect against this is made by dissolving 2 to 5 tablespoons of polyvinyl acetate chips in 1 pint of toluene. Alternatively, a cement glue can be dissolved in toluene until the mixture is thin enough to be painted on. Toluene is inflammable and probably not healthy to breathe, so this solution should only be used out of doors. A clear plastic aerosol spray sold in cans is useful when you are collecting certain fossils preserved as carbon films. If the rock is moist when collected, these fossils may flake off as the rock dries. Let the specimen dry away from wind, and then use a plastic spray on it, or carry away the moist shale in a tightly closed plastic bag to prevent evaporation until you get the specimen home.

Preparing a Fossil: The appearance of most fossils can be improved if they are prepared at home, with the proper equipment. Many fossils need only enough washing to remove loose soil, rock, or mud. But proceed cautiously, since some rock and some fossils disintegrate when wet. Often fossils are trapped in rock. If the rock is shale, it can sometimes be broken down to free calcitic or siliceous fossils. A strong detergent and very hot water are sometimes all that is necessary. On shale, many professional paleontologists use a special formula called Quaternary-O, obtainable from scientific supply houses.

Using Acids: Acids are commonly used to dissolve a limestone matrix around silicified fossils. But remember that most fossils are of calcium carbonate, and will also dissolve in acid. Experiment with expendable fossils first. The safest and easiest acid to employ is the acetic acid contained in white vinegar, which can be used straight from the bottle. Photography supply stores also sell acetic acid, but it must be diluted. Hydrochloric acid is stronger and more destructive. A commercial grade suitable for fossil work is called muriatic acid. To dissolve limestone, add about 1 part muriatic acid to 9 parts water—and watch carefully.

Removing Matrix: Probably the most common kind of preparation is the removal of bits of rock matrix attached to a fossil. This can be done with a fine chisel, pick, or needle. Because there is almost always a plane of weakness between the fossil and the matrix, the matrix is not scraped off, but flaked off. Power tools can help with this job; a small rotary tool with exchangeable bits will grind down matrix, or can be used with cutting wheels to trim rock margins from around a fossil. Some collectors use dentist's drills. But even in laboratories, most preparation is still done by hand. Many of the fossils illustrated in this book were prepared by hand 100 years ago, long before the advent of power tools.

Casting from Fossil Molds:
Casts reproducing the original form of fossils that are molds can often be made. This is most easily done with liquid latex, sold in hobby shops. The latex is poured into the fossil mold, allowed to harden, and pulled out. It is also quite easy to make molds of fossils in order to duplicate the fossil in plaster. Here the latex is painted over the fossil in thin layers. When the latex is dry, the mold is peeled off. This mold is supported with aluminum foil or in a box of sand, and casting plaster is then poured into it. When the plaster is hard, the mold can be peeled off and used again. The plaster cast can be painted to resemble the original fossil.

Storing Fossils:
The best system for storing fossils safely and accessibly is a set of wide, shallow drawers in which each fossil is kept in a separate cardboard box. Before putting your fossils away, give each a number and a label with its scientific name, if known; exactly where it was collected and in which rock formation; the age of the formation; the name of the collector; and the date of collection. Many collectors also keep a notebook or card file, with an entry or card for each specimen, giving the same information.

Identifying a Fossil:
Once your fossil has been carefully cleaned, numbered, and labeled, you can try to identify it. This field guide will act as your initial resource for locating and identifying fossils. But since there are thousands of fossil genera in North America, you will find various reference books helpful. One of the most important is *Index Fossils of North America* by Hervey W. Shimer and Robert R. Shrock, published in 1944 in New York by John Wiley &

Sons. This work contains photographs of about 2500 fossil genera and thousands of common species. Most college and university libraries have a copy. Because this volume is almost 40 years old, many of the genus names have been changed. For the current name of an invertebrate genus, check the *Treatise of Invertebrate Paleontology,* a set of more than 20 volumes published by the University of Kansas Press beginning in 1953. When complete, the *Treatise* will describe all valid invertebrate fossil genera. In the index of the appropriate volume, find the genus name that you think refers to your fossil. You may discover that this name is no longer in use, but has become synonymous with another genus name. If your fossil fits a description in the *Treatise,* then it can be identified, at least to genus. In certain groups, particularly the corals and bryozoans, identification will depend on making sections through the fossil to see its internal structure—a process that inevitably damages the fossil. At this point, it might be wise to seek expert help, through a fossil-collecting club or perhaps an open house at a local museum. You can send photographs of your specimens to the curator of invertebrate fossils at a museum near the area where the fossil was collected; include a self-addressed stamped envelope for their return. From a good photograph, the paleontologist may be able to identify its genus or species. Do not mail the fossils themselves unless invited to do so. Remember, finally, that many fossils can never be identified, either because they are too poorly preserved or because they would be destroyed in the process.

Discovering New Fossils: Amateur fossil collectors have made significant contributions to paleontology by discovering new species—many of which were then named for their discoverers. If you locate an unusual fossil, such as a rare vertebrate skeleton, be sure to report it to the nearest museum of natural history. The paleontologist will probably want merely to note its occurrence. If it is an important specimen, however, you may wish to donate it to the museum's collection. Science benefits from work done at every level, and you—as a fossil collector—can share in the adventure of exploring our planet's past.

GLOSSARY

Abdomen In crustaceans, chelicerates, and insects, the body segments posterior to the thorax.

Ambulacral groove In echinoderms, a groove on the surface of the body for moving food to the mouth.

Anal plate One of the radial plates on the posterior side of the dorsal cup of crinoids, often bearing the anal tube.

Anal series In crinoids, plates between the ray plates, lying below the anus or forming the anal tube.

Anal tube A tube on the upper surface of the theca of a crinoid, ending at the anus.

Angle change In gastropods, a ridge running along the outside of the whorls, where 2 surfaces meet at an angle.

Antenna (pl. antennae) In arthropods and annelids, a sensory appendage on the head.

Anterior Located at or toward the front end of the body.

Anterior margin In brachiopods, the margin of the shell opposite the hinge.

Anus The posterior termination of the digestive tract.

Aperture An opening in the skeleton of a gastropod, cephalopod, or bryozoan through which the soft parts protrude.

Apex In gastropods, the first-formed part of the shell, generally pointed. In brachiopods, the first-formed part of a valve, around which the shell has subsequently grown.

Appendage A jointed extension of an arthropod segment, usually a leg but sometimes modified for various other functions such as feeding or reproducing.

Aragonite A form of calcium carbonate with orthorhombic crystals, common in invertebrate shells.

Arm In some echinoderms, 1 of several long, plated extensions from the body.

Arm facet In some echinoderms, a depressed or ridged area on the body where an arm is attached.

Articulation A movable connection between 2 skeletal plates or bones.

Attachment scar An area on the outside of a shell where it was attached to a substrate; may be concave or rough.

Auricle In bivalves, an earlike extension of the dorsal margin; in brachiopods, a similar extension of the posterior margin.

Axis (pl. axes) An imaginary line in corals, running down the center of the corallite; in coiled gastropods, running through the shell apex around which the whorls are coiled; in bilaterally symmetrical coiled shells, passing through the center at right angles to the plane of symmetry.

Axial sculpture: In gastropods, ornamentation running parallel to the shell's axis, or from spire to base.

Basal plate One of a ring of plates in the dorsal cup of crinoids, just below the radial plates.

Bedding plane In sedimentary rocks, a former surface on which sediment was accumulating; sedimentary rocks tend to split along bedding planes; also called a bedding surface.

Bilateral symmetry A body plan in which the 2 halves of an organism, 1 on each side of an anterior-posterior plane, are mirror images of each other.

Bivalved Having a shell composed of 2 parts, or valves.

Body chamber In shelled cephalopods, the space between the aperture and the first septum, holding the soft parts of the animal.

Body whorl In gastropods, the outermost and largest whorl of the shell, containing most of the soft parts.

Brachial Pertaining to the lophophore of a brachiopod, or to the valve to which the lophophore is attached.

Brachiophore In some brachiopods, a blade of shell on the inside of the dorsal valve that supports the lophophore.

Brackish Having a salinity lower than that of sea water and higher than that of fresh water.

Brood chamber In bryozoans, a structure consisting either of a specialized individual or of a globular swelling attached to a feeding individual, for holding larvae.

Byssal gape In bivalves, an opening between the margins of the valves for the passage of the byssus.

Byssal notch In many bivalves with auricles, an indentation below the right front auricle for passage of the byssus or for protrusion of the foot.

Byssal thread In bivalves, 1 of the fine strands that make up the byssus.

Byssate Having a byssus.

Byssus A bundle of fine strands by which some bivalves can make a temporary attachment to some object.

Calcareous Composed mainly of calcium carbonate.

Calcified Impregnated with calcium carbonate.

Calcite A form of calcium carbonate with hexagonal crystals, common in invertebrate shells.

Calcium carbonate A white chemical salt consisting of calcium, carbon, and oxygen, found in many minerals and bones and in the shells of many animals.

Calice In corals, the cavity that held the coral animal, or polyp, when the coral was alive.

Callus In gastropods, a thickened shell deposit on the inner lip, sometimes covering the umbilicus.

Carbonaceous Composed mainly of the element carbon.

Carboniferous A period of geologic time recognized in Europe, but in North America usually divided into the Mississippian (Lower Carboniferous) and Pennsylvanian (Upper Carboniferous) periods.

Cardinal process In brachiopods, a blade or bump of shell situated centrally inside the dorsal valve near the posterior margin, serving for the attachment of the muscles that open the valves.

Cast The natural filling of a mold that was left when a fossil dissolved after being enclosed by rock.

Cementation scar An area on a shell where it was cemented to a substrate; sometimes concave or rough.

Channel In gastropods and bivalves, a groove inside the aperture for conducting water in or out of the shell.

Chela (pl. chelae) In some arthropods, a pincer borne at the end of an appendage.

Chelicera (pl. chelicerae) In chelicerate arthropods, 1 of a pair of appendages located just in front of the mouth, each bearing a chela, and used for grasping food.

Chitin A durable organic substance forming all or part of the skeleton of arthropods and certain other invertebrates.

Cilium (pl. cilia) A microscopic, hairlike structure on the surface of a cell that beats in coordination with others to produce a current of water.

Cirrus (pl. cirri) In echinoderms, a jointed, sensory or grasping extension of the stem.

Collar cell In sponges, 1 of the cells that line water channels and cause water to move by the beat of tiny whiplike hairs, or flagella.

Colony An assemblage of connected animals that have all budded from a single parent individual.

Columella In gastropods, the part of the spiral shell surrounding the axis, about which the shell coils, and formed by the inner sides of the whorls.

Columellar fold In gastropods, a spiral ridge on the columella that projects into the cavity of the shell.

Commissure In bivalves and brachiopods, the line along which the 2 valves come together.

Concentric sculpture Sculpture that is oriented parallel to the growth lines on a shell.

Connecting ring In cephalopods, a delicate tube extending from a septal neck to the next septum.

Corallite The skeleton of an individual coral polyp.

Cross-bedding A pattern of deposit in a sedimentary rock in which thin or narrow layers lie at an angle to the thicker ones, indicating that the sediment was deposited by current or wind.

Cross section The surface exposed when a fossil is cut along a plane perpendicular to the long axis of the organism; in colonial animals, the plane is parallel to the surface of the colony.

Crown In cystoids, blastoids, and crinoids, a collective term for the theca and arms.

Dental plate In brachiopods, one of several plates on the inside of the ventral valve that supports the hinge teeth.

Denticle A small ridge that resembles or functions as a tooth.

Deposit feeder An animal that eats sediment and detritus.

Detritus Decaying organic material and the microorganisms decomposing it.

Diaphragm In brachiopods, a thin, crescent-shaped sheet of shell on the inside of the dorsal valve, serving to restrict entry to the visceral cavity.

Dorsal cup The part of a crinoid theca below the arm facets.

Dorsal surface In coiled cephalopods, the inner, concave side of a whorl, closest to the axis of coiling; in straight or curved cephalopods, the side farthest from the siphuncle or from a sinus in the aperture.

Dorsal valve In brachiopods, the valve bearing the lophophore; also called the brachial valve.

Epifaunal Living on, rather than under, the bottom of a body of water.

Escutcheon In bivalves, a dorsal, lens-shaped, depressed area extending to the rear from the umbones.

Exhalant channel In gastropods and bivalves, a channel at the junction of the outer and inner margins of the aperture, occupied in life by a fold of the mantle that conducted water out of the mantle cavity.

External mold An impression in rock of the outside of an organism.

Facial suture In trilobites, a very narrow, uncalcified band across both sides of the head shield, where the skeleton split apart during molting.

Filter feeder An animal that feeds by filtering particles from water.

Flank In brachiopods, the part of a valve farthest from the anterior-posterior axis. In cephalopods, a side of the shell between the dorsal and ventral surfaces.

Fluted septum In cephalopods, a septum that has been folded near its contact with the shell to produce a complex suture pattern.

Food groove In echinoderms, a groove for moving food to the mouth; also called an ambulacral groove.

Foraminiferan A single-celled animal with a shell, related to amoebas.

Form genus A genus containing many species that are very similar to one another, but probably not all closely related.

Frill A high, thin element of concentric ornamentation on the surface of a shell.

Funnel In cephalopods, a tubular, ventral extension of the mantle, where water is forced out of the mantle cavity; important in locomotion.

Fusuline A member of a group of foraminiferans with shells coiled like scrolls and shaped like rice grains; abundant in the late Paleozoic.

Genus (pl. genera) A group of closely related species.

Gill A feathery or threadlike extension of the body wall, functioning for respiration, and sometimes used for feeding as well.

Glabella In trilobites, the raised central portion of the head shield.

Gonopore An opening for the release of eggs or sperm.

Growth line A line marking a former position of the aperture or valve margin in a skeleton; present in mollusks, barnacles, brachiopods, and many tubes.

Guard The heavy, solid, rear portion of a belemnoid shell; the part most commonly fossilized; also called a rostrum.

Guide fossil   A fossil that is widespread geographically but restricted in time, and therefore useful in correlating rocks of the same age in different areas; also called an index fossil.

Hinge   In bivalves, brachiopods, and certain arthropods, the area along the margin where the 2 valves are joined together.

Hinge line   In bivalves and brachiopods, the line along which 2 valves are hinged.

Hinge margin   In bivalves, the edge of the shell closest to the hinge line.

Hinge plate   In bivalves, an internal calcareous platform bearing hinge teeth and located below the umbones and in the same plane as the commissure.

Hinge tooth   In bivalves and brachiopods, a projection on the hinge line of the ventral valve, between the dental plates.

Homeomorphy   Similarity in external form between genera or species that are not closely related; very common in brachiopods.

Hypostoma   In trilobites, a small plate on the lower side of the head, in front of the mouth or covering it.

Inflated   In bivalves, with the valves expanded laterally; in brachiopods, with the valves expanded ventrally or dorsally.

Inhalant channel: In gastropods and bivalves, a channel at the junction of the outer and inner margins of the aperture, occupied in life by a fold of the mantle that conducted water into the mantle cavity.

Inner lip   In gastropods, the inner side of the aperture, closest to the columella.

Interarea   In articulate brachiopods, a flattened zone extending down the middle of the external surface of a valve, from the umbo to the posterior margin of the valve.

Internal mold   An impression in rock of the inside of a skeleton.

Ironstone nodule A rounded, hard stone formed as a concretion of iron carbonate (siderite) in softer rock, and often containing a fossil.

Larva (pl. larvae) An early stage in the life cycle of an animal, different from the adult and often planktonic.

Ligament In bivalves, the horny, elastic structure that joins the valves together dorsally and acts as a spring to force them open when the closing muscles relax.

Ligament pit A relatively broad depression in the area between the umbones of a bivalve, for holding the ligament.

Ligament plate In bivalves, a hollowed-out plate for the attachment of an internal ligament.

Lobe In cephalopods, the part of the suture that is concave toward the aperture.

Longitudinal ornamentation Raised lines on the surface of a shell, running from the apex to the aperture or margin of a conical tube or shell, at right angles to the growth lines.

Longitudinal section The surface exposed when a fossil is cut along a plane parallel to the long (usually anterior-posterior) axis of the body; in colonial animals, the plane is perpendicular to the surface of the colony.

Lophophore In brachiopods, a feeding organ made up of filaments and channels, symmetrically disposed about the mouth and attached to the inside of the dorsal valve.

Lunule In bivalves, a crescent- or heart-shaped depression in front of the umbones.

Mantle A thin sheet or lobe of tissue that surrounds the soft parts of a mollusk or brachiopod and secretes the shell, if there is one.

Mantle cavity In mollusks, the space enclosed by the mantle.

Matrix (pl. matrices) The rock or sediment in which a fossil is embedded.

Median ridge   In some brachiopods, a ridge along the interior of the dorsal or ventral valve, in the plane of shell symmetry.

Median septum   In some brachiopods, a calcified partition extending partway down the middle of the ventral valve interior.

Monticule   In bryozoans, a small raised area on the surface of a colony, containing enlarged or modified zooecia.

Muscle-attachment plate   In brachiopods, a shelly plate on the inside of a valve, for the attachment of a muscle.

Muscle field   In brachiopods, the area on the inside of a valve occupied by the muscle scars.

Muscle scar   The area on a skeleton where a muscle was attached; usually roughened and sometimes raised or depressed.

Nacreous   Consisting of thin layers of aragonite and having a pearly luster.

Operculum (pl. opercula)   A lid that closes off an aperture or opening.

Ornamentation   Ribs, ridges, bumps, and spines on the outside of the shells of mollusks, tubes, brachiopods, and other shelled animals.

Orthorhombic crystal   A crystal with 3 unequal axes, all at right angles to one another.

Outer lip   In gastropods, the side of the aperture away from the columella, extending from the suture to the tip of the columella.

Paleontology   The study of ancient life.

Pallial line   In bivalves, a line or narrow band on the inside of the valves near the margin, marking the area where the mantle was attached to the shell.

Pallial sinus   In some bivalves, an indentation in the pallial line near the rear end of the shell, where muscles for the retraction of the siphons were attached.

Parietal area   In gastropods, the area just inside and outside the inner lip of the aperture.

Parietal deposit Shell material deposited over the parietal area of a gastropod.

Parietal fold A spiral ridge on the parietal area, projecting into the shell interior.

Pedicellaria (pl. pedicellariae) A small, nipperlike structure on the surface of asteroids and echinoids.

Pedicle In brachiopods, the fleshy stalk that passes through the ventral, or pedicle, valve and attaches the animal to the substrate.

Pedicle groove In inarticulate brachiopods, a groove down the center of the pseudointerarea, for passage of the stalk or pedicle.

Pedicle valve In brachiopods, the valve not bearing the lophophore, but containing the stalk or pedicle; also called the ventral valve.

Peel A replica of the flat surface of a fossil made by etching the surface, flooding it with acetone, and applying an acetate sheet; when dry, the acetate can be lifted off, bringing a thin layer of the fossil with it.

Pelagic Inhabiting the open ocean; free-swimming.

Petal A petal-shaped area on the shell of an echinoid, outlined by pores or slits.

Phragmocone In belemnoid cephalopods, the chambered portion of the shell.

Pincer In some arthropods, the nipperlike tip of an appendage; the chela.

Pincher The nipperlike appendage of certain bryozoans or echinoderms.

Pinnule A small branch on the arm of a crinoid.

Plane of commissure In bivalves and brachiopods, a plane passing through the hinge and the valve margins when the valves are closed.

Planktonic Drifting in the sea, rather than swimming freely or living attached or buried.

Pleuron (pl. pleura)   In trilobites, the lateral portion of a thoracic or tail segment.

Posterior   Located at or toward the rear end of the body.

Posterior margin   In brachiopods, the margin of the shell near the hinge.

Pseudointerarea   In inarticulate brachiopods, a flattened area between the umbones and the hinge.

Pseudopunctate   In brachiopods, having the inside of the valve covered with small bumps.

Pteropod   One of a group of gastropods, with or without shells, which are members of the plankton and abundant in some oceanic sediments.

Pulmonate   In some gastropods, having the mantle cavity modified as a lung.

Punctate   In brachiopods, having tiny pits on the inside of the valves.

Pyrite   A mineral of iron disulfide common in sedimentary rock and sometimes forming fossil casts.

Radial plate   One of a ring of 5 plates on the dorsal cup of a crinoid theca in a direct line with the arms.

Radial sculpture   Ribs or lines that radiate from the umbones on the outer surface of bivalves or brachiopods.

Radial symmetry   A body plan in which repeated body parts are arranged around a central point, like the spokes of a wheel.

Radula   In gastropods, a small, tonguelike structure bearing teeth used for rasping.

Ray   The arm of an asteroid, or 1 element in the starlike pattern of an edrioasteroid.

Ray plate   A plate above a radial plate in crinoids, lying between 2 arms or supporting an arm.

Reconstruction A drawing or model of how a fossil is supposed to have looked in life, or when it was intact.

Resilifer In bivalves, an internal depression in the hinge area for holding the internal ligament.

Ripple marks Ridges and troughs in sedimentary rock made by current or wind on a bedding plane before the sediment hardened.

Root tuft In a sponge, the part of the skeleton that extends into the sediment to anchor the animal.

Saddle In cephalopods, the curved part of the suture that is convex toward the aperture.

Septal neck In cephalopods, the portion of a septum that is bent forward or backward around the siphuncle.

Septum (pl. septa) In corals, the small, longitudinal, calcareous plates that extend inward from the sides of a calice toward its center; in cephalopods, 1 of a series of partitions inside the shell, dividing the interior into chambers.

Sessile Living fixed in 1 place; sedentary.

Shell The external skeleton of a mollusk or brachiopod.

Shoulder In gastropods, a flattened, horizontal part of the shell below and parallel to the suture, or the angle change marking the outer edge of this flattened area.

Siliceous Composed mainly of silicon dioxide.

Sinus A shallow incision in the margin of the aperture of a mollusk shell.

Siphon In mollusks, a tubular extension of the mantle used to conduct water in or out of the mantle cavity.

Siphonal canal In gastropods, a tubular or troughlike extension of the lower part of the aperture, containing the siphon; also called the siphonal cavity.

Siphuncle In cephalopods, a long tube extending through all the chambers, and consisting of septal necks, connecting rings, and calcareous deposits.

Skeleton The hardened internal or external part of an animal, providing support or protection.

Slit In gastropods, a parallel-sided incision in the margin of the aperture.

Slit-band In gastropods, a narrow, raised or depressed band of crescent-shaped growth lines running along the surface of the whorls and generated by a narrow sinus or slit in the margin of the aperture.

Slope change In gastropods, a change in the slope of the surface of a whorl.

Socket A depression inside the dorsal margin of a bivalve or the posterior margin of a brachiopod, for reception of hinge teeth.

Species (pl. species) A population of animals or plants whose members are at least potentially able to interbreed, but which are reproductively isolated from other populations.

Spicule One of the small, usually needlelike elements that generally compose the skeletons of sponges, some soft corals, and a few echinoderms.

Spiracle An external opening of the water vascular system of some echinoderms.

Spiral ridges, threads, etc. In coiled mollusks, ornamentation that runs along the surface of the whorls, and is perpendicular to the growth lines.

Spire In gastropods, all the whorls except for the body whorl.

Steinkern An internal mold of a shell, free from the surrounding matrix.

Stolon In bryozoans, a slender, creeping tube from which new individuals develop.

Subgenus · A group of species within a large genus that are more closely related among themselves than they are to other species in the genus.

Suture · The junction of 2 or more parts of a skeleton; in gastropods, a continuous line on the shell surface where whorls adjoin. In cephalopods, the junction of a septum with the inner surface of the shell wall.

Symbiotic algae · Single-celled algae that live in the tissues of many marine invertebrates, especially some corals.

Tegmen · The ventral surface of the crown of a cystoid or crinoid.

Test · The skeleton of an echinoid, consisting of rows of tightly fused plates.

Theca · The hard part of a cystoid, blastoid, or crinoid, exclusive of the arms and stem.

Thoracic · Of or pertaining to the thorax.

Thorax · In trilobites, crustaceans, and chelicerates, the segments between the head and tail.

Trail · In brachiopods, an extension of either valve beyond the mantle cavity.

Transverse furrow · In trilobites, a groove across the glabella more or less at right angles to the anterior-posterior axis of the body.

Transverse ornamentation · Ribs, ridges, or threads on the surface of the whorls of gastropods, parallel to the outer lip.

Tube foot · A small water-filled appendage of some echinoderms, typically arranged in bands or along arms, and used for locomotion, respiration, and sensory reception.

Umbilicus · A cavity around the axis of coiled whorls. In gastropods, there is generally only 1 umbilicus, at the lower end, surrounded by the body whorl; in coiled cephalopods, there are generally 2, 1 on each side of the coiled shell.

Umbo (pl. umbones) In bivalves, branchiopods, and brachiopods, the earliest-formed part of the shell, usually near the hinge and often extended out into a beak.

Umbonal cavity In bivalves and brachiopods, the area inside the shell and under the umbo.

Valve In bivalves, brachiopods, and some arthropods, 1 of the 2 separate parts of the shell.

Varix (pl. varices) In cephalopods, a line or ridge on the shell that marks a former position of the aperture.

Venter The ventral surface of a cephalopod shell.

Ventral surface In coiled cephalopods, the outer side of the whorls; in straight or curved cephalopods, the side of the cone bearing a sinus in the aperture, or closest to the siphuncle.

Ventral valve In brachiopods, the valve that contains an opening for the pedicle; also called the pedicle valve.

Viscera The internal organs of an animal.

Visceral cavity In brachiopods, the cavity that contains the internal organs.

Volution One complete turn of a coil.

Whorl One complete turn of a coiled shell.

Zooecium (pl. zooecia) In bryozoans, a calcified or chitinous sac, chamber, or tube containing the soft parts of the animal.

Zooxanthella (pl. zooxanthellae) A single-celled, yellow-green alga living symbiotically in the tissues of some invertebrates.

LIST OF MAJOR FOSSIL COLLECTIONS

United States

Alabama

Tuscaloosa   University of Alabama Museum of Natural History

Alaska

Juneau   Alaska Historical Library and Museum

Arizona

Flagstaff   Museum of Northern Arizona
Holbrook   Rainbow Forest Museum, Petrified Forest National Park
Tucson   University of Arizona Mineralogy Museum

Arkansas

Fayetteville   University of Arkansas Museum
State College   Arkansas State College Museum

California

Berkeley   University of California Museum of Paleontology
Calistoga   Petrified Forest
Claremont   Raymond M. Alf Museum, Webb School
Los Angeles   Los Angeles County Museum of Natural History
George Page Museum of LaBrea Discoveries
San Diego   San Diego Natural History Museum, Balboa Park
San Francisco   California Academy of Sciences, Golden Gate Park

Santa Barbara	Santa Barbara Museum of Natural History

Colorado
Bayfield	Gem Village Museum
Boulder	University of Colorado Museum
Canon City	Canon City Municipal Museum
Denver	Denver Museum of Natural History
Florissant	Florissant Fossil Beds National Monument

Connecticut
Hartford	Trinity College Museum
New Haven	Peabody Museum of Natural History, Yale University
Rocky Hill	Dinosaur State Park
Stamford	Stamford Museum and Nature Center

District of Columbia
Washington, D.C.	National Museum of Natural History, Smithsonian Institution

Florida
Bartow	Phosphate Valley Exposition
Gainesville	University of Florida State Museum
Orlando	Central Florida Museum

Georgia
Atlanta	Fernbank Science Center
	Georgia State Museum of Science and Industry

Idaho
Pocatello	Idaho State University Museum

Illinois
Chicago	Chicago Academy of Sciences
	Field Museum of Natural History
Normal	Illinois State University Museum
Rockford	Natural History Museum
Springfield	Illinois State Museum of Natural History and Art
Urbana	University of Illinois Museum of Natural History

Indiana
Bloomington	Department of Geology and Botany, Indiana University
Greencastle	Department of Geology, DePauw University
Indianapolis	Children's Museum of Indianapolis
	Indiana State Museum

Notre Dame Department of Geology, University of
Notre Dame
Richmond Joseph Moore Museum, Earlham
College

Iowa
Des Moines Iowa State Museum
Waterloo Museum of History and Science

Kansas
Hays Sternberg Memorial Museum, Fort
Hays Kansas State University
Lawrence University of Kansas Museum of
Natural History

Kentucky
Bowling Green Kentucky Museum
Louisville Department of Geology, University of
Louisville
Louisville Museum
Union Big Bone Lick State Park

Louisiana
Baton Rouge Louisiana State University Department
of Geology Museum

Maine
Caribou Nylander Museum

Maryland
Baltimore Maryland Academy of Sciences
Natural History Society of Maryland
Solomons Calvert Marine Museum

Massachusetts
Amherst Pratt Museum, Amherst College
University of Massachusetts Scientific
Collection
Boston Boston Museum of Science
Cambridge Museum of Comparative Zoology,
Harvard University

Michigan
Ann Arbor University of Michigan Museum of
Paleontology
Battle Creek Kingman Museum of Natural History
East Lansing Michigan State University Museum
Grand Rapids Grand Rapids Public Museum
Traverse City Great Lakes Area Paleontological
Museum

Minnesota
Minneapolis Museum of Natural History, University
of Minnesota

St. Paul   St. Paul Institute Science Museum

Mississippi
Jackson   Mississippi Museum of Natural Science
State College   Dunn-Seiler Geological Museum,
Mississippi State University

Missouri
Jefferson   State Museum of Natural Resources

Montana
Bozeman   Museum of the Rockies
Chinook   Blaine County Museum
Ekalaka   Carter County Museum
Fort Peck   Fort Peck Project
Missoula   Montana State University Museum

Nebraska
Agate   Cook Museum of Natural History
Fort Robinson   University of Nebraska Trailside
Museum
Lincoln   University of Nebraska State Museum

Nevada
Nye County   Ichthyosaur State Park

New Hampshire
Hanover   Dartmouth College Museum

New Jersey
New Brunswick   Rutgers University Geological Museum
Newark   Newark Museum
Princeton   Princeton University Museum
Trenton   New Jersey State Museum

New Mexico
Abiquiu   Ghost Ranch Museum
Albuquerque   Museum of Natural History
Carlsbad   Natural History Museum, Carlsbad
Caverns National Park

New York
Albany   New York State Museum
Buffalo   Buffalo Museum of Science
New York City   American Museum of Natural History
Rochester   Rochester Museum of Arts and Sciences

North Carolina
Aurora   Aurora Fossil Museum
Chapel Hill   Research Laboratories of Anthropology,
Univeristy of North Carolina
Greensboro   Natural Science Center
Raleigh   North Carolina State Museum

Ohio

Cincinnati	Cincinnati Museum of Natural History
	University of Cincinnati Museum
Cleveland	Cleveland Metropolitan District Trailside Museum
	Cleveland Museum of Natural History
Columbus	Ohio State University Geological Museum

Oklahoma

Norman	Stovall Museum, University of Oklahoma

Oregon

Eugene	University of Oregon Museum of Natural History
John Day	John Day Fossil Beds National Monument
Portland	Oregon Museum of Science and Industry

Pennsylvania

Harrisburg	William Penn Memorial Museum
Philadelphia	Academy of Natural Sciences of Philadelphia
Pittsburgh	Carnegie Museum

South Dakota

Mission	Zeitner Geological Museum
Rapid City	South Dakota School of Mines and Technology Museum of Geology

Texas

Alpine	Big Bend Historical Museum, Sul Ross State College
Austin	Texas Memorial Museum
Canyon	Panhandle Plains Historical Museum
Dallas	Dallas Museum of Natural History
	Department of Geology, Southern Methodist University
El Paso	El Paso Centennial Museum, Texas Western College
Houston	University of Houston Geological Museum
Lubbock	West Texas Museum, Texas Tech University

Utah

Jensen	Dinosaur Quarry Visitor Center, Dinosaur National Monument
Provo	Earth Sciences Museum, Brigham Young University

Salt Lake City   University of Utah Museum of Natural History
Vernal   Utah Field House of Natural History

Washington
Seattle   Thomas Burke Memorial Museum, University of Washington
Vantage   Ginkgo Petrified Forest State Park

Wisconsin
Milwaukee   Thomas A. Greene Memorial Museum, Milwaukee-Downer College
Milwaukee Public Museum

Wyoming
Cheyenne   Wyoming State Museum
Kemmerer   Fossil Butte National Monument
Laramie   University of Wyoming Geological Museum

Canada

Alberta
Calgary   Geological Survey of Canada, Institute of Sedimentary and Petroleum Geology
Drumheller   Paleontological Museum and Research Institute
Edmonton   Provincial Museum of Alberta
Department of Geology, University of Alberta
Patricia   Dinosaur Provincial Park

British Columbia
Vancouver   Geological Museum, University of British Columbia
Victoria   British Columbia Provincial Museum

Manitoba
Winnipeg   Manitoba Department of Resources
Manitoba Museum of Man and Nature

New Brunswick
Fredericton   Department of Geology, University of New Brunswick

Newfoundland
St. John's   Department of Geology, Memorial University

Nova Scotia

Halifax Geological Survey of Canada, Atlantic Geoscience Center
Nova Scotia Museum

Ontario

Kingston Department of Geology, Queens's University
Ottawa Geological Survey of Canada
National Museum of Natural Sciences
Sudbury Department of Geology, Laurentian University
Toronto Royal Ontario Museum
Windsor Department of Geology, University of Windsor

Quebec

Miguasha Musée d'Histoire Naturelle
Montréal Département de Géologie, Université du Québec
Départements de Géographie et Géologie, Université de Montréal
Department of Geology, McGill University
Redpath Museum, McGill University
Québec Département de Géologie, Université Laval
Trois Rivières Département de Géologie, Université du Québec

Saskatchewan

Regina Saskatchewan Museum of Natural History

GEOLOGICAL SURVEY OFFICES IN THE UNITED STATES AND CANADA

Detailed information about fossil collecting sites and the laws regulating collecting may be obtained from the state and regional offices of geological surveys. These offices provide maps, lists of areas where fossils may be found, names of books about fossils, and dates of organized field trips. They also publish bulletins containing articles on fossils. Listed here are the geological surveys of the continental United States and Canada.

United States

National Cartographic
Information Center
U.S. Geological Survey
Mail Stop 507,
National Center,
Reston, VA 22092

Alabama Geological Survey
Drawer O,
University, AL 35486

Alaska Division of Geological
& Geophysical Surveys
3001 Porcupine Drive,
Anchorage, AK 99501

Arizona Bureau of Geology &
Mineral Technology
845 North Park Ave.,
Tucson, AZ 85719

Arkansas Geological
Commission
Vardelle Parham Geology
Center,
3815 West Roosevelt Road,
Little Rock, AR 72204

California Division of
Mines & Geology
1416 Ninth St.,
Room 1341,
Sacramento, CA 95814

Colorado Geological Survey
1313 Sherman St., Room 715,
Denver, CO 80203

Connecticut Geological &
Natural History Survey
Room 553, State Office Bldg.,
165 Capitol Ave.,
Hartford, CN 06115

Delaware Geological Survey
University of Delaware,
Newark, DE 19711

Florida Bureau of Geology
903 West Tennessee St.,
Tallahassee, FL 32304

Georgia Department of
Natural Resources
Earth & Water Division,
Room 402,
19 Martin Luther
King, Jr. Drive SW,
Atlanta, GA 30334

Idaho Bureau of Mines &
Geology
Morrill Hall, Room 332,
Moscow, ID 83843

Illinois State Geological Survey
Natural Resources Bldg.,
615 East Peabody Drive,
Champaign, IL 61820

Indiana Geological Survey
611 North Walnut Grove,
Bloomington, IN 47405

Iowa Geological Survey
123 North Capitol St.,
Iowa City, IA 52242

Kansas Geological Survey
1930 Avenue A,
Campus West,
University of Kansas,
Lawrence, KS 66044

Kentucky Geological Survey
311 Breckinridge Hall,
University of Kentucky,
Lexington, KY 40506

Louisiana Geological Survey
Box G, University Station,
Baton Rouge, LA 70803

Maine Geological Survey
Dept. of Conservation,
State House Station #22,
Augusta, ME 04333

Maryland Geological Survey
Merryman Hall,
Johns Hopkins University,
Baltimore, MD 21218

Massachusetts Department of
Environmental Quality
Engineering
Division of Waterways,
1 Winter St.,
7th Floor,
Boston, MA 02108

Michigan Department of
Natural Resources,
Geological Survey Division
Box 30028,
Lansing, MI 48909

Minnesota Geological Survey
1633 Eustis St.,
St. Paul, MN 55108

Mississippi Bureau of Geology
P.O. Box 5348,
Jackson, MS 39216

Missouri Division of Geology
& Land Survey
Box 250,
Rolla, MO 65401

Montana Bureau of
Mines & Geology
Montana College of Mineral
Science & Technology,
Butte, MT 59701

Nebraska Conservation & Survey
Division
University of Nebraska,
113 Nebraska Hall,
Lincoln, NE 68588-0517

Nevada Bureau of Mines &
Geology
University of Nevada,
Reno, NV 89557-0088

New Hampshire Department of
Resources & Economic
Development
James Hall,
University of New Hampshire,
Durham, NH 03824

New Jersey Geological Survey
CN-029,
Trenton, NJ 08625

New Mexico Bureau of Mines
& Mineral Resources
Socorro, NM 87801

New York State Geological
Survey
Cultural Education Center,
Room 3140,
Albany, NY 12230

North Carolina Department of
Natural Resources &
Community Development
Geological Survey Section,
Box 27687,
Raleigh, NC 27611

North Dakota Geological
Survey
University Station,
Grand Forks, ND 58202

Ohio Division of Geological
Survey
Fountain Square, Bldg. B.,
Columbus, OH 43224

Oklahoma Geological Survey
830 Van Vleet Oval,
Room 163,
Norman, OK 73019

Oregon Department of
Geology & Mineral Industries
1005 State Office Bldg.,
1400 SW Fifth Ave.,
Portland, OR 97201

Pennsylvania Bureau of
Topographic & Geologic
Survey
Department of Environmental
Resources,
Box 2357,
Harrisburg, PA 17120

Rhode Island Department of
Economic Development
7 Jackson Walkway,
Providence, RI 02903

South Carolina Geological
Survey
Harbison Forest Road,
Columbia, SC 29210

South Dakota Geological
Survey
Science Center,
University of South Dakota,
Vermillion, SD 57069

Tennessee Division of Geology
G-5 State Office Bldg.,
Nashville, TN 37219

Texas Bureau of Economic
Geology
University of Texas,
University Station, Box X,
Austin, TX 78712

Utah Geological & Mineral
Survey
606 Black Hawk Way,
Salt Lake City, UT 84108

Vermont Geological Survey
Agency of Environmental
Conservation,
Heritage II Office Bldg.,
Montpelier, VT 05602

Virginia Division of
Mineral Resources
Box 3667,
Charlottesville, VA 22903

Washington Division of
Geology & Earth Resources
Dept. of Natural Resources,
Mail Stop PY-12,
Olympia, WA 98504

West Virginia Geological &
Economic Survey
Box 879,
Morgantown, WV 26507-0879

Wisconsin Geological &
Natural History Survey
1815 University Ave.,
Madison, WI 53706

Wyoming Geological Survey
Box 3008, University Station,
University of Wyoming,
Laramie, WY 82071

Canada

The Geological Survey of Canada has branches in Calgary, Alberta; Dartmouth, Nova Scotia; and Vancouver, British Columbia.

Geological Survey of Canada
Department of Energy,
Mines & Resources,
601 Booth St.,
Ottawa, Ontario K1A 0E8

Alberta Department of Energy & Natural Resources
Petroleum Plaza,
South Tower 9915,
108th St.,
Edmonton, Alberta T5K 2C9

British Columbia Ministry of Energy, Mines & Petroleum Resources
Mineral Resource Division,
Parliament Bldg.,
Victoria, British Columbia V8V 1X4

Manitoba Department of Energy & Mines
Geological Branch,
989 Century St.,
Winnipeg, Manitoba R3H 0W4

New Brunswick Department of Natural Resources
Box 6000,
Fredericton, New Brunswick E3B 5H1

Newfoundland Department of Mines & Energy
Government of Newfoundland Labrador,
Box 4750,
St. John's, Newfoundland A1C 5T7

Northwest Territories
Department of Indian Affairs & Northern Development
Geology, Box 1500,
Yellowknife,
Northwest Territories X1A 2R3

Nova Scotia Department of Mines & Energy
1690 Hollis St., Box 1087,
Halifax, Nova Scotia B3J 2X1

Ontario Ministry of Natural Resources
The Whitney Block,
99 Wellesley St. West,
Room 1640,
Toronto, Ontario M7A 1W3

Prince Edward Island
Department of Tourism,
Industry & Energy
Energy Branch, Box 2000,
Charlottetown,
Prince Edward Island C1A 7N8

Quebec Geological
Services Offices
Dept. of Natural Resources,
1620 Blvd. de L'Entente,
Quebec G1S 4N6

Saskatchewan Department of Mineral Resources
Geology & Mines Branch,
1914 Hamilton St.,
Regina, Saskatchewan S4P 4V4

Yukon Northern Affairs
Program
Geology Department,
200 Range Road,
Whitehorse, Yukon Y1A 3V1

SPECIMEN LOCALITIES

This list indicates the place of origin of each fossil specimen illustrated. The numbers correspond to the plate numbers in the photograph captions, and begin with number 34, the plate number of the first fossil. The state, when known, is listed first by postal code abbreviation, and followed, where possible, by a more detailed location. If a specimen is part of a museum collection, that information is included next. Specimens in the Smithsonian Institution, Washington, DC are indicated by the initials SI; in the Peabody Museum of Yale University by YPM; in the Princeton University Museum by PUM; in the Museum of Pennsylvania State University by PS; and in the University of Utah's Museum by UUT.

34 KS (Turner's Station, Wyandotte Co.); YPM

35 AK (Kanaga Island, Aleutian Islands); SI

36 VA (Yorktown, York Co.); PUM

37 TX (Jacksboro, Jack Co.); SI

38 TX (Mary's Creek, Benbrook, Tarrant Co.); YPM

39 OH (Oxford, Butler Co.); PUM

40 Appalachian Mts.; YPM

41 SD (s. fork of Cheyenne River; southwestern); YPM

42 MD (Mattawoman Swamp, Charles Co.); SI

43 OK (Allen, Pontotoc Co.); YPM

44 AL (Alabama River; southwestern); YPM

45 MD (southern or eastern); PUM

46 VA (York River; eastern); PUM

47 FL (Ft. Denaud, near La Belle, Hendry Co.); PUM

48 TX (Jacksboro, Jack Co.); SI

49 IN (Weisburg, Dearborn Co. [left], Versailles, Ripley Co. [right]); YPM

50 Ontario (between Thedford and Port Frank); YPM

51 OH (Maysville, Clermont Co.); YPM

52 IL (Peoria Co.); YPM

53 NY (Moscow, Livingston Co.); YPM

54 OH (Sylvania, Lucas Co.); SI

55 OH (Cincinnati, Hamilton Co.); YPM

56 TX (15 m. n.e. of Cisco, Eastland Co.); YPM

57 IA (Rockford, Floyd Co.); YPM

58 MO (Knobnoster, Johnson Co.); YPM

59 TX (n. of Cisco, Eastland Co.); YPM

60 TX (15 m. n.e. of Cisco, Eastland Co.); YPM

61 TN (western); YPM

62 IL (Hamilton, Hancock Co.); YPM

63 NY (western); YPM

64 IA (Cedar River, s.w. of Brandon, Buchanan Co.); YPM

65 IL (Dixon, Lee Co.); YPM

66 IA (Clayton Co.); SI

67 IN (Richmond, Wayne Co.); PUM

68 OH (Cincinnati, Hamilton Co.); SI

69 NY (Moscow, Livingston Co.); YPM

70 NY (Avery's Gully, near Buffalo, Erie Co.); YPM

71 NY (Lockport, Niagara Co.); YPM

72 IL (Dixon, Lee Co.); YPM

73 MO (Cyrene, Pike Co.); YPM

74 IN (Charlestown, Speed, and Sellersburg, Clark Co.); YPM

75 NY (Schoharie, Schoharie Co.); PUM

76 IN (Vevay, Switzerland Co., or Versailles, Ripley Co.); YPM

77 IA (Cerro Gordo Co.); YPM

78 OH (Oxford, Butler Co.); PUM

79 MO (Flat River, St. Francois Co.); YPM

80 NY (Westford, Otsego Co.); YPM

81 OH (Cincinnati, Hamilton Co.); PUM

82 IN (Wabash River; western or northern); YPM

83 NY (West Hill, Schoharie, Schoharie Co.); PUM

84 NY (Schoharie, Schoharie Co.); YPM

85 MN (Cannon Falls, Goodhue Co.); SI

86 NY (Tully, Onondaga Co.); YPM

87 TN (Coon Creek, McNairy Co.); YPM

88 NY (near Clarksville, Albany Co.); YPM

89 MN (St. Paul, Ramsey Co.); YPM

90 MO (St. Joseph, Buchanan Co.); YPM

91 IN (Waldron, Shelby Co.); YPM

92 TX (Glass Mountains; western); YPM

93 IL (Evansville, Randolph Co.); YPM

94 MO (Callaway Co.); YPM

95 NY (Clarksville, Albany Co.); YPM

96 MO (St. Joseph, Buchanan Co.); YPM

97 NY (western); YPM

98 IN (Waldron, Shelby Co.); SI

99 AL (Old Canton Landing, Alabama River, Wilcox Co.); SI

100 TX (Bell Co.); PUM

101 WY (Warm Springs Canyon, Fremont Co.); PUM

102 NY (Ft. Hunter; Little Falls, Herkimer Co.); YPM

103 MO (Weston, Platte Co.); YPM

104 MN (Winona, Winona Co.); YPM

105 NJ (near New Egypt, Ocean Co.); SI

106 IA (Davenport, Scott Co.); SI

107 TX (Mt. Bonell, Austin, Travis Co.); YPM

108 IA (Johnson Co.); YPM

109 NY (Stafford, Genesee Co.); YPM

110 NY (18 Mile Creek, Erie Co.); YPM

111 MO (Fulton, Callaway Co.); SI

112 MT (Three Forks Quad., Gallatin Co.); YPM

113 British Columbia (Mt. Bosworth); YPM

114 MO (Pleasant Hill, Cass Co.); YPM

115 NY (Troy, Rensselaer Co.); YPM

116 NY (Cranes Corner, Herkimer Co.); YPM

117 NJ (Weehauken, Hudson Co.); YPM

118 WY (Johnson Creek, Bighorn Mountains; north central); YPM

119 NY (Rochester, Monroe Co.); YPM

120 KY (Covington, Kenton Co.); YPM

121 NY (Canandaigua Lake, Ontario Co.); YPM

122 NY (Pompey Hills, Onondaga Co.); YPM

123 IA (Boonsboro, Boon Co.); YPM

124 TN (Coon Creek, McNairy Co.); YPM

125 AL (Claiborne, Monroe Co.); YPM

126 VA (Suffolk, Nansemond Co.); YPM

127 IN (Charlestown, Clark Co.); YPM

128 TN (Coon Creek, McNairy Co.); YPM

129 SD (Trail City, Dewey Co.); YPM

130 NJ (Hudson River; northern); YPM

131 CA (La Jolla, San Diego Co.); YPM

132 AL (Baldwin Co.); PUM

133 KY (Breckinridge Co.); YPM

134 TX (Cisco, Eastland Co.); YPM

135 VA (Yorktown, York Co.); YPM

136 SC (southern or central); YPM

137 AL (Claiborne, Monroe Co.); YPM

138 DE (Chesapeake and Delaware Canal, New Castle Co.); PUM

139 MD (Plum Point, Calvert Co.); YPM

140 TX (Jacksboro, Jack Co.); YPM

141 Quebec (Montréal); YPM

142 MD (Patuxent River); YPM

143 MS (Jackson, Hinds Co.); YPM

144 FL (Shell Creek, DeSoto Co.); YPM

145 MS (Tippah Co.); YPM

146 NY (Dresden, Yates Co.); YPM

147 KY (Ohio Co.); YPM

148 NY (Otisco Lake, Onondaga Co.); PUM

149 NY (Georgetown, Madison Co.); YPM

150 PA (Tyrone, Blair Co.); YPM

151 Quebec (Hull); SI

152 OH (Berea, Cuyahoga Co.); YPM

153 WY (Torchlight, Oregon Basin, Big Horn Co.); YPM

154 OK (southern or western); YPM

155 KS (Turner, Wyandotte Co.); YPM

156 NY (Berkshire, Tioga Co.); SI

157 SD (Fox Hills; northwestern); YPM

158 NE (Milford, Milford Co.); YPM

159 CO (Fossil Ridge, Greeley, Weld Co.); PUM

160 VA (Nansemond River; southeastern); SI

161 IA (Rockford, Floyd Co.); SI

162 WY (Chugwater Creek, Platte Co.); YPM

163 TX (Bell Co.); PUM

164 NJ (Birmingham, Burlington Co.); PUM

165 TX (McKewfeld Trail); YPM

166 TX (Brownwood State Park, Brown Co.); YPM

167 MI (northern); YPM

168 Manitoba (Garson or Lockport); PUM

169 OK (Copan, Washington Co.); YPM

170 AL (Alabama River; southwestern); YPM

171 MO (Webb City, Jasper Co.); YPM

172 AR (Smithville, Lawrence Co.); YPM

173 MO (Bridger, Carbon Co.); PUM

174 TN (Trenton, near Nashville, Gibson Co.); SI

175 FL (Ft. Denaud, near La Belle, Hendry Co.); PUM

176 TN (Cypress Creek Crest, 2 m. s.e. of Camden, Benton Co.); YPM

177 MO (central); PUM

178 Newfoundland (Greenhead, Port au Port); YPM

179 NY (Valcour Island, Lake Champlain, Clinton Co.); YPM

180 NY (Palatine Bridge; east central); YPM

181 Mexico (Valle de las Delicias, Coahuila); YPM

182 TX (Fort Worth, Tarrant Co.); PUM

183 TX (Jacksboro, Jack Co.); YPM

184 IN (south central); PUM

185 ID (5 m. e. of John Gray's Lake; southeastern); YPM

186 SD (15 mi. w. of Orlrichs, Fall River Co.); YPM

187 KY (Floyd Co.); YPM

188 KY (Floyd Co.); SI

189 SD (mouth of Belle Fourche River, Meade Co.); YPM

190 IN (south central); PUM

191 IA (Rockford, Floyd Co.); SI

192 Ontario (Arkona); YPM

193 SD (Sage Creek, Stanley Co.); YPM

194 SD (Belle Fourche, Butte Co.); YPM

195 OH (Marblehead, Ottawa Co.); SI

196 SD (northwestern); YPM

197 SD (Sage Creek, Stanley Co.); YPM

198 SD (north central); YPM

199 SD (Belle Fourche, Butte Co.); YPM

200 IN (Cloverland, Clay Co.); SI

201 SD (Belle Fourche, Butte Co.); YPM

202 MS (northeastern); YPM

203 TX (Fort Worth, Tarrant Co.); PUM

204 IN (Wabash, Wabash Co.); PUM

205 NJ (Vincentown, Burlington Co.); PUM

206 AL (Plymouth Bluff, Tombigbee River; western); YPM

207 IA (Lyons, Greene Co.); SI

208 NJ (Watertown); SI

209 WY (Beaver Creek, Weston Co.); PUM

210 TX (Chatfield, Navarro Co.); SI

211 DE (Chesapeake and Delaware Canal, New Castle Co.); PUM

212 TX (Washita, Bell Co.); PUM

213 TX (Salado, Bell Co.); PUM

214 AL (Claiborne, Monroe Co.); YPM

215 NY (Schoharie, Schoharie Co.); YPM

216 IN (Crawfordsville, Montgomery Co.); PUM

217 TX (Stephens Co.); YPM

218 TX (Hudspeth Co.); PUM

219 TX (western); PUM

220 AL (Claiborne, Monroe Co.); YPM

221 Atlantic or Gulf Coast; PUM

222 MT (Yellowstone River; southeastern); YPM

223 KY (s. of Highbridge; north central); YPM

224 MO (Henry Co.); SI

225 MN (St. Anthony Falls, Hennepin Co.); YPM

226 OH (Waynesville, Warren Co.); PUM

227 IN (Montgomery Co.); SI

228 MN (Deming, Luna Co.); SI

229 FL (north central); PUM

230 VA (Old Moore House); PUM

231 FL (Ft. Denaud, near La Belle, Hendry Co.); PUM

232 TX (north central); YPM
233 OK (n.e. of Copan, Washington Co.); YPM
234 TN (middle); YPM
235 IA (Monmouth, Jackson Co.); SI
236 Newfoundland (Conception Bay); SI
237 AL (Claiborne, Monroe Co.); YPM
238 AR (Ravendens Springs, Randolph Co.); PUM
239 IL (1 m. n. of Teutopolis, Effingham Co.); SI
240 TX (Elrath Co.); YPM
241 MD (Calvert Beach, Calvert Co.); YPM
242 AL (Claiborne, Monroe Co.); YPM
243 MD (St. Mary's Co.); PUM
244 VA (James River; eastern or central); PUM
245 FL (Caloosahatchee River; southwestern); PUM
246 FL (Ft. Denaud, near La Belle, Hendry Co.); PUM
247 TX (Glass Mountains; western); SI
248 FL (Harmby Canal, Lake Okeechobee); PUM
249 FL (De Sota Co.); YPM
250 FL (Ft. Denaud, near La Belle, Hendry Co.); PUM
251 FL (Caloosahatchee River; southwestern); PUM
252 AL (Claiborne, Monroe Co.); YPM
253 TN (McNairy Co.); YPM
254 TN (Coon Creek); YPM
255 AL (Prairie Bluff, Alabama River, Wilcox Co.); YPM
256 TX (central); YPM
257 FL (Ft. Denaud, near La Belle, Hendry Co.); PUM
258 MT (near Harlowton, Wheatland Co.); PUM

259 NY (Pratt's Falls, Madison Co.); YPM
260 KY (Moreland, Lincoln Co.); PUM
261 TX (Stephens Co.); YPM
262 AL (Claiborne, Monroe Co.); YPM
263 MS (Warren Co.); YPM
264 TX (near Bryan, Brazos Co.); YPM
265 FL (Harmby Canal, Lake Okeechobee); YPM
266 AL (Claiborne, Monroe Co.); YPM
267 NC (Duplin Co.); YPM
268 UT (House Range, Millard Co.); PUM
269 OK (south Criner Hills, southwest of Ardmore, Carter Co.); PUM
270 NY (Lockport, Niagara Co.); PUM
271 OK (Ardmore, Carter Co.); PUM
272 MO (Glenn Park, Jefferson Co.); PUM
273 NY (Madison Co.); PUM
274 AL (Piney, Cherokee Co.); SI
275 OH (Cincinnati, Hamilton Co.); PUM
276 IN (Hamilton Co.); SI
277 IN (Crawfordsville, Montgomery Co.); SI
278 OH (Silica, Lucas Co.); SI
279 AL (Cedar Bluff, Cherokee Co.); PUM
280 IL (Grafton, Jersey Co.); PUM
281 UT (North Canyon, Millard Co.); PUM
282 MT (Three Forks Quad., Gallatin Co.); YPM
283 British Columbia (Mt. Stephen, above Field); PUM
284 British Columbia (n.w. slope, Mt. Stephen, above Field); PUM

285 British Columbia (n.w. slope, Mt. Stephen, above Field); PUM

286 NY (Rome, Oneida Co.); PUM

287 Sweden (Shane); YPM

288 PA (Fruitville, Lancaster Co.); SI

289 NY (Springbrook, Erie Co.); PUM

290 NY (Lockport, Niagara Co.); SI

291 Alberta (Popes Creek); YPM

292 VT (Georgia, Franklin Co.); YPM

293 PA (Fruitville, Lancaster Co.); PUM

294 Newfoundland (Conception Bay); SI

295 NY (Trenton Falls, Oneida Co.); YPM

296 VT (Parker Hill, Franklin Co.); PUM

297 AL (Cedar Bluff, Coosa River, Cherokee Co.); SI

298 PA (Swatara Gap, Lebanon Co.); SI

299 NY (Black River, Watertown, Jefferson Co.); YPM

300 NY (Lockport, Niagara Co.); PUM

301 Ontario (Collingwood); PUM

302 MN (Stillwater, Washington Co.); YPM

303 UT (Millard Co.); PUM

304 IL (Mazon Creek, Grundy Co.); YPM

305 OK (s. of Ada, Pontotoc Co.); PUM

306 NY (Buffalo, Erie Co.); PUM

307 OH (Waynesville, Warren Co.); SI

308 NJ (Holmdel, Monmouth Co.); PUM

309 IL (Mazon Creek, Grundy Co.); YPM

310 CO (northwestern)

311 CO (Rio Blanco Co.)

312 CO (Rio Blanco Co.)

313 CO (Rio Blanco Co.)

314 CO (northwestern)

315 CO (Rio Blanco Co.)

316 CA (southern); PUM

317 VA (Chesapeake Bay); PUM

318 IN (Spergen Hill, Washington Co.); YPM

319 IA (Burlington, Des Moines Co.); PUM

320 OH (Cedarville, Greene Co.); SI

321 MO (Kansas City, Jackson and Clay Co.); PUM

322 IN (Crawfordsville, Montgomery Co.); PUM

323 VT (South Island); PUM

324 IN (Crawfordsville, Montgomery Co.); PUM

325 IN (Crawfordsville, Montgomery Co.); PUM

326 IN (Crawfordsville, Montgomery Co.); PUM

327 IN (Crawfordsville, Montgomery Co.); PUM

328 KS (western); YPM

329 IA (Burlington, Des Moines Co.); PUM

330 IN (Waldron, Shelby Co.); PUM

331 OH (Cincinnati, Hamilton Co.); PUM

332 NY (Lewiston, Niagara Co.); PUM

333 IL (Chester, Chester Co.); PUM

334 TN (Decatur Co.); YPM

335 OH (Cedarville, Greene Co.); PUM

336 IN (Waldron, Shelby Co.); YPM

337 Ontario (Arkona); YPM

338 SD (Sage Creek, Ft. Pierre, Stanley Co.); YPM

339 NJ (Brewers Ditch, near Hornerstown, Monmouth Co.); YPM

340 AL (Claiborne, Monroe Co.); YPM

341 Newfoundland (Isle La Motte); SI

342 NY (Watertown, Jefferson Co.); YPM

343 NY (Belmont, Allegany Co.); YPM

344 OH (Yellow Springs, Greene Co.); PUM

345 IN (Richmond, Wayne Co.); PUM

346 NY (Jerusalem Hill, Herkimer Co.); YPM

347 WI (Colfax, Dunn Co.); SI

348 OH (Cincinnati, Hamilton Co.); PUM

349 OH (Waynesville, Warren Co.); PUM

350 KY (Falls of the Ohio River; northern); PUM

351 KY (Falls of the Ohio River; northern); SI

352 KY (Moreland, Lincoln Co.); PUM

353 LA (Montgomery, Grant Co.); SI

354 MS (Jackson, Hinds Co.); SI

355 TX (Smithville, Bastrop Co.); PUM

356 MO (St. Louis, St. Louis Co.); PUM

357 OH (Caydraga Gorge, Akron, Summit Co.); SI

358 AL (Claiborne, Monroe Co.); PUM

359 AL (Claiborne, Monroe Co.); PUM

360 IN (Clarke Co.); PUM

361 TN (Perry Co.); PUM

362 TX (Bell Co.); PUM

363 TX (Sierra Blanca Quad., Hudspeth Co.); SI

364 GA (Perry, Houston Co.); SI

365 FL (Citrus and Levy Cos.); SI

366 NC (Pender Co.); SI

367 TX (15 m. w. of Ft. Worth, Tarrant Co.); SI

368 CA (Kettleman Hills, Kings and Kern Cos.); SI

369 TN (Tippah Co.); SI

370 AL (Cosa Valley, Cherokee Co.); PUM

371 NY (Trenton Falls, Oneida Co.); SI

372 TX (Everman, Tarrant Co.); PUM

373 TX (Hudspeth Co.); SI

374 TX (Hudspeth Co.); SI

375 KY (Bigby); YPM

376 NY (Schoharie, Schoharie Co.); PUM

377 IA (Iowa City, Johnson Co.); PUM

378 KY (Kings Mountain); SI

379 IN (Charlestown, Clark Co.); PUM

380 NY (LeRoy, Genesee Co.); PUM

381 KS (Manhattan, Riley Co.); PS

382 NC (Wilmington, New Hanover Co.); PS

383 FL (Miami, Dade Co.); PS

384 MN (St. Paul, Ramsey Co.); YPM

385 IN (Putnamville, Putnam Co.); PUM

386 NM (Gallup, McKinley Co.); UUT

387 OH (Sylvania, Lucas Co.); PS

388 NY (Ithaca, Tompkins Co.); PUM

389 MI (Emmet Co.); SI

390 NY (LeRoy, Genesee Co.); PUM

391 IN (Versailles, Ripley Co.); PUM

392 IN (Richmond, Wayne Co.); PUM

393 OH (Cincinnati, Hamilton Co.); YPM

394 KS (Manhattan, Riley Co.); PS

395 NC (Wilmington, New Hanover Co.); PS

396 NC (Wilmington, New Hanover Co.); PS

397 KS (Council Grove, Morris Co.); PS

398 NJ (Vincentown, Burlington Co.); YPM

399 KS (Council Grove, Morris Co.); PS

400 IN (Brookville, Franklin Co.); PS

401 MO (Sedalia, Pettis Co.); SI

402 KY (Jeffersontown, Jefferson Co.); SI

403 NC (Colerain Beach, Bertie Co.); PS

404 Bahamas (Joulters Cays); PS

405 KS (Manhattan, Riley Co.); PS

406 PA (Huntingdon, Huntingdon Co.); PS

407 TX (Joplin, Jack Co.); PUM

408 PA (probably Perry Co.); PUM

409 NY (Schoharie, Schoharie Co.); PUM

410 TN (Henry Co.); YPM

411 MT (Philipsburg, Granite Co.); PUM

412 OH (Camden, Preble Co.); PS

413 OH (Sylvania, Lucas Co.); PS

414 NC (Wilmington, New Hanover Co.); PS

415 NC (Wilmington, New Hanover Co.); PS

416 IN (Waldron, Shelby Co.); YPM

417 NJ (Allentown, Monmouth Co.); PUM

418 NY (Cape Vincent, Jefferson Co.); YPM

419 KY (Jeffersontown, Jefferson Co.); SI

420 IA (Iowa City, Johnson Co.); PUM

421 IN (Utica, Clark Co.); PUM

422 OH (Oxford, Butler Co.); PUM

423 OH (Cincinnati, Hamilton Co.); YPM

424 TN (Nashville, Davidson Co.); SI

425 FL (Ft. Denaud, near La Belle, Hendry Co.); PUM

426 NC; PUM

427 VA (James River); PUM

428 IA (Iowa City, Johnson Co.); PUM

429 United States (exact location unknown); PUM

430 OH; SI

431 IN (Harrison Co.); PUM

432 MI (Alpena, Alpena Co.)

433 KY (Falls of the Ohio River; northern); SI

434 NY (Chazy, Clinton Co.); PUM

435 NY (Moscow, Livingston Co.); PUM

436 MS (Vicksburg, Warren Co.); PUM

437 VA (Yorktown, York Co.); PUM

438 AL (Claiborne, Monroe Co.); PUM

439 Middle Basin; SI

440 Sweden (Gothland); PUM

441 Ontario (near Thedford); PUM

442 FL (La Belle, Hendry Co.); PUM

443 IA (Nora Springs, Floyd Co.); PUM

444 NY (Clarksville, Allegany Co.); YPM

445 NY (Cohocton, Steuben Co.); YPM

446 OH or KY; PUM

447 Ontario (Thedford and Bartlett Mills); SI

448 OH (Cincinnati, Hamilton Co.); PS

449 NY (East Bethany, Genesee Co.); YPM

450 NC (Wilmington, New Hanover Co.); YPM

451 NC (Wilmington, New Hanover Co.); PS

452 TN (Perry Co.); PUM

453 ID (Butte Co.); PS

454 OH (Linton); PUM

455 KY (near Hardinsburg, Breckinridge Co.); YPM

456 IL (Warsaw, Hancock Co.); PUM

457 NY (Oxtinga Creek near Fort Plain, Montgomery Co.); YPM

458 NY (Trenton, Oneida Co.); YPM

459 NY (Kenwood, Madison Co.); YPM

460 NY; YPM

461 Quebec (Pt. Lévis); YPM

462 NY (Kenwood, Madison Co.); YPM

463 NY (Rochester, Monroe Co.); YPM

464 NY (Rochester, Monroe Co.); YPM

465 ME (Penobscot River, Penobscot Co.); PUM

466 Newfoundland (Kelly Island); PUM

467 NY (Fort Ann Village, Washington Co.); PUM

468 Quebec (Percé, Gaspé); PUM

469 TX (Clifton, Bosque Co.); PUM

470 MN (Taylor's Falls, Chicago Co.); PUM

471 KS (Manhattan, Riley Co.); PS

472 NJ (Roseland, Essex Co.); PUM

473 CT (Rocky Hill, Hartford Co.); PUM

474 MA (Horse Race, Mintagne); YPM

475 SD (Big Badlands, Pennington Co.); PUM

476 SD (Big Badlands, Pennington Co.); PUM

477 SD (Big Badlands, Pennington Co.); PUM

478 SC (Beaufort, Beaufort Co.); PUM

479 NY (Orange Co.); PUM

480 WY (Newcastle, Weston Co.); PUM

481 SD (White River, Mellette Co.); PUM

482 NE (Sand Creek, Crawford, Dawes Co.); PUM

483 PA (Canonsburg, Washington Co.); PUM

484 UT (Emery Co.); PUM

485 DE (Bridge, New Castle Co.); PUM

486 NJ (Boonton, Morris Co.)

487 WA (Vantage, Kittitas Co.); PUM

488 MT (Glacier Nat. Park); PUM

489 Nova Scotia (S. Jogeins); YPM

490 IL (Jo Daviess Co.); PUM

491 PA; PUM

492 NY (Gilboa, Schoharie Co.); YPM

493 PA (eastern); PUM

494 NY (Gilboa, Schoharie Co.); YPM

495 MA (Horse Run, Gill, Franklin Co.); YPM

496 Alberta; PUM

497 AK (southwestern); PUM

498 Alberta; PUM

499 IL (Mazon Creek, Grundy Co.); PUM

500 IL (Mazon Creek, Grundy
   Co.); YPM
501 PA (St. Clair); PUM
502 IL (near Coal City,
   Grundy Co.); YPM
503 NY (Gilboa, Schoharie
   Co.); YPM
504 PA (Cannelton, Beaver
   Co.); PUM
505 MT (Glendive, Dawson
   Co.); PUM
506 OR (western); PUM
507 WY (Yellowstone Park
   plateau); PUM

PICTURE CREDITS

All photographs were taken by Townsend P. Dickinson, with the exception of those listed below.

The numbers in parentheses are plate numbers. Agency names appear in boldface. Photographers hold the copyright to their works.

INDEX

Numbers in boldface type refer to color plates. Numbers in italics refer to pages.

A

Abatocrinus, 697

Acanthocladia, **394,** *614*

Acanthoclymenia, 519

Acanthodians, *762*

Acanthotelson, **309,** *591*

Acrocyathus, **431,** *366*

Acrothele, **113,** *630*

Acrothoracican barnacle burrows, **471,** *590*

Actinastrea, 371

Actinoceras, **342,** *504*

Actinoceratoidea, *504*

Actinocerida, *504*

Actinostroma, **443,** *348*

Actinostromaria, 349

Aeora, 489

Agathiceras, 521

Agelacrinites, 711

Agerostrea, **164,** *473*

Agnatha, *758*

Agnathans, *758*

Agnostida, *551*

Agoniatites, 519

Albertella, **291,** *563*

Alectryonia, 474

Alethopteris, **501,** *746*

Algae
blue-green, *741*
coralline, *743*
green, *742*
red, *743*

Alveolites, **441,** *383*

Amblysiphonella, 347

Ambocoelia, **97,** *678*

Ambonychia, **39,** *458*

Ambonychiacea, *458*

Ammonitida, *528*

Ammonoidea, *516*

Amphibia, *762*

Amphibians, *762*

Amphigenia, 683

Anadara, **47,** *453*

Anarcestida, *517*

Anastomopora, 614

Anchisauripus, **474**

Anchura, **255**, *427*

Angiosperms, *749*

Annelida, *537*

Annularia, **493**, *745*

Annuliconcha, *463*

Anomia, **124**, *467*

Anomiacea, *467*

Anoria, *561*

Anthophyta, *749*

Anthophytes, *749*

Anthozoa, *353, 356*

Antiarchs, *758, 759*

Antirhynchonella, *665*

Apheoorthis, *637*

Aphera, *438*

Aphrodina, *489*

Arachnids, *754*

Arcacea, *452*

Archaeogastropoda, *393*

Archaeopteryx, *766*

Archimedes, **455, 456,** *612*

Archohelia, *372*

Archosaurs, *765*

Argopecten, *465*

Armenoceras, *505*

Arthrodires, *758*

Arthropoda, *547, 751*

Articulata, *635*

Articulata, *705*

Ash, 21, 24, *92*
volcanic, *92*

Astarte, *481*

Asterobillingsa, *363*

Asteroidea, *707*

Astraeospongia, *347*

Astraeospongium, **452,** *346*

Astrangia, **437,** *371*
danae, *372*

Astrhelia, **426,** *372*

Astrocoenia, *371*

Astrodaspis, *717*

Astylospongia, **361,** *341*

Athyris, **110,** *675*

Atrina, *457*

Atrypa, *672, 673*

Atrypina, *672*

Augustoceras, *510*

Aulacera, **446,** *351*

Aulocaulis, *388*

Aulocystis, *388*

Aulopora, **389,** *387*

Auloporida, *386*

Aves, *765*

Aviculopecten, **34,** *463*

B

Babylonites, *403*

Bactrites, **337,** *518*

Bactropora, *615*

Baculites, **338,** *527*

Badlands, *91*

Balanophyllia, *377*

Balanus, **316, 317,** *589*

Barbatia, **46,** *452*

Barrandella, *665*

Barrandeoceras, *513*

Barroisella, *628*

Barycrinus, 702

Baryphyllum, 362

Bathynotus, **296,** 557

Bathytormus, **143,** 482

Bathyuriscus, **282,** 561

Batocrinus, **318, 325,** 696

Batostoma, 608

Batostomella, 607

Baylea, **239,** 405

Beachrock, 96

Beecheria, **103,** 684

Belemnitella, **339,** 536

Belemnoidea, 535

Bellerophon, **218,** 396

Bellerophontacea, 393

Bellerophontina, 394

Beloitoceras, 510

Bembexia, **167,** 405

Bickmorites, **204,** 512

Bicorbula, 495

Billingsaria, 380

Billingsella, **101,** 636

Bilobites, 646

Birds, 765

Bivalvia, 445

Blastoidea, 687, 690

Bothriolepis, 759

Brachiopoda, 625

Brachiospongia, **375,** 345

Brachydontes, 456

Branchiopoda, 586, 588

Briscoia, 567

Brontosaurs, 764

Brooksella, **370,** 354

Bryozoa, 597

Bucanella, **176,** 394

Bucania, 396

Bucanopsis, 396

Buccinacea, 434

Buchiola, **40,** 451

Bulimorpha, 422

Bulla, **266,** 442

Bumastus, **272,** 573

Busycon, **250,** 435

Bythopora, 608

C

Calamites, **489,** 745

Calcarea, 346

Calcispongia, 346

Callavia, 556

Callianassa, **308,** 591

Calliostoma, **229, 230,** 416

Calopora, 610

Calymene, **280,** 580

Calyptraeacea, 428

Calyptraphorus, **252,** 426

Camarotoechia, 668

Camerata, 695

Cameroceras, 503

Campylorthis, 640

Cancellaria, **241,** 437

Caninia, 365

Carcharodon, **478,** 761
carcharias, 761

Cardiacea, 482

Cardiidae, 483

Cardiola, 452

Cardiomorpha, 497

Carditacea, 480

Cardium, 483

Carinaropsis, 396

Carneyella, 710

Caryocrinites, **332,** 688

Caryomanon, 341

Caryophylliida, 373

Cassiduloida, 718

Cedaria, **281,** 569

Celleporaria, **404,** 622

Celtis, **487,** 750

Centipedes, 751

Centroceras, **195,** 515

Cephalaspis, 758

Cephalopoda, 501

Ceramopora, **406,** 603

Ceramoporella, 603

"*Cerastoderma,*" **45,** 483

Ceratitida, 525

Ceratopea, **238,** 410

Ceraurus, **295,** 578

Ceriopora, **403,** 600

Cerithiacea, 423

Cerithium, **262,** 424

Chaetetes, **401,** 378

Chaetetida, 378

Chalk, **11,** 96

Cheilostomata, 617, 619

Cheirurus, **305,** 577

Chelicerata, 547, 592

Chelonia, 764

Chesapecten, **36,** 465

Chione, 490
californiensis, 491
cancellata, 491
grus, 491

Chlamys, **35,** 465

Chlorophyta, 742

Chlorophytes, 742

Chondrichthyes, 759

Chondrites, **464,** 734

Chonetinella, 658

Chonostegites, 388

Chordata, 723, 755

Cidaroidea, 713

Cirripedia, 586, 588

Cladopora, **411,** 382

Claibornites, 478

Clam
Asiatic, 487
Atlantic Surf, 485

Clathrodictyon, 350

Clathrodrillia, **257,** 441

Clathrospira, 407

Cleiothyridina, **133,** 676

Clementia, 490

Climacograptus, **458,** 728

Cliona, **417,** 340

Clionoides, 341

Clionolithes, 340

Clypeasteroida, 715

Cnidaria, 353

Coal, **31,** 90

Codakia, **125,** 478

Coelocaulus, 419

Coenostroma, 350

Coleoidea, 535

Coleopteran, 314

Collenia, **488,** 741

Composita, **134,** 677

Conacea, 440

Conchicolites, 546

Conchidium, **82,** 664
alaskense, 665

Conchostraca, 588

Concretions, 92

Coniferophyta, 748

Coniferophytes, 748

Conifers, 748

Conocoryphe, 565

Conodiscus, 631

Conopeum, **382,** 620

Conotreta, 630

Constellaria, **423,** 604

"*Conularia,*" **356, 357,** 355

Conulariida, 355

Conus, **249,** 440

Cooperidiscus, 711

Cooperoceras, **247,** 513

Coprolite(s), shark, **483,**
761

Coquina, 15, 27, 96

Coral, Northern Stony, 372

Corbicula, **129,** 487
fluminea, 487

Corbiculacea, 487

Corbula, **139,** 494

Cordaites, 748

Cordaites, **504,** 748

Cornulitella, 546

Cornulites, **345,** 545

Cornulitidae, 545

Corynexochida, 558

Coscinopleura, **398,** 621

Costispirifer, **75,** 681

Cranaena, **108,** 685

Crania, 633

Crassatella, 482

Crassatellacea, 481

Crassostrea, **157,** 472

Crawling traces, 732

Crepicephalus, 566

Crepidula, **214,** 428

Cribrilaria, **415,** 621

Cribrilina, 622

Crinoidea, 693

Crinoid stems, **323, 343,**
694

Crossopterygian, 762, 763

Crustacea, 547, 586

Cruziana, **466,** 732

Cryptoblastus, 693

Cryptolithoides, 576

Cryptolithus, **298,** 575

Cryptostomata, 598, 610

Ctenodonta, **151,** 447

Ctenodontacea, 447

Ctenostomata, 617, 618

Cucullaea, **42,** 454

Cupularostrum, **80,** 668

Cyanophyta, 741

Cyanophytes, *741*

Cyathocrinites, **326,** *701*

Cycadophyta, *747*

Cycads, *747*

Cyclobathmus, 412

Cyclonema, **226,** *414*
bilix, 414

Cyclophoracea, *420*

Cyclostomata, *598*

Cylichna, 443

Cymatonota, 498

Cymbophora, 485

Cypraea, **267,** *429*

Cypraeacea, *429*

Cyprimeria, 490

Cyrtina, **94,** *678*

Cyrtograptus, 729

Cyrtospirifer, 680

Cystihalysites, 386

Cystiphyllida, *358*

Cystiphylloides, **447,** *358*

Cystiphyllum, 359

Cystoidea, *687, 688*

Cystoporata, *598, 602*

Cyzicus, **117,** *588*

D

Dactylogonia, 652

Dalmanella, **73,** *644*
edgewoodensis, 645

Dalmanites, **290,** *583*

Dawsonoceras, **344,** *508*

Delocrinus, **321,** *702*

Demospongea, *340*

Dendraster, **368,** *717*

Dendrophylliida, *376*

Dentalium, **340,** *500*

Derbyia, **56,** *656*

Desmoceras, **182,** *530*

Desmorthis, 641

Desquamatia, **74,** *672*

Devonochonetes, **54,** *656*

Diaphragmus, **93,** *659*

Dicaelosia, 646

Dicellograptus, **462,** *727*

Dicellomus, **118,** *628*

Dichograptus, 726

Dichotrypa, 617

Dicoelosia, **95,** *646*

Dicranograptus, 726

Dictyoclostus, 662

Dictyonema, 634

Dictyospongia, 345

Didymograptus, **459,** *725*

Dikelocephalus, **302,** *567*

Dinichthys, 759

Dinorthis, **67,** *639*

Dinosaurs, *764*

Diodora, 413

Diparelasma, 641

Dipleura, **273,** *582*

Diplodonta, 478

Diploporaria, 615

Discosorida, *511*

Doleroides, 642

Dolerorthis, **71,** *637*

Dolomite, **12,** *96*

Dorycrinus, **319**, *695*

Dosinia, **126**, *489*

Dosiniopsis, *489*

Douvillina, **64**, *653*

Douvillinaria, *654*

Drepanophycus, *492*, *745*

Dunbarella, *462*

Dunkleosteus, *759*

Dwelling structures, *732*

Dystactospongia, *391*, *342*

E

Eatonia, **88**, *670*

Ecculiomphalus, **180**, *401*

Echinarachnius, *718*

Echinoconchus, **62**, *660*

Echinodermata, *687*

Echinoidea, *711*

Echinosphaerites, *689*

Ecphora, **244**, *433*

Ectomaria, *419*

Edmondia, **123**, *496*

Edmondiacea, *496*

Edrioasteroidea, *687*, *710*

Eggs, bird, *481*, *482*, *766*

Ehmania, *565*

Elasmonema, *414*

Elrathia, **283**, *565*

Elrathina, *565*

Emarginula, **237**, *412*

Emmonsia, *381*

Encope, *716*

Encrinurus, **276**, **300**, *579*

Endelocrinus, *703*

Endoceras, **341**, *503*

Endoceratoidea, *502*

Endocerida, *503*

Endopachys, **438**, *377*

Ensis, *498*

Entalophora, *599*

Enteletes, **96**, *643*

Enterolasma, *361*

Eoorthis, **79**, *636*

Eospirifer, **98**, *677*

Eotomaria, *406*

Equisetum, *745*

Eretmocrinus, *698*

Eristalis, **313**

Eubrontes, *473*

Eucalyptocrinites, **330**, **336**, *699*

Euchondria, *464*

Eucladocrinus, *701*

Euomphalacea, *400*

Eupatagus, **365**, *721*

Euphemites, **219**, *394*

Euproops, **304**, *593*

Eurypterida, *595*

Eurypterus, **306**, *595*

Euryzone, *409*

Eutaxocrinus, *704*

Eutrephoceras, **189**, *516*

Eutrochocrinus, **329**, *697*

Exogyra, **211**, *470*

F

Fasciculiconcha, *463*

Fasciolaria, 242, 436

Favia, **439**, 369

Faviida, *368*

Favistella, *360*

Favistina, **430**, 359

Favosites, **377**, **420**, 381

Favositida, *380*

Feeding structures, *734*

Fenestella, **378**, **385**, 611

Fenestralia, 612

Fenestrapora, 612

Fenestrellina, 612

Ferns
seed, 746
true, 745

Ficopsis, 432

Ficus, **251**, 432

Finkelnburgia, **102**, 640

Fishes
bony, 761
jawless, 758
lobe-fin, 763

Fissurellacea, *412*

Fistulipora, **413**, **416**, 603

Flabellum, **354**, 376

Flexibilia, *703*

Flexicalymene, **275**, 581

Foerstephyllum, **434**, 380

Fungia, *368*

Fungiida, *367*

G

Gastrioceras, **187**, **188**, 524

Gastropoda, *391*

Ginkgo, **505**, 747

Ginkgophyta, 747

Ginkgos, 747

Girtycoelia, *347*

Girtyocoelia, **407**, 347

Glabrocingulum, **232**, 406

Glossograptus, 728

Glossopleura, 561

Glycymeris, **128**, 455

Glyptocrinus, **331**, 698

Glyptorthis, **66**, 639

Goniatites, 523

Goniatitida, *519*

Goniobasis, **258**, 425

Gonioceras, **299**, 506

Goodhallites, **203**, 532

Grammysia, **122**, 498

Grammysioidea, 497

Granocardium, 483

Graptolite Fragments, **460**, 724

Graptolithina, *723*, 724

Greenops, **289**, 585

Greensand, **10**, **14**, 93

Grewingkia, **348**, **349**, 361

Griffithides, **277**, 574

Gryphaea, 469

Gymnolaemata, 597, 617

Gypidula, **106**, 665

Gyrodes, **170**, 429

H

Hadrophyllum, **360**, 362

Haimesastraea, 373

Halimeda, 742

Hallopora, 610

Halysites, 440, 385

Halysitida, 385

Hamites, 527

Hamulus, 206, 539

Hardouinia, 369, 718

Hebertella, 76, 642

Hederella, 449, 602

Helcionella, 391

Helicoceras, 528

Helicotoma, 172, 177, 401

Heliolites, 424, 384

Heliolitida, 383

Heliophyllum, 350, 435, 364

Hemiaster, 367, 720

Hemichordata, 723

Hemicystites, 711

Hemiphragma, 608

Hesperonomia, 638

Hesperornis, 766

Hesperorthis, 72, 638

Heteraster, 721

Heterocoelia, 347

Heterophrentis, 351, 352, 360

Heteropora, 601

Heteroschisma, 691

Hexactinellida, 343

Hexagonaria, 364

Hiatella, 494

Hiatellacea, 495

Hindia, 408, 409, 410, 341

Hippocardia, 43, 86, 444

Hirudinia, 537

Holectypoida, 714

Holectypus, 362, 363, 714

Holmia, 288, 555

Holocystites, 335, 689

Holopea, 223, 413

Holosteans, 761

Holothuroidea, 687

Homotelus, 269, 271, 307, 570

Hoploscaphites, 196, 529

Hormotoma, 419

Hornera, 396, 600

Hudsonaster, 371, 707

Hughmilleria, 596

Hustedia, 90, 673

Hyalospongea, 343

Hydnoceras, 445, 344

Hyolithes, 347, 543

Hyolithida, 543

I

Ianthinopsis, 256, 421

Ichthyostega, 763

Idmidronea, 395, 599

Idonearca, 455

Illaenus, 270, 573

Ilymatogyra, 212, 213, 471

Imitoceras, 191, 520

Inadunata, 701

Inarticulata, 627

Inoceramus, 158, 162, 461

Insecta, 751

Insects, *751*

Isonema, 414

Isoteloides, 571

Isotelus, 571

J

Juresania, **60,** *661*

K

Kingena, 686

Kionoceras, 508

Knightites, **217,** *397*

Kootenia, 559

Kouphichnium, 733

L

Labechia, 351

Laevicardium, 483

Lasiograptus, 728

Lecanospira, **178,** *399*

Leioproductus, 658

Leiopteria, 460

Leperditia, **116,** *587*

Lepidodendron, **491,** *744*

Lepidophyllum, 744

Lepidostrobus, 744

Lepisosteus, 762

Leptaena, **61,** *651*

Leptellina, 649

Leptobolus, 629

Leptodesma, **156,** *460*

Leptograptus, 727

Leptostrophia, 654

Lichenopora, **451,** *601*

Lima, **38,** *468*

Limacea, *468*

Limestone(s), 16, 28, 94, 95
fossiliferous, *13, 19, 95*
siliceous, *6, 96*

Limpet, Keyhole, *413*

Limulus, 593, 594

Lindstroemella, 632

Lingula, **152,** *627*

Lingulella, **104,** *629*

Lingulepis, 629

Linoproductus, **59,** *663*

Linthia, 721

Lirodiscus, **137,** *481*

Lirophora, **136,** *491*

Lithostrotion, 367

Lithostrotionella, 367

Littorina, 418

Loculipora, 612

Lopha, **163,** *474*

Lophophyllidium, 365

Lophophyllum, 365

Loxonema, **259,** *423*

Loxonematacea, *422*

Loxoplocus, **234,** *407*

Lucina, 478

Lucinacea, *477*

Lunatia, 431

Lycophyta, *744*

Lycophytes, *744*

Lycopodium, 744

Lycopods, *744*

Lyriopecten, 463

Lyropecten, 466

Lytoceratida, *526*

Lytospira, 402

M

Maclurea, 400

Macluritacea, *398*

Macluritella, 402

Maclurites, **168, 179,** *399*

Macropleura, 677

Mactracea, *484*

Malacostraca, *586, 590*

Mammalia, *766*

Mammals, *766*

Mammut, **479,** *768*

Mangelia, 426, 442

Manticoceras, 518

Margarites, 417

Marginella, 437

Marjumia, 559

Mastodon, *768*

Mataxa, 438

Meandrina, 442, 373

Mediospirifer, 681

Meekella, **92,** *654*

Meekoceras, **185,** *525*

Meekospira, **261,** *422*

Melania, 426

Membranipora, **386,** *619*

Mercenaria, **135,** *492*
campechiensis, 493
mercenaria, 493
tridacnoides, 492

Merista, 675

Meristella, **109,** *674*

Meristina, 675

Merostomata, *592, 593*

Merriamaster, 717

Merycoidodon, **477,** *768*

Mesogastropoda, *420*

Mesohippus, **475,** *767*

Mesolobus, **58,** *657*

Metablastus, 691

Metacoceras, 515

Metasequoia, **496,** *748*

Metoicoceras, **201,** *533*

Michelia, 419

Michelinia, 382

Michelinoceras, **320,** *507*

Micrabacia, 374

Microcyclus, 362

Microdoma, **240,** *416*

Microdomatacea, *415*

Micromitra, **112,** *634*

Millepedes, *751*

Modiolopsis, **150,** *475*

Modiolus, 457

Modiomorpha, 475

Modiomorphacea, *475*

Mollusca, *389*

Monograptus, **463,** *729*

Monoplacophora, *389, 390*

Montastrea, 370

Monticulipora, **392, 412,**
606

Monticuloporella, 606

Mortonella, 716

Mortoniceras, 532

Mosses, club, *744*

Mourlonia, 406

Mucrospirifer, **50,** *679*

Mudstone, **3,** *91*

Muensteroceras, **184, 190,** *523*

Multicostella, 640

Murchisonia, **260,** *419*

Murchisoniacea, *418*

Murex, **246,** *433*

Muricacea, *432*

Mya, **130, 141,** *493*

Myacea, *493*

Myalina, **165,** *459*

Mytilacea, *456*

Mytilus, 456

N

Natica, 431

Naticacea, *430*

Naticopsis, **224,** *418*

Nautilida, *513*

Nautiloidea, *506*

Nautilus, **501,** *516, 535*

Neilo, 449

Nemagraptus, 730

Nematoceran, **310**

Nemodon, 454

Neochonetes, 657

Neogastropoda, *432*

Neopilina, 390

Neospirifer, **52,** *682*

Neptunea, **243,** *435*

Nerinea, 425

Nerinella, 425

Neritacea, *417*

Neuropteris, **503,** *746*

Nilssonia, **498,** *747*

Nisusia, 637

Nucula, **145,** *448*

Nuculacea, *448*

Nuculana, **154,** *450*

Nuculanacea, *449*

Nuculoidea, 448

Nuculopsis, 449

O

Obolella, **115,** *633*

Oculina, **436,** *372*

Odontocephalus, 585

Odontochile, 585

Oepikina, 651

Ogygopsis, **284,** *560*

Oldhamia, **465,** *734*

Olenellus, **292,** *553*

Oleneothyris, **105,** *684*

Olenoides, **274, 279,** *558*

Olenus, **287,** *568*

Oligochaeta, *537*

Oliva, 437

Olivella, **265,** *437*

Omphalotrochus, **228,** *403*

Oncoceras, **208,** *510*

Oncocerida, *509*

Onniella, **68,** *645*

Onychaster, 709

Onychocrinus, **324,** *705*

Ophioglypha, 709

Ophiura, **372,** *708*

Ophiuroidea, *708*

Opisthobranchia, *392, 442*

Orbiculoidea, **114,** *632*

Ornithischia, *764*

Orria, **303,** *562*

Orthacanthus, **454,** *760*

Orthida, *635*

Orthis umbonata, *678*

Orthoceras, *508*

Orthocerida, *507*

Orthograptus, **457,** *727*

Orthonema, *424*

Orthonota, **148,** *498*

Osteichthyes, *761*

Ostracoda, *586, 587*

Ostracoderms, *758*

Ostrea, **160,** *473*

Ostreacea, *469*

Owenella, *395*

Oxybeloceras, **209,** *526*

Oxytoma, **41,** *460*

P

Pachyphyllum, **429,** *362*

Paedeumias, **293,** *554*

Pagiophyllum, **494, 495,** *748*

Palaeaster, *708*

Palaeoneilo, **146,** *449*

Palaeostylus, *416*

Paleofavosites, *381*

Paleolimulus, *594*

Panopea, **142,** *496*

Paracyclas, **127,** *479*

Paradoxides, **294,** *556*

Parallelodon, **149,** *453*

Parallelopora, **376,** *350*

Paranomia, *467*

Parasmilia, **353,** *375*

Paraspirifer, *681*

Parazyga, *674*

Parvohallopora, **393,** *609*
rugosa, *610*

Paterina, *634*

Peachella, *553*

Pecopteris, **499, 500, 502,** *746*

Pecten, *465, 466*

Pectinacea, *462*

Penicillus, *742*

Penniretepora, **453,** *614*

Pentamerida, *663*

Pentamerus, **119,** *664*

Pentremites, **333,** *692*

Periarchus, **364,** *715*

Periglyptocrinus, *699*

Perissodactyla, *767*

Perissolax, *436*

Perrinites, *523*

Petrocrania, **78,** *632*

Phacopida, *577*

Phacops, **278,** *582*

Phanocrinus, *703*

Pharkidonotus, *397*

Phillipsia, *575*

Pholadomya, 153, *499*

Pholadomyacea, *497*

Phragmoceras, 207, *511*

Phylactolaemata, *597*

Phymatopleura, 407

Pinna, 159, *457*

Pinnacea, *457*

Pitar, 131, *488*

Placenticeras, 193, 197, *531*

Placodermi, *758*

Placoderms, *758, 762*

Plaesiomys, 640

Plagioglypta, 501

Plankton, *742*

Planolites, 467, 468, *735*

Planorbis, 173, 175, *443*

Plants, *738*

Plasmopora, 384

Platanophyllum, 507, *749*

Platyceras, 215, 216, *415*

Platyceratacea, *413*

Platycrinites, 327, *700*

Platyorthis, 647

Platystrophia, 51, *642*

Plecia, 315

Plectodonta, 649

Plectorthis, 641

Plethorhyncha, 669

Pleuracanthodii, *760*

Pleurodictyum, 433, *382*

Pleurorthoceras, 508

Pleurotomaria, 409

Pleurotomariacea, *404*

Polhemia, 401

Polinices, 221, *430*

Polychaeta, *537, 538*

Polyplacophora, *389*

Polypora, 380, 381, *613*

Porifera, *339*

Praecardiacea, *451*

Praecardium, 452

Prasopora, 418, *607*

Prasoporina, 607

Prionocyclus, 186, *533*

Priscoficus, 432

Prismatophyllum, 428, 432, *363*

Prismodictya, 444, *344*

Prodentalium, 501

Productella, 111, *658*

Prohysteroceras, 532

Promytilus, 155, *456*

Pronemobius, 311

Properrinites, 523

Proplina, 391

Propora, 384

Prosobranchia, *392*

Protaraea, 422, *385*

Protista, *741*

Protists, *741, 742*

Protocardia, 100, *484*

Protomedusae, *354*

Protoscutella, 716

Protospongia, 343

Prototreta, 630

Protozyga, 671

Pseudogygites, **301**, *572*

Pseudomonotis, **37**, *464*

Pseudoparalegoceras, *524*

Pseudozygopleura, **416**, *423*

Pteriacea, *459*

Pteridophyta, *745*

Pteridophytes, *745*

Pteridospermophyta, *746*

Pteridospermophytes, *746*

Pterinopecten, **121**, *462*

Pterosaurs, *766*

Pterygotus, *596*

Ptilodictya, **388**, *616*

Ptilopora, *615*

Ptychagnostus, **268**, *552*

Ptychoceras, *526*

Ptychodus, **480**, *760*

Ptychopariida, *564*

Ptychopteria, *460*

Ptylopora, *615*

Pulmonata, *392*, *443*

Pycnodonte, **138**, *469*

Pyropsis, *436*

Q

Quercus, **506**, *749*

R

Rafinesquina, **65**, *650*

Rangia, **132**, *485*

Rastellum, *473*

Receptaculitid(s), **490**, *742*

Redlichiida, *552*

Redwoods, dawn, *748*

Reedops, *583*

Rensselaeria, **83**, *683*

Reptaria, *602*

Reptiles, *763*

Reptilia, *763*

Resserella, *645*

Reticulatia, **48**, *662*

Retrorsirostra, *640*

Rhinidictya, **384**, *616*

Rhipidomella, **70**, *647*

Rhodophyta, *743*

Rhodophytes, *743*

Rhombopora, **397**, *615*

Rhombotrypa, **400**, *609*

Rhyncholampas, **366**, *719*

Rhynchonellida, *666*

Rhynchotrema, **85**, **89**, *667*

Rhynchotreta, **91**, *666*

Roemerella, *632*

Ropalonaria, **405**, *618*

Rostroconchia, **389**, *444*

Rotularia, **205**, *540*

Rugosa, *353*, *357*

S

Sandstone(s), **17**, **18**, **20**, **22**, **25**, **26**, **31**, *90*, *92*, *93*, *94*
beach, **8**, *94*
fossiliferous, **5**, *93*
freshwater, *94*
with glauconite, **7**, *93*

Sarcinulida, *380*

Saurischia, *764*

Saxicava, *494*

Scabrotrigonia, **87**, *476*

Scaphites, **194**, **199**, *529*

Scaphopoda, *389*, *500*

Scenella, **236**, *390*

Schellwienella, *655*

Schizoblastus, *692*

Schizocrania, **120**, *631*

Schizodus, **140**, *476*

Schizophoria, **77**, *644*

Schizoporella, **383**, **414**, *622*

Schuchertella, **57**, *655*

Scleractinia, *353*, *367*

Scyphozoa, *354*

Selachii, *760*

Semicoscinium, *612*

Semionotus, **486**, *761*

Septastrea, **427**, *370*

Sequoia, **497**, *749*

Serpula, **469**, *538*

Serpulidae, *540*

Shale(s), **20**, **31**, **32**, **33**, **89**, *90*
black, **1**, *90*
calcareous, **2**, *91*
freshwater, **21**, **22**, **23**, **24**, **26**, *90, 92*
lake, **29**, *90*
paper, **4**, *91*
slaty, **30**, *91*

Shansiella, **166**, *409*

Sharks, *759*

Shumardites, **183**, *521*

Siderastrea, **425**, *368*

Sigillaria, *745*

Siltstone, **9**, **24**, *91*

Sinuites, *395*

Sinum, **220**, *431*

Siphonophrentis, *366*

Skolithos, **470**, *732*

Small Tubes of Uncertain Affinities, *541*

Solemya, **147**, *451*

Solemyacea, *450*

Solenoceras, *526*

Solenopora, *743*

Sowerbyella, **49**, *649*

Spatangoida, *720*

Sphenodiscus, **198**, *535*

Sphenophyta, *745*

Sphenophytes, *745*

Spinatrypa, **161**, *672*

Spinocyrtia, **53**, *680*

Spirifer, *642*, *680*

Spiriferida, *670*

Spirophyton, *736*

Spirorbis, *539*

Spisula, **470**, *485*
solidissima, *485*

Spondylospira, *679*

Stauriida, *359*

Stegocoelia, *419*

Stelleroidea, *706*

Stellipora, *605*

Stenolaemata, *597*, *598*

Stereocidaris, **373**, **374**, *713*

Stigmaria, *744*

Stomatopora, **448**, **450**, *598*

Straparollus, **171**, *402*

Straparolus, 403

Streblotrypa, 615

Streptelasma, 361

Striatopora, 383

Stricklandia, 664

Stromatolites, *741*

Stromatopora, **421,** *349*

Stromatoporoidea, *347*

Strombacea, *426*

Strombus, 436

Stropheodonta, 652

Strophodonta, **63,** 652

Strophomena, **55,** 649

Strophomenida, *648*

Strophonella, 652

Strophonelloides, 653

Stylonurus, 596

Subhyracodon, **476,** 768

Subulitacea, *421*

Subulites, 422

Sulcoretepora, 387, 617

Synphoria, 585

Syringopora, **379, 390,** *388*

T

Tabulata, *353, 377*

Tabulipora, **399,** 608

Taeniopora, 617

Tarphycerida, *512*

Taxocrinus, **322,** 704

Tellina, **144,** 486

Tellinacea, *486*

Temnocheilus, **200,** 514

Tentaculites, **346,** *542*

Tentaculitida, *542*

Terebra, **263,** *441*

Terebratalia, 685

Terebratulida, *682*

Terebratulina, **99,** 685

Tetradiida, *379*

Tetradium, **402, 419,** *379*

Tetragraptus, **461,** *725*

Texanites, **202,** 534

Texigryphaea, 470

Thamniscus, 614

Tiarasmilia, 375

Tipula, **312**

Tonnacea, *431*

Tornoceras, **192,** 520

Trace Fossils, *731*

Trachydomia, 418

Trematocystis, 690

Trematospira, 674

Trepospira, **169, 227,** *404*

Trepostomata, *598, 605*

Tretaspis, 576

Triarthrus, **286,** 568

Tricrepicephalus, **297,** 566

Trigonia, 477

Trigoniacea, *476*

Trigoniidae, *477*

Trigonoglossa, 628

Trilobita, *549*

Trilobitomorpha, *549*

Trimerus, 582

Trimerus (*Dipleura*), 582

Trionyx, **485**, *763*

Trochacea, *416*

Trochidae, *417*

Trochocyathus, **358**, **359**, *374*

Trochonema, **225**, **235**, *411*

Trochonematacea, *411*

Troosticrinus, **334**, *690*

Tropidodiscus, **174**, *395*

Tropidoleptus, **69**, *647*

Tubipora, *388*

Turbinolia, **355**, *374*

Turrilites, **210**, *528*

Turritella, **264**, *424*

Typhis, **248**, *434*

U

Uintacrinus, **328**, *706*

Uncinulus, **84**, *669*

Uniramia, *547*

Uperocrinus, *698*

Ursirivus, *495*
pyriformis, *495*

V

Vaginoceras, *504*

Valcourea, *640*

Vasum, **245**, *439*

Veneracea, *488*

Venericardia, **44**, *480*

Venus
Common Californian, *491*
Cross-barred, *491*
Gray Pygmy, *491*

Vertebrates, *755*

Vinella, *618*

Virgiana, *664*

Viviparus, **222**, *420*

Volutacea, *437*

Volutoderma, **254**, *439*

Volutomorpha, **253**, *438*

W

Waagenoceras, **181**, *522*

Waagenoconcha, *660*

Waconella, **107**, *686*
wacoensis, *686*

Wanneria, *554*

Worthenia, **233**, *408*

X

Xenophora, **231**, *427*

Xenorthis, *636*

Xiphosura, *593*

Y

Yoldia, *450*

Yunnania, *409*

Z

Zacanthoides, **285**, *563*

Zittelella, *342*

Zoophycos, *736*

Zygospira, **81**, *671*

NATIONAL AUDUBON SOCIETY
FIELD GUIDE SERIES

Also available in this unique all-color,
all-photographic format:

African Wildlife • **Birds** (*Eastern Region*) • **Birds**
(*Western Region*) • **Butterflies** • **Fishes** • **Insects
and Spiders** • **Mammals** • **Mushrooms** • **Night Sky**
• **Reptiles and Amphibians** • **Rocks and Minerals**
• **Seashells** • **Seashore Creatures** • **Trees** (*Eastern
Region*) • **Trees** (*Western Region*) • **Tropical Marine
Fishes** • **Weather** • **Wildflowers** (*Eastern Region*)
• **Wildflowers** (*Western Region*)

Prepared and produced by Chanticleer Press, Inc.

Founding Publisher: Paul Steiner
Publisher: Andrew Stewart

Staff for this book:

Editor-in-Chief: Gudrun Buettner
Executive Editor: Susan Costello
Managing Editor: Jane Opper
Project Editor: Carol Anne Slatkin
Natural Science Editor: John Farrand, Jr.
Picture Editor: Townsend P. Dickinson
Art Director: Carol Nehring
Art Associate: Ayn Svoboda
Production Manager: Helga Lose
Picture Library: Edward Douglas
Maps and Symbols: Paul Singer
Drawings: Patrick J. Lynch, Dolores R. Santoliquido

Original series design by Massimo Vignelli

All editorial inquiries should be addressed to:
Field Guides
P.O. Box 479
Ascutney, VT 05030
editors@thefieldguideproject.com

To purchase this book or other National Audubon Society
illustrated nature books, please contact:
Alfred A. Knopf
1745 Broadway
New York, NY 10019
(800) 733-3000
www.aaknopf.com